T0138974

Materials Science in Construction: An Introduction

Materials Science in Construction: An Introduction explains the science behind the properties and behaviour of construction's most fundamental materials (metals, cement and concrete, polymers, timber, bricks and blocks, glass and plaster). In particular, the critical factors affecting in situ materials are examined, such as deterioration and the behaviour and durability of materials under performance. An accessible, easy-to-follow approach makes this book ideal for all diploma and undergraduate students on construction-related courses taking a module in construction materials.

Ash Ahmed is a senior lecturer in construction materials science and module leader of several undergraduate and postgraduate materials science modules at the School of the Built Environment and Engineering at Leeds Beckett University. His research specialises in the evaluation of the mechanical and physical properties of commercial materials as well as novel sustainable materials in civil engineering.

John Sturges is a visiting professor at the School of the Built Environment and Engineering at Leeds Beckett University. His research interests include the environmental impact of materials, the energy efficiency of buildings and the whole area of sustainability and its impact on UK industry.

Materials Science in Construction: An Introduction

Ash Ahmed and John Sturges

Routledge
Taylor & Francis Group

LONDON AND NEW YORK

First published 2015
by Routledge
2 Park Square, Milton Park, Abingdon, Oxon OX14 4RN

and by Routledge
711 Third Avenue, New York, NY 10017

Routledge is an imprint of the Taylor & Francis Group, an informa business

© 2015 Ash Ahmed and John Sturges; individual chapters, the contributors

The right of Ash Ahmed and John Sturges to be identified as authors of this work has been asserted by them in accordance with sections 77 and 78 of the Copyright, Designs and Patents Act 1988.

All rights reserved. No part of this book may be reprinted or reproduced or utilised in any form or by any electronic, mechanical, or other means, now known or hereafter invented, including photocopying and recording, or in any information storage or retrieval system, without permission in writing from the publishers.

Trademark notice: Product or corporate names may be trademarks or registered trademarks, and are used only for identification and explanation without intent to infringe.

Disclaimer: Every effort has been made to contact and acknowledge copyright holders. The authors and publishers would be grateful to hear from any copyright holder who is not acknowledged here and will undertake to rectify any errors or omissions in future printings or editions of the book.

British Library Cataloguing-in-Publication Data
A catalogue record for this book is available from the British Library

Library of Congress Cataloging-in-Publication Data
Ahmed, Ash.
   Materials science in construction / Ash Ahmed and John Sturges.
      pages cm
   Includes bibliographical references and index.
   1. Building materials.   I. Sturges, John (Construction engineer)
   II. Title.
   TA403.A34 2015
   691—dc23   2014007765

ISBN: 978-1-85617-688-0 (pbk)
ISBN: 978-0-08-095850-7 (ebk)

Typeset in Bembo
by Keystroke, Station Road, Codsall, Wolverhampton

Contents

Contents

1

Introduction

This chapter provides an overview of mankind's use of materials, comparing the use of materials in various industrial sectors, and making clear the point that construction is the world's largest consumer of materials. It goes on to examine the importance of material properties in their selection for use, and outlines the various types and classes of materials and their importance in construction. A discussion of the service behaviour of materials and the problems of degradation and failure follows. Finally, in view of the current importance of sustainability, their environmental impact is stressed, and an outline of the book's contents brings the chapter to a close.

Contents

1.1 The industrial use of materials

The science and use of materials is central to all branches of industry, and as such is a subject of enormous importance. The range of materials we have at our disposal is enormous, and is being added to as the results of research and development are put to use week by week and month by month. The industries using materials include construction, aerospace, automobile, shipbuilding, white goods, electronics, railways, etc. Each industry has its own particular concerns about materials; in the aircraft business, the over-riding concern is with lightness and weight-saving, with cost being secondary to this. This is well illustrated at the present time with the advent of the Boeing 787 Dreamliner, which is pioneering the use of fibre-reinforced composite materials instead of the usual aluminium alloys for the construction of the airframe. Although aluminium is already a light metal, its use is being abandoned in favour of fibre-reinforced polymeric materials which are lighter still.

In the automobile industry, the luxury car sector currently is turning to the use of aluminium alloys for body structures in place of the traditional steel, and again the driver for this is weight-saving, with the consequent saving in running (fuel) costs. This may well be the precursor for the wider, progressive replacement of steel with light metals or reinforced polymers. In the construction industry, the *leitmotif* is most often low cost, i.e. weight-saving is not usually an issue, whereas keeping costs down to a minimum is very important. Having said this, in the construction of very tall buildings, engineers do take steps to save weight, usually in the construction of the upper floors of skyscrapers, in an effort to reduce the enormous loads that need to be borne by their foundations. Steel and concrete are such popular materials in construction; steel possesses high stiffness, high tensile and compressive strength and good ductility, with prices starting at around £500 per tonne, and concrete represents the cheapest way to buy one unit of compressive strength. Both are excellent value for money and this is most important in construction.

Each industry has its own preoccupations with the types of materials used. In most cases the products made will be created in a well-regulated, factory environment. In construction, on the other hand, the product is created on site in a less well-regulated environment, and this factor must be borne in mind. In most industries materials can be tested before they are used, to ascertain their quality and fitness for purpose. In construction, however, concrete falls outside this rule as it is made and used on site in one operation. If serious mistakes are made, the defective piece of concrete may have to be broken out and replaced. Concrete is the most widely used material of all, with over 12 billion tonnes being used worldwide each year, and yet it differs from all other important materials in not being able to be tested before it is used.

Construction also uses a lot of timber, a traditional construction material which has been used for centuries. We often lose sight of the fact that timber is a 'smart' material when it is growing as part of a tree. In a growing tree, timber can sense where compressive stresses are increasing due to weight increase brought about by the growth of new timber, and is able to respond by increasing the size of branches that are bearing the increased weight. So far engineers have not been able to produce such a remarkable material.

Finally, we must remember that whatever the industry using and specifying materials, what they are really doing is *specifying desirable properties*. We shall return to this fact later. It has been estimated that we currently have between 40,000 and 80,000 different materials (Ashby, 1992) at our disposal, if we count separately all the different alloy steels, all the different polymers, species of timber, types of glass, types of composites, etc. Making the correct selection can be a complex matter. Furthermore, for certain applications we cannot always meet our requirements from single materials among the 40,000–80,000 available. Sometimes none of these have the particular combination of properties we need; in such cases we may have to make recourse to *composite materials*; we shall look in more detail at these later in the book.

1.2 Importance of construction materials

In the UK, the construction industry is one of the largest, employing 1.0–1.5 million people (Harvey & Ashworth, 1997), and it rivals the NHS in size. It is responsible for at least 8 per cent of the UK's gross domestic product, currently being worth in excess of £60 billion per year. Roughly half of the industry's work involves new build, while the other half is maintenance, repairs and refurbishment. In addition, there is the UK building materials industry: brick-making, cement production, steel-making, as well as the industries that produce glass, plastics, gypsum plaster products, timber products, paints, fasteners, etc.

The construction industry in the UK consists of over 170,000 individual companies, the overwhelming preponderance of which are very small. Fewer than 50 firms employ more than 1,200 people.

Only 100 have more than 600 employees, so construction is often called a fragmented industry, and this is the situation in most of the countries of the world. These firms are located all over the UK; everyone has their local builders, plumbers, joiners, etc. One of the main reasons for the large number of small firms is that the barriers to entry to the industry are so low as to be virtually non-existent. By this we mean that little capital is required to begin. A skilled (or in some cases, unskilled!) bricklayer or roofer can set himself up in business very easily. The result is that every week in the construction industry in the UK, scores of firms cease trading, while new firms are started every week.

Another reason for the presence everywhere of construction firms is that unlike the products of the manufacturing industry, buildings are erected in a particular place; they cannot, in general, be built and transported. Because of this everyone needs their local builder.

Because the industry is so large, it is therefore a huge consumer of materials, both in the UK and worldwide. In fact, it is the largest consumer of materials in the UK and worldwide by a considerable margin. The total weight of materials consumed by all other industries combined is barely a quarter of that used in construction. The consequence of this is twofold: first, our use of construction materials has a major impact on our environment; and second, it is of the utmost importance that these materials are used correctly and as efficiently as possible.

Construction uses a wider range of materials than any other. Materials used include cement and aggregates to make concrete, metals – primarily steel, but with significant amounts of copper, copper alloys and aluminium alloys – timber, fired clay products, glass, gypsum products, polymers, bituminous materials, etc. The global consumption of the principal materials is shown in Table 1.1.

The table shows the proportion of these materials going to construction. It does not include fired clay products, which are widely used in construction, because global figures for this material are difficult to obtain. Such materials are not used in significant amounts by the other industrial sectors.

For comparison, the automotive industry consumes just 15 per cent of the world's steel output, a total of just over 200,000,000 tonnes. Steel is the principal material of the car makers, and the other materials it uses are 40 per cent of the world's rubber and 25 per cent of the world's glass output. So automobile production uses a tiny fraction of the quantities used in construction, and the aerospace business uses less still. The two major constructors of large passenger aircraft, Boeing (US) and Airbus Industrie (Europe), each deliver between 300 and 500 planes per year, depending upon the economic climate. If the average weight of each plane is 250 tonnes, this gives a total material consumption of 250,000 tonnes in a good year. Globally, the annual consumption of materials in the aircraft business is therefore under 1,000,000 tonnes. However, these materials will be very high value, with aero-engine materials in particular costing up to £1,000,000 per tonne.

The shipbuilding industry is another large consumer of material; steel is the one used in the greatest amounts. The present size of the world's merchant fleet stands at 1.0 billion tonnes dead-weight, and comprises just under 43,000 vessels. For the purposes of these statistics, a merchant ship must be over

Table 1.1 World production of principal materials, and the approximate proportion going to the construction industry

Material	Annual world production (tonnes)	% of world production used in construction
Cement	2,400,000,000	95–100
Aggregates	12,000,000,000	95–100
Steel	1,450,000,000	Up to 50
Timber	1,000,000,000	c.60
Polymers	150,000,000	c.20–25
Total	17,000,000,000	

Table 1.2 Approximate global annual consumption of materials by construction, shipbuilding, automobile and aircraft production

Industry	Annual consumption of materials (tonnes)
Construction	At least 10,000,000,000
Shipbuilding	At least 500,000,000
Automobiles	Around 400,000,000
Aircraft	Under 1,000,000

70 m long (about 230 feet). This tonnage is being added to at the rate of nearly 100,000,000 tonnes per year, with China now being the largest constructor. These figures do not include naval construction, or the construction of smaller vessels of less than 70 m length. The consumption of steel in shipbuilding may therefore be taken as between 150,000,000 and 200,000,000 tonnes per annum. Table 1.2 compares the approximate amounts of material consumed annually by construction, shipbuilding, automobile manufacture and aircraft production.

Of course, steel is also used in agriculture, white goods manufacture, machine tools, etc. But Table 1.2 shows the predominance of construction in material consumption.

Finally, all industries have used more and more materials as the years go by, i.e. the production and consumption graphs are all climbing with time. However, in the developed world, more materials are going into buildings as a proportion of the total than ever before (see Figure 1.1). Figure 1.1 gives data for the United States, but the pattern is the same all over the developed world.

The graphs shown in Figure 1.1 are very interesting. We can see that consumption of all classes of materials increased during the 100 years from 1900 to 2000. The peaks caused by the World Wars, and the

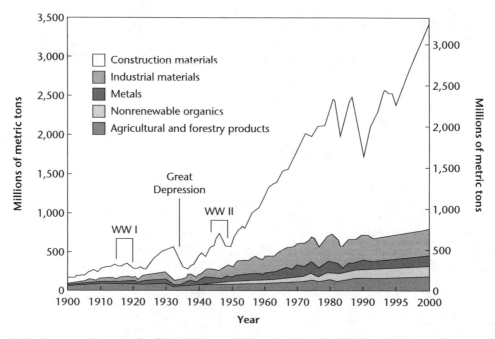

Figure 1.1 Consumption trends of various materials by the United States for the twentieth century

(*Source:* United States Geological Survey)

troughs occasioned by the Great Depression of the early 1930s, and the mid 1970s and 1980 oil crises are clearly visible. However, two features are noteworthy: (1) the increase in consumption of materials for construction far outstrips all other classes of materials, and (2) economic recessions have a disproportionately large effect on the construction industry. The data are for the United States, but the same trends are widely observed across the rest of the world. In short, this graph illustrates the pre-eminent importance of construction and its consumption of materials, and this theme will be returned to in the final chapter of this book.

1.2.1 Brief history of building materials

The finding and provision of shelter is one of the most basic human needs. When *homo sapiens* first appeared on Earth, they existed as hunter-gatherers, and would find shelter in caves and other convenient natural features. Around 8000 BC, however, mankind began to make the transition from hunter-gatherer to farmer, and men ceased to be nomadic and settled on their farmland. The need to build permanent settlements on their land became a major concern, and this step initiated the development of man as a building constructor. At this time, the population of the Earth would be perhaps 20,000,000 people in all. Once the human population density reached more than two people per square mile, the hunter-gatherer lifestyle was no longer sustainable, and more intensive methods for providing food became imperative.

In building his shelter man would utilise the materials that came to hand locally, such as timber, stone, animal skins and bone, etc. The mastery and use of fire led to the discoveries of ceramics, including fired clay, glass and also the smelting of metals. As millennium followed millennium, men discovered how to utilise a gradually increasing array of materials. So important were the materials used by men that the ages of mankind's development were named after the materials used, i.e. Stone Age, Bronze Age, Iron Age, and so on.

The Industrial Revolution, beginning at the start of the eighteenth century, initiated an acceleration in the pace of discovery and technological development, so that by the end of the nineteenth century, man had perhaps 100 or so materials at his disposal to meet all of his needs. The twentieth century saw the acceleration become an explosion in the number of available materials; Ashby (1992) has suggested that we now have between 40,000 and 80,000 different materials from which to choose.

The balance between types of materials has changed dramatically over time. In early historical times (10000 BC onwards) ceramics and glasses were important, together with the use of natural polymers and elastomers, and early composites such as straw-reinforced bricks and paper. The use of metals was known, but only a few metals had been identified – gold and copper being two of the earliest. By the middle of the twentieth century, metals had become the single most important class of materials. Since that time discoveries in the fields of polymer and ceramics have redressed that balance, and developments of engineering composites have also had a major impact. Figure 1.2 illustrates the balance between the various material types over time. It is important to recognise that the figure shows relative importance of the various material types and not absolute amounts. For example, at the start of the twentieth century, the total annual consumption of materials was well under one billion tonnes, whereas now it runs at over 16–17 billion tonnes.

However, the need for shelter is just as important as it ever was; this is true for all peoples in all countries. The construction industry has grown and developed to meet this need, and as a result it is the largest industry in the UK and in the rest of the world. This industry is the largest consumer of materials by far, as well as using a much wider range of material types than any other. The manufacture and use of all this material has an enormous impact on our world, which is our natural environment. This whole area of environmental impact and sustainability is a matter of increasing concern; we shall need to say something about this, and this discussion will be found in the last chapter of this book.

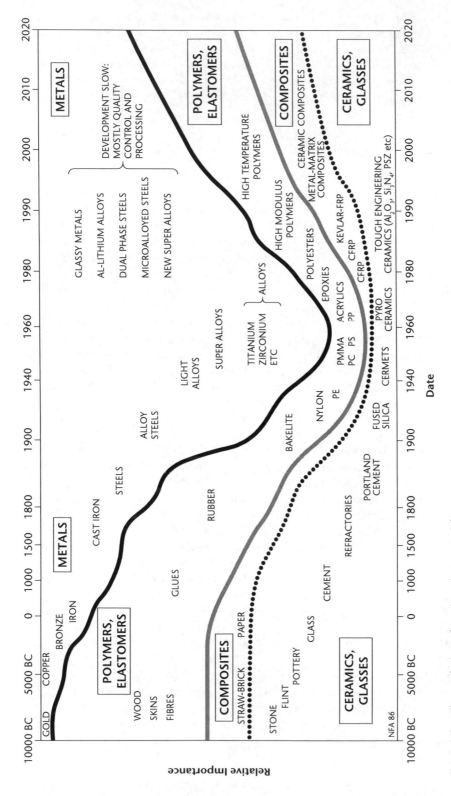

Figure 1.2   The evolution of engineering materials
(After Ashby, 1992)

1.2.2 The materials of construction

A wider range of materials are used in building construction than any other branch of industry. This range includes steel and certain other non-ferrous metals, cement, concrete, plaster, clay bricks and tiles, timber, glass, polymers, bituminous materials, natural stone, etc. For convenience, we shall classify these materials in three groupings:

1   metals, ferrous and non-ferrous
2   ceramics and other inorganic materials
3   polymers and natural organic materials.

It will be necessary to spend a little time on some of the underlying scientific principles governing the behaviour of materials. We need to appreciate the reasons why metals used in construction such as steel, lead and copper are ductile, and concrete and bricks, for example, are not.

1.3 Properties of materials

When a builder, architect or engineer specifies a material, he or she is really specifying a property or combination of properties. Materials are used for the properties that they possess, whether it be compressive strength, thermal insulation, high electrical conductivity, appearance, low cost or whatever. The properties that materials possess derive from their structures, i.e. the way that their component atoms and molecules are put together. This book does not set out to be a physics, chemistry or engineering text, but we require a little insight into the structures of materials if we are to understand how they perform in service.

It is worth bearing in mind that when we use a material, as was pointed out earlier, we are really making use of its properties. For this reason it will be a valuable exercise to look next at the process of *selecting* materials. This is not something usually covered in books on construction materials; the topic has rather been the subject of texts produced for engineers. However, materials selection will be dealt with as part of Chapter 18, under the heading of rational selection methods.

However, it will be useful to take a preliminary look at the range of properties we have at our disposal, and it is illuminating to consider properties by material classes such as ceramics, metals, polymers and composites. Such is the very wide range of property values that the values have to be plotted on a logarithmic scale. Figure 1.3 compares values of yield strength σ_y for ceramic, metallic, polymeric and composite materials. We can immediately see that ceramics have very much higher strengths than polymers, while metals have a much wider range of strengths, from alloy steels down to some very soft, pure metals.

Similar wide variations are seen for stiffness values (E), in Figure 1.4, and for density (ρ), in Figure 1.5.

Note the very wide variation in properties shown in Figure 1.3. The strength covers six orders of magnitude, i.e. the highest value (diamond) is nearly one million times stronger than the lowest (foamed polymer).

In Figure 1.4, again note the wide variation in stiffness values, with the data spanning six orders of magnitude. The stiffest material (diamond) is a million times stiffer than the least stiff material (foamed polymer).

Density values shown in Figure 1.5 span three orders of magnitude, with the densest metals (platinum and tungsten) being about 1000 times more dense than the lightest (foamed polymers).

1.4 Behaviour of materials in service

We shall consider the conditions under which materials serve in buildings, including the loadings to which they are subjected and the environmental influences which surround them. This is very important because we need to be as economical with materials, which are precious resources, as we can.

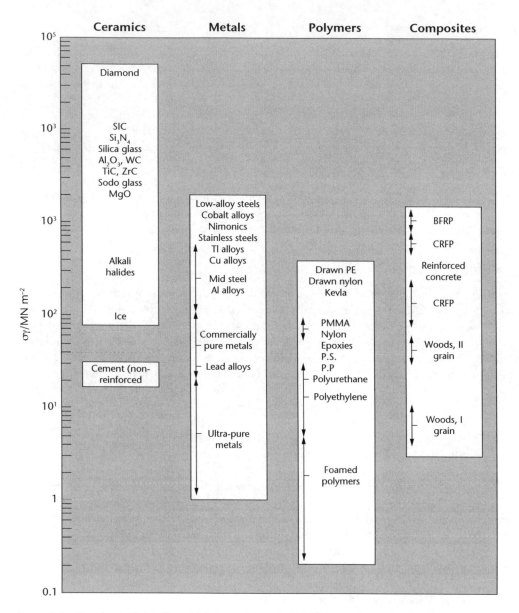

Figure 1.3 Bar chart of data for yield strength, σ_y (MN/m²)

(After Ashby & Jones, 1980)

In service, we are nearly always concerned with the durability of our materials. Unlike motor vehicles, which have a life of perhaps a decade, buildings are usually expected to last considerably longer, and massive repair and maintenance costs are not welcome to those who are responsible for them. Durability is the term used to describe the robustness of materials in the face of the service conditions that they endure; in simple terms, how long they last.

In Victorian times, mankind had perhaps 100 or so different materials which had to meet all of our needs, and which were somewhat less than ideal for their application. Fortunately, most of those materials

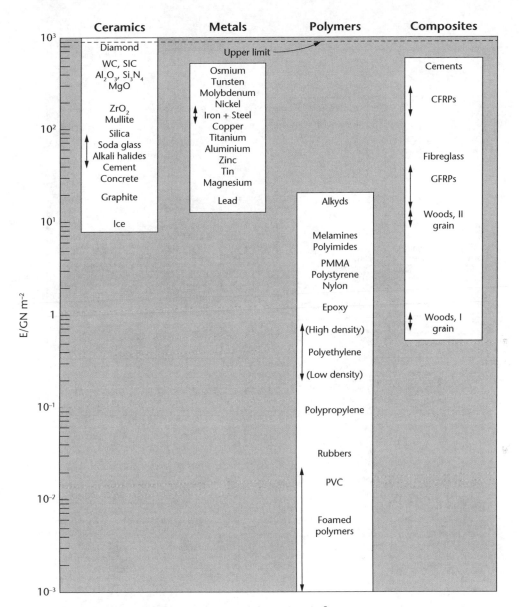

Figure 1.4 Bar chart of data for Young's modulus, E (GN/m^2)

(After Ashby & Jones, 1980)

were tolerant of abuse, and while not totally suited to their use nevertheless performed adequately. We are much more fortunate today, in the twenty-first century, in having a very much greater number of available materials, including composites, so that we can select materials having optimal properties for the particular application that we wish to fulfil.

An important part of this book will be to deal with how the various materials perform in service, together with the ways in which they can fail. We need to appreciate the environmental conditions in which the materials serve, as well as the types of events they may encounter or endure during their service lives.

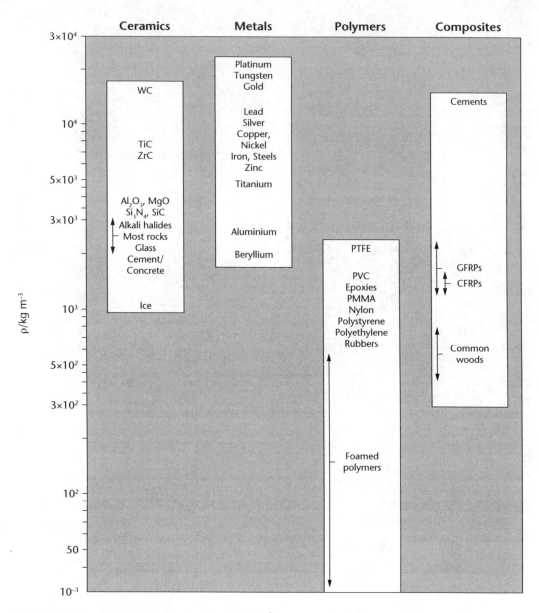

Figure 1.5 Bar chart of data for density, ρ (kg/m^3)

(After Ashby & Jones, 1980)

In the first place, buildings stand, apparently doing nothing, and are subject to the elements of the weather, and local meteorological conditions. On a day-to-day basis, this includes variations in temperature, precipitation and humidity, and wind conditions. If the building is in an urban location, vibrations from vehicular traffic may be a factor. The temperature may fall below 0 °C from time to time, and this can pose serious problems if water has been absorbed into cracks in structures or into individual materials. On freezing, water undergoes a 9 per cent volume increase, and this can cause stressing and cracking of those material which are inherently brittle.

This fact raises another very important factor in material performance, and that is the subject of porosity in materials. Some materials are porous, such as clay bricks, concrete and timber. Others are fully dense and impermeable to water. Such materials include metals, sheet plastics, glass, etc. The porosity will determine whether water will be absorbed by materials during service or not. Porosity will also be a determinant of the mechanical properties of materials, as we shall see. The phenomenon of capillarity means that porous materials can absorb water when they are exposed to it, and also tend to retain it even when the surplus water has drained from their surfaces. The water does not always have to be in liquid form, porous materials containing moisture will equilibrate with their surroundings and absorb moisture from the atmosphere during times when the weather is wet or humid. Similarly, they will then dry out when the weather is dry or less humid. These effects cause expansion and contraction effects in addition to those caused by temperature variations.

Another factor that is sometimes overlooked is that although buildings are static structures, they are always under stress. A very large building will possess an enormous weight, and this load is carried by the structural elements and foundations. For example, the Empire State Building in New York weighs over 300,000 tonnes. The stresses induced will be mainly compressive and monotonic. However, we need to bear in mind that while the weight will be responsible for the so-called dead loads on the structure, there will also be the live loads, i.e. those that are continuously varying from hour to hour and day to day. The live loads will include loads due to wind pressure, varying occupancy, etc. In bridge structures some of the stresses can be tensile in nature.

If the stresses vary in a cyclical way with time, they can lead to fatigue in metallic structural elements. Such conditions can occur in bridge structures, and several historically famous bridge failures have occurred because of fatigue.

However, even simple monotonic compressive stress can cause problems with some materials. Under such conditions, the phenomenon of *creep* can occur. Creep can occur in many types of materials including metals, concrete, polymers, masonry, glass and timber. Some of the expensive mistakes made in the construction of multi-storey tower blocks in the 1960s arose because creep and other relative movement effects were not taken into account. We shall examine this later.

In addition to these routine environmental variations, we must also consider other events such as fires, explosions and earthquakes. In the UK, fire is the commonest of these hazards. Explosions also occur, though less frequently than fires. One of the most common causes of explosions is gas leaks in domestic properties. Such explosions can be very destructive, often resulting in the partial or complete demolition of the house in which the leak occurred. During such events, the materials from which the building structure is made are subjected to rapid, dynamic loading, and the response of materials to such loading can be markedly different from their response to gradually or more slowly applied loads. This is also true for seismic activity. Fortunately, earthquakes are infrequent and relatively minor events in the UK. In other parts of the world, building design codes are written to take account of earthquakes, and the dynamic loading that they give rise to. In the UK, such earthquakes as do occur are mid-plate phenomena, like the one that occurred in January 2008, which measured 5.4 on the Richter scale. This was sufficient to cause only very minor damage to some buildings.

1.4.1 The use of materials and their impact on the environment

Because at least three-quarters of the materials used on planet Earth go into buildings and infrastructure, and because this is such a huge quantity, the manufacture, use and disposal of building materials has an enormous impact on our environment. The consequences of their use include energy consumption, pollution effects in air, water and soil, despoliation of landscapes, resource depletion, etc., on a correspondingly large scale. Unless energetic steps are taken to minimise and mitigate these effects, there is a real danger that we shall leave a degraded world to our successors and descendants rather than one enhanced for its occupants.

The impact that the use of materials has is not simple, and can have several ramifications. For example, extraction of raw materials can lead to spoiling of the landscape, as also can the deposit of waste materials from the production process. The production process can give rise to the emission of dust and gases into the atmosphere, as well as waste liquids and other solids. We need ways of measuring or quantifying these effects if we are to control these adverse impacts, and this question will be re-visited in the last chapter of the book. This is a topic of considerable current importance, with concerns increasingly being expressed about how sustainable our present mode of life will be in the long term. Unfortunately, the word sustainable is now very widely used, and not always by people who understand what it might mean. In the serious academic community, its meaning is still being debated and clarified. This topic will be dealt with in the final chapter of this book.

1.5 The contents of this textbook

The construction industry uses a very wide range of materials, wider in fact than that used by any other industry. Metals, ceramics, organics – natural and man-made – are used in enormous quantities. It is important that these materials are used economically and efficiently and not wasted, as construction materials in the past unfortunately have been (Anon, 1987). Waste is the hallmark of the present age. Archaeologists and anthropologists learn a good deal about ancient civilisations by excavating their middens and waste dumps. Archaeologists of the future (if they are still around) will be amazed at what our current civilisation throws away! But in today's increasingly environmentally conscious world, waste is belatedly being seen for what it is – gross mismanagement of our planet's resources. This theme will be taken up in the final chapter of the book.

As far as the main body of the book is concerned, it falls into three sections:

1. *Basic principles*. First, the book will attempt to cover the basic science of the materials of construction. The aim will be to give the minimum coverage to the principles governing the behaviour and properties of these materials. This first section will also include a chapter on the basic principles of structures, since many materials are used to build structures.
2. *Individual materials and classes of materials*. The second section will then deal in detail with the individual types of materials and how they perform in service. It will therefore have something to say about how construction materials in buildings degrade and fail. This section will build on the basic science of the first section and will move on to deal with the individual types of material in turn.
3. *Materials in service, durability and failure*. The third section will deal with those issues arising when materials are put into service, including different modes of failure, the effects of corrosion and solar irradiation, the effects of stress and types of fracture, and the effects of fire, etc.

Finally, since construction is by far the largest industry globally and the largest consumer of materials, and given the current preoccupation with sustainability, the enormous impact that construction has on our environment will be dealt with in the concluding chapter.

1.6 Critical thinking

The aim of this section is to provide the student with the opportunity to reflect on what he/she has learned, and to think about some of the main ideas outlined in the chapter. Questions for thought will be outlined in a critical thinking box, and other questions will be set out for students to work through, to help their understanding of important sections of the text.

1.7 Concept review

1   Construction materials are, in the main, inexpensive and low-tech. Explain why they are considered to be so important in the world of the twenty-first century.
2. Why are steel and concrete two of the most important materials of construction?
3. List the principal materials of construction. Which of these materials are porous and which are impermeable?
4. Why is porosity such an important factor in determining the behaviour of construction materials?
5. What, in general, is the link between the cost of a material and how much of that material is used?

1.8 References and further reading

ANON. (1987), *Materials for Construction and Building in the UK*, The Materials Forum and The Institution of Civil Engineers, The Institute of Metals, London.

ASHBY, M.F. (1992), *Materials Selection in Mechanical Design*, Pergamon Press, Oxford.

ASHBY, M.F. and JONES. D.R.H. (1980), *Engineering Materials: An Introduction to their Properties and Applications*, Pergamon Press, Oxford.

COLE, R.J. (1999), Building Environmental Assessment Methods: Clarifying Intentions, *Building Research & Information*, Vol. 27 (4/5), pp. 230–246.

GREENMAN, D. (ed.) (2008) *Jane's Merchant Ships 2008*, Jane's Information Group

HARVEY, R,C. and ASHWORTH, A. (1997), *The Construction Industry of Great Britain*, 2nd edition, Laxton's, Oxford.

McKINNEY, M.L., SCHOCH, R.M. and YONAVJAK, L. (2007), *Environmental Science: Systems and Solutions*, 4th edition, Jones and Bartlett Publishers, Sudbury, MA.

Part I
Basic principles: material structures and properties

2

Bonding and structures

Contents

2.1 Fundamentals and the structure of the atom

The idea that matter is made up of small, discrete particles is a very old one. Such a scheme was described by Democritus (460–370 BC) over 2,000 years ago. The modern notion of the atom was put forward by the English chemist John Dalton in 1800. He envisaged atoms as small, indivisible particles, with the atom being the smallest quantity of an element obtainable that retains the properties of that element. We now know that atoms are not indivisible, but they consist of smaller sub-atomic particles, protons, neutrons and electrons. Indeed, modern research in high-energy physics has shown that these particles can be broken down even further, but for our purposes an atom is most easily visualised as a nucleus surrounded by orbiting electrons. The nucleus contains most of the mass of the atom, and consists of neutrons (large, electrically neutral particles) and protons (large, positively charged particles). The orbiting electrons are negatively charged, and are tiny compared to the nuclear particles. The number of protons in the nucleus defines which element it is, and this number is called the atomic number Z. Table 2.1 shows the relative masses and charges of these three types of particle.

Table 2.1 Relative masses and charges carried by atomic component particles

Particle	Mass	Charge
Neutron	1,840	Zero
Proton	1,836	+1
Electron	1	−1

In a stable atom, the number of protons and electrons will be equal, and so overall the atom will carry no charge. However, as we shall see later, atoms can both lose and gain electrons. If they do this they are said to become ionised; gaining an electron will make them into a negatively charged ion (also known as an anion), losing an electron will result in them becoming a positively charged ion (also known as a cation).

2.2 The periodic table

In 1869 the Russian chemist Dmitri Mendeleev first noted that the chemical elements exhibited a 'periodicity of properties'. He had tried to organise the chemical elements according to their increasing atomic weights. He had assumed that their properties would progressively change as their atomic weights increased, but he found that their properties changed and then seemed to be repeated at sudden distinct steps, so that they could be arranged or grouped into distinct periods. One of his particular insights was that in 1869 there were elements that remained undiscovered, and which would, when found, occupy the missing places in his periodic scheme. This insight enabled him to predict accurately many of the properties that that an element was found to possess when it was isolated later. For example, he gave the name ekasilicon to the element germanium, which had not yet been discovered in 1869, and he successfully predicted several of its properties.

The modern periodic table of the elements is shown in Figure 2.1, and is based upon Mendeleev's ideas. It is organised by atomic number Z, and not by atomic weight. As we move from left to right along a row or period, the properties of the elements gradually change. The last element in each row is chemically inert, i.e. helium, neon, argon, krypton, xenon, radon – these are the inert gases. Those to their immediate left are the very reactive halogens, i.e. fluorine, chlorine, bromine, iodine, astatine. Their reactivity or inertness results from their outer electronic structures, as we shall see later. Therefore the elements in any column (called groups) tend to possess similar chemical properties. The periodic table has enormous significance in understanding the chemical behaviour of the elements, because it is rooted in their atomic numbers and therefore in their electronic structures.

At this point it is appropriate to point out that of the 100 or more elements in the periodic table, three-quarters are metals. Some of these, such as the rare earth metals, are found in nature in tiny amounts. In this modern age of electronic goods, some of these metals have become of great technological importance, despite being used in tiny amounts. The world annual production of many of these metals is often only a few hundred tonnes in total.

2.3 Bonding

We have looked briefly at the properties of atoms, but in practice our materials are made up of assemblies of atoms arranged in a myriad different ways, and we must now examine the various ways in which they can be bonded together. We shall see that it is the outer electronic structure of the atoms that is responsible for the bonding, and not the nuclear cores. We shall also see that it is the qualities of these bonds that determine the properties of our materials. By properties we mean principally the mechanical, thermal and electrical properties of materials.

The periodic table of the elements

1	2											13	14	15	16	17	18
hydrogen 1 **H** 1.0079																	helium 2 **He** 4.0026
lithium 3 **Li** 6.941	beryllium 4 **Be** 9.0122											boron 5 **B** 10.811	carbon 6 **C** 12.011	nitrogen 7 **N** 14.007	oxygen 8 **O** 15.999	fluorine 9 **F** 18.998	neon 10 **Ne** 20.180
sodium 11 **Na** 22.990	magnesium 12 **Mg** 24.305											aluminium 13 **Al** 26.982	silicon 14 **Si** 28.086	phosphorus 15 **P** 30.974	sulfur 16 **S** 32.065	chlorine 17 **Cl** 35.453	argon 18 **Ar** 39.948
potassium 19 **K** 39.098	calcium 20 **Ca** 40.078	scandium 21 **Sc** 44.956	titanium 22 **Ti** 47.867	vanadium 23 **V** 50.942	chromium 24 **Cr** 51.996	manganese 25 **Mn** 54.938	iron 26 **Fe** 55.845	cobalt 27 **Co** 58.933	nickel 28 **Ni** 58.693	copper 29 **Cu** 63.546	zinc 30 **Zn** 65.38	gallium 31 **Ga** 69.723	germanium 32 **Ge** 72.64	arsenic 33 **As** 74.922	selenium 34 **Se** 78.96	bromine 35 **Br** 79.904	krypton 36 **Kr** 83.798
rubidium 37 **Rb** 85.468	strontium 38 **Sr** 87.62	yttrium 39 **Y** 88.906	zirconium 40 **Zr** 91.224	niobium 41 **Nb** 92.906	molybdenum 42 **Mo** 95.96	technetium 43 **Tc** [98]	ruthenium 44 **Ru** 101.07	rhodium 45 **Rh** 102.91	palladium 46 **Pd** 106.42	silver 47 **Ag** 107.87	cadmium 48 **Cd** 112.41	indium 49 **In** 114.82	tin 50 **Sn** 118.71	antimony 51 **Sb** 121.76	tellurium 52 **Te** 127.60	iodine 53 **I** 126.90	xenon 54 **Xe** 131.29
caesium 55 **Cs** 132.91	barium 56 **Ba** 137.33	lutetium 71 **Lu** 174.97	hafnium 72 **Hf** 178.49	tantalum 73 **Ta** 180.95	tungsten 74 **W** 183.84	rhenium 75 **Re** 186.21	osmium 76 **Os** 190.23	iridium 77 **Ir** 192.22	platinum 78 **Pt** 195.08	gold 79 **Au** 196.97	mercury 80 **Hg** 200.59	thallium 81 **Tl** 204.38	lead 82 **Pb** 207.2	bismuth 83 **Bi** 208.98	polonium 84 **Po** [209]	astatine 85 **At** [210]	radon 86 **Rn** [222]
francium 87 **Fr** [223]	radium 88 **Ra** [226]	lawrencium 103 **Lr** [262]	rutherfordium 104 **Rf** [261]	dubnium 105 **Db** [262]	seaborgium 106 **Sg** [266]	bohrium 107 **Bh** [264]	hassium 108 **Hs** [277]	meitnerium 109 **Mt** [268]	ununnilium 110 **Uun** [271]	unununium 111 **Uuu** [272]	ununbium 112 **Uub** [277]		ununquadium 114 **Uuq** [289]				

lanthanum 57 **La** 138.91	cerium 58 **Ce** 140.12	praseodymium 59 **Pr** 140.91	neodymium 60 **Nd** 144.24	promethium 61 **Pm** [145]	samarium 62 **Sm** 150.36	europium 63 **Eu** 151.96	gadolinium 64 **Gd** 157.25	terbium 65 **Tb** 158.93	dysprosium 66 **Dy** 162.50	holmium 67 **Ho** 164.93	erbium 68 **Er** 167.26	thulium 69 **Tm** 168.93	ytterbium 70 **Yb** 173.05
actinium 89 **Ac** [227]	thorium 90 **Th** 232.04	protactinium 91 **Pa** 231.04	uranium 92 **U** 238.03	neptunium 93 **Np** [237]	plutonium 94 **Pu** [244]	americium 95 **Am** [243]	curium 96 **Cm** [247]	berkelium 97 **Bk** [247]	californium 98 **Cf** [251]	einsteinium 99 **Es** [252]	fermium 100 **Fm** [257]	mendelevium 101 **Md** [258]	nobelium 102 **No** [259]

Figure 2.1 The periodic table of the elements

Bonding is the name given to the mechanism by which two (or more) atoms join together to form compounds. There are several types of primary bond that can be formed, and these are all determined by the extra-nuclear or electronic make-up of the elements. It is the electrons, and not the nuclei, that form the various types of bond. Furthermore, these bond types can vary widely in their strengths; the energy required to separate two bonded atoms is the bond energy, and this governs not only mechanical strength but also other properties such as melting temperature. An element or compound with a high bond strength will be more difficult to melt, i.e. it will tend to have a higher melting temperature. It will also have a higher value of stiffness (Young's modulus of elasticity, E).

Before we look in more detail at the various kinds of chemical bonding that are commonly found, it will be profitable to consider the cohesive forces that hold atoms together and the binding energies involved. These considerations apply to the forces of attraction between any pairs of atoms or molecules, be they in the gaseous, liquid or solid states. In Chapter 6 we shall consider thermal properties, but it is worth pointing out here that heat is atoms and molecules in motion. In solids, the atoms or molecules vibrate or 'wriggle' about a mean position; in liquids they move randomly around inside the liquid, colliding with each other from time to time; in gases the atoms and molecules move with greatest speed, and move around randomly within whatever space is available to them.

In any group of atoms or molecules, be they gas, liquid or solid, there will be short-range forces of attraction operating. These can be represented by curve 1 in Figure 2.2. There will also be strong, short-range forces of repulsion that operate to resist compression when the atoms or molecules come into close proximity, and these are represented by curve 2 in Figure 2.2. If these curves (forces) are added we obtain a resultant force, curve 3. We can see that this resultant gives a position of stability at a spacing of a_0; any deviation from this spacing will be opposed by a restoring force, either tensile of compressive. Curve 4 shows that the system has minimum energy with this zero-force spacing, and will therefore be in

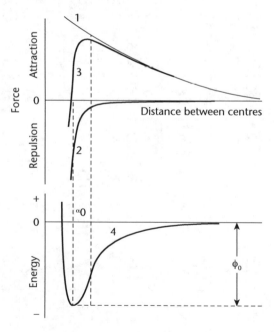

Figure 2.2 Forces and energy of interaction between particles
(After Cottrell, 1964)

Table 2.1 Showing values of strength, stiffness (Young's modulus) and melting temperature for a range of elements

Element	Yield strength (MPa)	Stiffness (E) (GPa)	Melting/softening temperature
Diamond	50,000	1,000	3,800
Tungsten	6,000	450–650	3,380
Steel 0.4%C	400	200	1,450
Iron	50	196	1,537
Copper	60	124	1,083
Aluminium	40	69	660
Lead	11	14	327

equilibrium. It can be seen that this resultant curve 3 is the same as the important curve shown in Figure 3.2 in Chapter 3.

2.3.1 Bond strength and material properties

As has been mentioned above, the strength of the bonds between atoms has a major influence on the properties of the solids made up of these atoms. This is quite logical, as the phenomena of elastic deformation, fracture and melting all involve the pulling apart of atoms in solids. So we would expect to see materials having high strengths also being difficult to melt (high melting temperatures) and having high stiffness values. In fact, there is such a link between these properties, and this is illustrated in Table 2.1. This table gives properties for diamond (a covalently bonded solid) and six metals, and from it we can see that ionic and metallic bonding can give rise to strong bonds.

The bonding mechanisms all involve electron donation from one atom to another, or electron sharing in some form, or electrostatic attraction brought about by electron gain or loss. There are a number of primary (strong) bond types, the most common being ionic, covalent and metallic. Secondary (weak) bonds include van der Waals and hydrogen bonds, which we shall now examine in turn.

2.3.2 Ionic bonding

This is the simplest type of bond. It has been pointed out above that halogen atoms are very reactive. This is due to the fact that when an atom has an outer set of electron orbits (known as a 'shell') which is full, bonding will not happen. If the atom has an outer shell that lacks one electron, then it will be keen to gain one more, and it is this fact that makes the halogens so reactive as a group.

On the left-hand side of the periodic table are the alkaline metals – lithium, sodium, potassium, etc. These metals have just one electron in their outer shell. Therefore, by donating an electron to a halogen, an ionic or electrovalent bond is formed. Consider the case of sodium chloride. The donation of an electron by the sodium atom produces a positively charged sodium ion (Na^+) while the donation of an electron to a chlorine atom produces a negatively charged chloride ion (Cl^-). So the charged ions form a so-called ionic bond.

2.3.3 Covalent bonding

This is a very common type of bond; it is found in most polymeric type materials, and it involves electron sharing. We have already looked at chlorine, one of the halogens. Chlorine forms a stable diatomic molecule

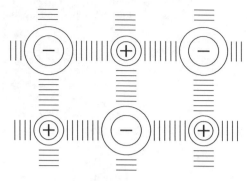

Figure 2.3   Schematic representation of ionic bonding

(After Smallman & Bishop, 1995)

(Cl_2) by each of the two chlorine atoms donating one electron to give a pair, which are then shared. This is the so-called covalent bond.

There is another variation on the covalent bond – the coordinate covalent bond. In this case, one of the atoms that are bonded together donates both of the shared electrons. One important characteristic of covalent bonds is that they can be very strong; the bonding between the carbon atoms in diamond is covalent. Diamond is a very strong solid, with a high stiffness (high value of Young's modulus E), and a high melting temperature. All of these properties result from a high value of bond strength. The bonding in many polymers is also covalent – mountain climbers trust their lives to nylon climbing ropes!

2.3.4 Metallic bonding

We have seen that metals are the largest group of elements in the periodic table, and metallic bonding is the name given to the bonds formed when metal atoms aggregate together to form solid pieces of metal. Metallic bonding is different from both ionic and covalent bonding. Metals have crystalline structures, and to form bonds each metal atom loses its outer electrons to form cations, i.e. positively charged ions. The free electrons from all the metal atoms thereby form a 'sea' of electrons, which can flow around these cations and through the lattice. These electrons are sometimes called 'de-localised electrons', because they are not confined to one place or freedom of mobility. This arrangement is illustrated in Figure 2.5.

The cations are all positively charged, and they will be subject to forces of attraction and also to forces of repulsion (their positive charges will form part of the repulsive force). The atoms will take up constant

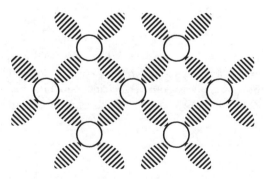

Figure 2.4   Schematic representation of covalent bonding

(After Smallman & Bishop, 1995)

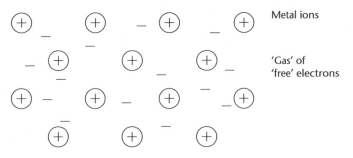

Figure 2.5 Schematic representation of metallic bonding

(After Ashby & Jones, 1980)

spacings from each other, and the distance apart will be the distance at which the attractive and repulsive forces are exactly equal. As we shall see, it is these forces that give the metal its *elastic* properties.

The 'sea' of electrons, coupled with the regular electrical periodicity, is what gives metals their excellent electrical conductivity properties. Finally, because all the atoms are lined up in a regular array, they are ideally positioned for the easy and rapid transmission of heat energy. So their crystalline structure confers on metals their excellent thermal conductivity properties as well.

2.3.5 van der Waals and hydrogen bonding

Whenever atoms are in close proximity to each other, they will attract each other by weak electrostatic forces. These forces are seen even between chemically inert atoms like the inert gases. They are called van der Waals forces and they are much weaker than ionic or covalent forces. They are very short-range forces, and they exist between all atoms and molecules, regardless of whatever other forces may be involved. These forces are caused by the fact that, even though the average electrical field of a neutral, spherical atom is zero, its instantaneous field is not zero. This is due to the fact that the electrons within the atom will move and cause the field to fluctuate. If the electrons in one atom move and leave the positively charged nucleus somewhat exposed to the electrons in a second atom, the atoms are able to correlate their electronic movements so that they are attracted to each other, and so form a weak van der Waals bond.

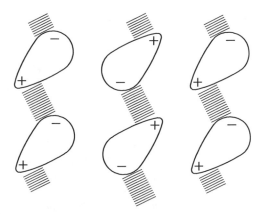

Figure 2.6 Schematic representation of van der Waals bonding

(After Smallman & Bishop, 1980)

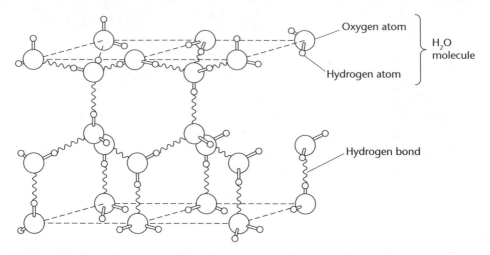

Figure 2.7 Arrangement of water molecules in ice, showing hydrogen bonds

(After Ashby & Jones, 1980)

Such van der Waals bonding causes inert gases to liquefy, and later in this book we shall examine the structures and behaviour of polymeric materials. Thermoplastics are of great technological importance, and we shall see that they consist of long carbon chain molecules where the bonding is strongly covalent. Strong, covalently bonded structures can be brittle. Nevertheless, thermoplastics are ductile. While the carbon–carbon bonds along each molecule are covalent, van der Waals bonds form between the chains. When thermoplastics are stressed to the point where they deform, these van der Waals bonds are broken and the carbon chains slide past each other, thereby allowing the materials to change shape.

2.4 Crystal structures

Solid materials can be classified according to the regularity or otherwise with which their atoms and ions are arranged relative to one another. Crystal structures are highly ordered; the state is characterised by a regular, periodic three-dimensional array of atoms, ions or molecules, and these crystalline solids have properties that result from this high degree of internal order. Non-crystalline structures, on the other hand, are those without such long-range atomic order. Such materials are variously described as amorphous, glassy or vitreous, and these terms are usually used synonymously.

Metals are a very important class of materials, and one of their characteristics is that they are *crystalline*, i.e. their atoms are arranged in very regular arrays. Metals therefore have some of the most highly ordered structures of all materials. While there are 14 different possible atomic arrangements with crystalline systems, most of the metals of common technological importance conform to one of the three simplest arrangements shown in Figure 2.8. These three common arrangements are:

1. body-centred cubic (BCC)
2. face-centred cubic (FCC)
3. hexagonal close-packed (HCP).

2.4.1 Body-centred cubic

This crystal structure type is found among some common metals, such as ferritic steel, pure iron, chromium, tungsten, etc. We can characterise the various crystal forms by their atomic packing factor (APF).

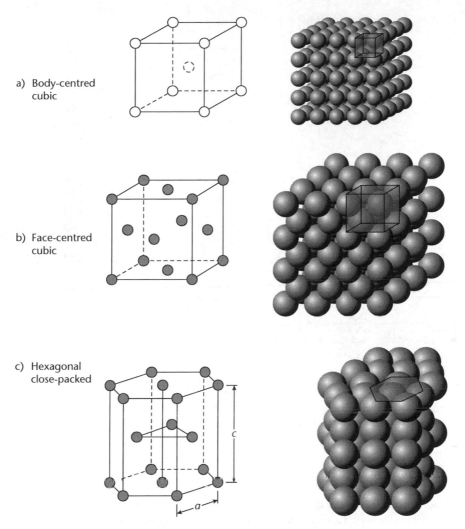

a) Body-centred cubic

b) Face-centred cubic

c) Hexagonal close-packed

Figure 2.8 The three main metallic crystal structures, (a) BCC, (b) FCC, (c) HCP
(After Callister, 1994)

$$\text{APF} = \frac{\text{Volume of atoms in a unit cell}}{\text{Total volume of unit cell}}$$

With BCC metals the atomic packing factor is 0.67, i.e. 67 per cent of the unit cell volume is taken up by the constituent atoms.

2.4.2 Face-centred cubic

This is the structure found in some of the common engineering metals such as austenitic steel, aluminium, copper, gold, silver, nickel, etc. (see Table 2.2). With FCC metals, the atomic packing factor is 0.74. This means that 74 per cent of the unit cell volume is taken up by the constituent atoms. We can see from the APF figures that FCC metals are more close-packed than BCC metals.

Table 2.2 Crystal structures for some commonly used metals

Metal	Crystal structure	Inter-atomic distance (nm)	Atomic radius (nm)
Aluminium	FCC	0.2862	0.1431
Cadmium	HCP	0.2978	0.1489
Chromium	BCC	0.2498	0.1249
Cobalt	HCP	0.2496	0.1248
Copper	FCC	0.1556	0.1278
Gold	FCC	0.2882	0.1441
Iron(α)	B.C.C	0.24824	0.12412
Iron(γ)	FCC	0.2540	0.1270
Lead	FCC	0.3499	0.1750
Magnesium	HCP	0.3209	0.1610
Molybdenum	BCC	0.2720	0.1360
Nickel	FCC	0.2491	0.1246
Platinum	F.C.C	0.2774	0.1387
Silver	FCC	0.2888	0.1444
Tantalum	BCC	0.2858	0.1429
Titanium	HCP/BCC	0.2876	0.1438
Tungsten	BCC	0.2738	0.1369
Zinc	HCP	0.2665	0.1390

2.4.3 Hexagonal close-packed

This is the third common structure type, and the common metals conforming to this type include zinc, magnesium, cobalt and cadmium. The atomic packing factor for HCP metals is 0.74, so it is a truly close-packed structure.

2.5 Polymorphism and allotropy

Polymorphism of a solid material refers to its ability to exist in more than one form of crystal structure. The particular form the material adopts will depend upon the local conditions of temperature and pressure. Polymorphism can occur in various types of substance, elements and compounds, and in organic and inorganic materials. In elements it is called allotropy.

One familiar example is found in carbon, where graphite is the stable allotrope at ambient conditions, and diamond is formed at extremely high temperatures and pressures. In metals, the allotropes of greatest technological significance are those occurring in iron. At room temperature, pure iron exists as a crystal with a BCC structure (α-iron or ferrite), and above 910 °C it transforms instantaneously to an FCC structure (γ-iron or austenite). If heated up to 1,394 °C, the austenite reverts instantaneously to a BCC structure (δ-iron), before melting at 1,538 °C. Because the FCC arrangement is more close-packed than the BCC arrangement, there is a slight volume reduction when the alpha to gamma transformation occurs, and a corresponding expansion when the FCC structure reverts to the BCC at 1,394 °C. The alpha to gamma transformation is of enormous technological significance because austenite will dissolve around 100 times more carbon than ferrite. If austenitic steel is rapidly cooled, there is no time or energy for the carbon atoms to diffuse, and so the carbon is trapped in solution, thus preventing the FCC structure from transforming to BCC ferrite. Instead, a body-centred tetragonal structure called martensite is produced. This is a non-equilibrium structure possessing very high hardness and strength, and it provides the basis for the heat-treatment of steels. That steel can be hardened to a remarkable degree by producing non-equilibrium martensite by rapid cooling has been known for centuries, but it has only been understood since the middle of the twentieth century. The ability to achieve an almost infinite variety of combinations of hardness, strength and ductility in what is essentially the cheapest industrial alloy is the reason for the fact that the various grades of steel comprise about 90 per cent of all metals and alloys used each year. It is because steel is a material of such technological importance and is used in such large quantities that a whole chapter is devoted to it later in this book (Chapter 8).

2.6 X-ray diffraction

Because X-rays have wavelengths of the same order of size as the inter-atomic spacing in metal crystals (10^{-10} m), metals will cause diffraction effects in beams of X-rays which impinge upon them. This effect was discovered by Friedrich and Knipping acting on the suggestion of Max von Laue at the end of the nineteenth century. Later work by W.H. Bragg and W.L. Bragg (1913) related the lattice parameter (atomic spacing) of a metal crystal to the wavelength of the X-rays used and the angle of diffraction in a simple equation:

$$n.\lambda = 2d.\sin\theta$$

where:

λ = wavelength of the X-ray beam
d = atomic spacing of the diffracting planes
θ = incident angle of X-ray beam to atomic plane
n = an integer

If a beam of X-rays strikes a plane of metal atoms at some angle θ, then there will be a path difference of $2d.\sin\theta$, if the planes of atoms are separated by distance d. If the path difference for that particular value of θ is equal to the wavelength of the X-radiation, then constructive interference will occur, and the radiation will be strongly reflected. For other values of θ, the path difference will not be a whole wavelength, and so destructive interference will occur and there will be no strong reflection.

Being crystalline, metals have some of the simplest structures, and X-ray diffraction techniques and the Bragg equation proved to be a very powerful tool in elucidating their structures, and measuring both their lattice parameters and crystal type. Much of this work on metals was done before and just after the Second World War. However, it was quickly realised that X-ray diffraction was a very powerful technique and could be developed and applied to determine the structures of many more complex non-metallic materials, including organic (both natural and man-made) materials and of various ceramic materials. The technique has indeed proved to be very effective; for example, 60 years ago it was used to help decipher the structure of DNA, the genetic material at the heart of all living cells.

The inter-atomic spacings quoted in Table 2.2 were all determined using X-ray diffraction.

2.7 Critical thinking and concept review

1. What is the difference between the atomic weight of an element and its atomic number?
2. Explain what is meant by an ionic bond, and give an example of an ionically bonded solid.
3. Explain what is meant by a covalent bond, and give an example of a covalently bonded solid.
4. Describe the nature of the metallic bond.
5. Produce a sketch of each of the three main crystal arrangements in metals, and give an example of a common metal having such a crystal structure, for each of the three main types.

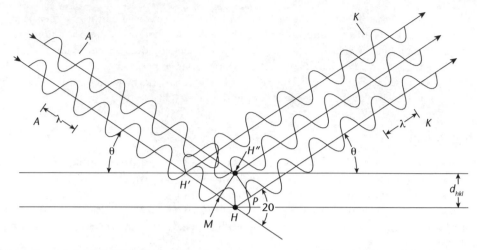

Figure 2.9 Bragg's law of diffraction: diffraction only occurs when the conditions of the Bragg equation are met

(After Van Vlack, 1974)

6. Explain why each metal crystal has its own unique lattice parameter (i.e. inter-atomic spacing).
7. What is meant by the term polymorphism?
8. A piece of pure iron is heated up to its melting temperature. Explain the changes to its crystal structure that occur as it is heated, and give the temperatures at which these changes occur.
9. Which crystal structure is the more close-packed, body-centred-cubic or face-centred-cubic?
10. In iron, which crystal form (BCC or FCC) will dissolve the most carbon?

2.8 References and further reading

ASNBY, M.F. and JONES, D.R.H. (1980) *Engineering Materials: An Introduction to their Properties and Applications.* Pergamon, Oxford.

CALLISTER, W.D. (1994), *Materials Science and Engineering: An Introduction*, 3rd Edition, John Wiley, New York.

COTTRELL, A.H. (1964), *The Mechanical Properties of Matter*, John Wiley & Sons Inc., New York and London.

FRIEDRICH, W. and KNIPPING, P. (1912), *Ann. Phys.*, 4, p. 971.

SMALLMAN, R.E. and BISHOP, R.J. (1995), *Metals and Materials*, Butterworth-Heinemann, Oxford.

TABOR, D. (1979), *Gases, Liquids and Solids*, Cambridge University Press, Cambridge.

VAN VLACK, L.H. (1974). *Materials Science for Engineers*, Addison-Wesley, Reading, MA.

3

Dislocations, imperfections, plastic flow and strengthening mechanisms in metals

This chapter provides an outline of the crystalline structure of metals and explains, in terms of this crystalline structure, how metals deform both elastically and plastically. Depending upon the temperature, metals will contain a population of crystal defects, and it is these defects that make them capable of plastic deformation. By controlling the number, size and type of these defects, the strength and ductility of metals can be controlled to meet the requirements of many design situations. The presence of these crystal defects also influences in a major way the final failure and fracture of metals.

Contents

3.1 Introduction

Metals are a group of materials of great technological importance in construction and in engineering generally. Since the advent of the Industrial Revolution, the application and use of metals has had an incalculable impact on our modern society. The most important and most used metal is steel, an alloy of iron and carbon. Other metals used in construction are copper, lead, zinc and a number of alloys such as brass, and small amounts of stainless steel. Metals have remarkable properties. For example, many of them can be shaped into an amazing variety of shapes and sections by being plastically deformed. This capacity for plastic deformation without suffering fracture is called *ductility*. A piece of steel 100 mm thick can be rolled down into a very thin strip 0.1 mm thick, without failure. This illustrates the amazing ductility possessed by metals as a group. If we tried to roll a piece of concrete or a brick in the same way, they would not deform, but suffer brittle fracture.

Metals also offer us *tensile strength*, which few of the traditional materials like natural stone or fired clay possess. Because of this remarkable combination of tensile strength and ductility, we can create all types of buildings and other structures that were not possible before the Industrial Revolution. Metals also offer *fracture toughness*, a resistance to brittle fracture not found in natural stone, clay brick or concrete, and this makes them indispensable in the design and construction of many large buildings.

Metals, and steel in particular, possess high values of *stiffness*, i.e. high values of elastic modulus, E. The high value of Young's modulus is one of the major advantages of steel as a structural material. A metal's elastic modulus is a fundamental property of the metal, and its physical basis lies in the crystalline structure of the metal, and in the nature and strength of the bonds between the atoms in the metal's crystal lattices.

Metals also possess excellent *electrical and thermal conductivities*, particularly FCC metals (see below) such as copper and aluminium. These are properties that we can make use of in heat exchangers, central heating systems and in the provision of lighting systems and numerous electrical goods used in buildings. We therefore need to gain an insight into the properties of metals and the reasons for them. We shall find that all these properties arise from the highly ordered, crystalline structures found in metals and alloys.

In the previous chapter we have examined the nature of atomic bonding, and we have seen how metals as a class have very ordered, crystal structures. In the next chapter we shall examine in more detail the mechanical properties of materials, and it is the task of this chapter to examine in simple terms how the mechanical and physical properties of metals, including those mentioned above, are a consequence of their crystalline nature.

3.2 Crystalline structure of metals

Metals are *crystalline*, that is to say their atoms are arranged in patterns with the highest degree of symmetry and order of any of the materials used by man. Their properties, including their strength and ductility, their excellent thermal and electrical conductivities all arise from their crystallinity. We shall therefore examine the crystal structures of the metals commonly used in construction. The bonding between the atoms that make up metal crystals is also special, and is known as the *metallic bond*. We have examined the various types of bonds between atoms in Chapter 2, and we have seen that chemical bonding involves the extra nuclear structure of atoms, i.e. the electronic structure. The atomic nucleus in all elements is orbited by electrons, the number of electrons depending on which element it is, i.e. what the *atomic number* of the element is. It is the various kinds of interactions between these electrons that enable atoms to bond together in different ways (ionic, covalent, metallic, etc.).

3.2.1 Crystal structures of the common metals

There are 14 types of crystal possible, with a crystalline atomic arrangement, but fortunately the metals of common interest conform to three of the simplest of these, and we shall confine our attention to these three arrangements. In Section 2.4 we saw that these three arrangements are:

1 body-centred cubic (BCC)
   * face-centred cubic (FCC)
   * hexagonal close-packed (HCP).

These atomic arrangements are shown again in Figure 3.1.

These crystal structures or atomic arrays are called *crystal lattices* by metallurgists and material scientists. Are there any differences between these three crystal types? Does the structure have a noticeable effect on the behaviour of the metal? The answer to these questions is yes, the structure does have an effect. Later in this chapter we shall examine the processes of plastic or permanent deformation in metals, and we shall see that deformation is accomplished by the movement of crystal defects called 'dislocations'. Dislocation movements are studied by crystallographers, and they have found that dislocations move by a process called slip on close-packed planes of atoms along close-packed directions. We shall not discuss this in detail here, but the crystal arrangement that has the most slip planes is the FCC structure. This being so, we would expect that metals possessing an FCC structure to be the most ductile, i.e. capable of the most plastic deformation before they fail. This is exactly what we find – the FCC metals are the most ductile. Aluminium, copper, lead, silver and gold are all very ductile metals and they all have the FCC-type structure.

The other two types are less ductile; in general, BCC metals are less ductile than the FCC type, and HCP metals possess lower ductility than the previous two. It is important to remember that all metals are ductile; FCC metals are outstandingly ductile, with the other two being less so.

Compared with the other materials of construction, these metal crystal structures are the simplest. For this reason, the metals were the first class of materials to have their structures investigated and to be well-understood in work that was mainly conducted before the Second World War. The techniques that were

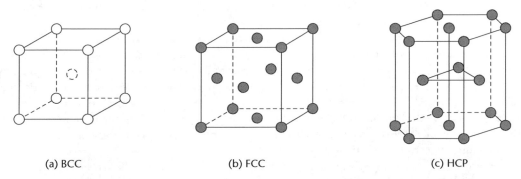

| (a) BCC | (b) FCC | (c) HCP |

Figure 3.1 (a) the body-centred cubic, (b) face-centred cubic and (c) hexagonal close-packed arrangements of atoms

(After Callister, 1994)

Table 3.1 Inter-atomic distances, atomic radii and crystal type for some common metals

Metal	Crystal structure type	Inter-atomic distance (nm)	Atomic radius (nm)
Aluminium	FCC	0.2862	0.1431
Copper	FCC	0.2556	0.1278
Iron (α)	BCC	0.24824	0.12412
Iron (γ)	FCC	0.2540	0.1270
Nickel	FCC	0.2491	0.1246
Zinc	HCP	0.2665	0.1390
Lead	FCC	0.3499	0.1750
Magnesium	HCP	0.3209	0.1610
Silver	FCC	0.2888	0.1444
Gold	FCC	0.2882	0.1441

used to investigate them, such as X-ray diffraction, were then used with others to elucidate the structures of the other classes of materials in research carried out since the Second World War. As we shall see, the properties of metals such as their excellent thermal and electrical conductivities, their ductility and fracture behaviour derive from their crystalline nature.

We may ask why the atoms arrange themselves in this way, and the detailed answer to this question lies in the field of solid state physics. However, the simple answer is that there are forces of both attraction and repulsion acting between the atoms. With a lattice at room temperature, the distance between the atoms is always the same. This is true for any piece of iron anywhere on planet Earth. The distance between the atoms in a piece of gold will be different, but again, constant for any other piece of gold, and so on. This inter-atomic spacing is sometimes called the *lattice parameter*, and values for all metals can be found in textbooks on crystallography. The lattice parameters for a few common metals are given in Table 3.1.

These inter-atomic spacing vales are constant because of the balance of forces existing between the atoms composing the crystals. We shall examine the nature and balance of these inter-atomic forces in the next section.

3.2.2 Forces between atoms in crystals

The atoms separate themselves at a distance at which the forces of attraction and repulsion between the atoms are equal. The inter-atomic spacings given above are the same for a piece of copper or iron wherever in the world they are produced. Figure 3.2 shows the relationship between the separating distance between two atoms in a crystal and the force existing between them.

The line shown in Figure 3.2 shows the resultant force acting between the atoms as a function of the separation distance between them. This line is the resultant of two other graphs; the graph showing the attractive force between atoms as a function of separation distance, and another showing the repulsive force as a function of separating distance. We can see that if we put the lattice into tension, we get a positive tensile resisting force, and if we compress the lattice we experience a negative compressive force of resistance. Note also that the graph has a virtually linear slope in the region of the neutral or strain-free position. This explains Hooke's Law: the load is proportional to extension in an elastic solid, i.e. the stress vs strain graph is linear in the elastic region.

If we heat the metal, its atoms will gain energy, and the repulsive forces will increase slightly, and this will result in the inter-atomic spacing increasing slightly, giving rise to the familiar thermal expansion effects observed in metals when they are heated. We shall look at this again in Chapter 6. Let us now consider the mechanical (stress–strain) properties of metals.

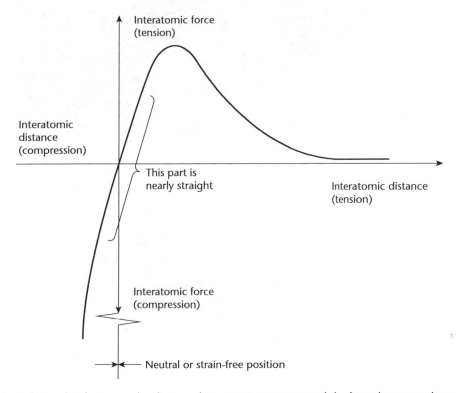

Figure 3.2 Relationship between the distance between two atoms and the force between them
(After Gordon, 1971)

3.3 Stress vs strain behaviour of metals

Metals, and particularly steel, are of vital importance in construction and civil engineering; indeed, the construction of most of the impressive buildings, bridges and other structures created during the twentieth century would not have been possible without steel. For this reason it will be very worthwhile to examine the load-bearing behaviour of steel.

The stress vs strain graph obtained by testing a metal such as plain carbon steel to destruction is shown in Figure 3.3. We can immediately see that the line consists of two regions; an initial linear portion, followed by a non-linear portion. The last point on the curve, marked by a small cross, is the point of fracture, i.e. this is the point at which the test piece can sustain no more strain and it fails. In this case, we can see that failure has occurred at a strain of 42.5 per cent. This means that the test piece gauge length was 42.5 per cent longer than it was at the start of the test. Furthermore, this 42.5 per cent extension was *permanent extension*. The 42.5 per cent was measured by putting the broken test piece ends together in an extensometer gauge and measuring the new length as at the point of fracture.

What happened to the steel as it was strained from zero to the point of fracture? What processes occurred inside the metal that resulted in the stress–strain graph being of the form that we observe? To answer these questions, we need to take a look at the structure of metals, i.e. at how the atoms of which they are composed are arranged.

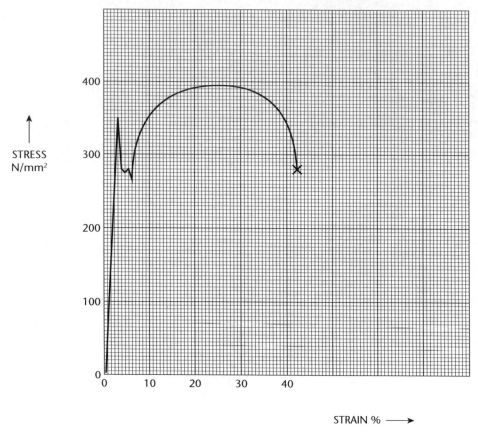

Figure 3.3 Stress vs strain graph for a 0.1 per cent carbon steel

3.3.1 Crystal structure of steel

Steel is iron with a small amount of carbon added to it. The addition is typically less than 1.0 per cent; many structural steels contain about 0.4 per cent carbon. So steel consists of iron crystals, with a small amount of iron carbide (Fe_3C) present in the microstructure. We need not consider the metallurgy of steel in any depth here. At room temperature iron crystals have the BCC-type structure, as shown in Figure 3.4.

3.3.2 Elastic behaviour

The spacing between the atoms is the separation at which *the force of attraction is exactly balanced by the force of repulsion*. This is logical, and it is the basis of elastic behaviour in metals. The result of these forces of attraction and repulsion is that they resist any applied loading, and as a result, we can model a metal crystal as a lattice where the atoms are attached to each other by springs, as shown in Figure 3.5.

This spring model can help us understand the elasticity of metals. By elastic deformation we mean temporary and recoverable deformation. We stretch a piece of rubber or a spring and it extends. When we release one end of the rubber or spring, it immediately 'springs back' to the original length it had before we stretched it. We see exactly the same behaviour in compression or torsion. Load, deform followed by

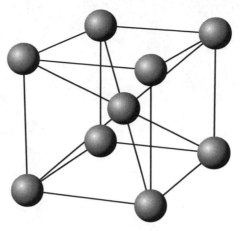

Figure 3.4   The arrangement of atoms in the BCC structure

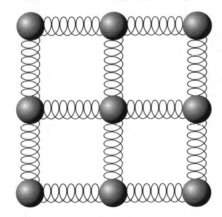

Figure 3.5   Material in the unstrained condition

(After Gordon, 1971)

spring-back when the load is released. This behaviour is illustrated in Figure 3.6. This shows both the tensile and compressive loading situations.

If we apply a tensile load to our metal, we stretch it, and it resists our stretching – we can feel the resistance. What is happening, and where does the resisting force come from? Remember, we are looking at *elastic* behaviour.

When we apply a tensile load to our metal, at the crystal level it is like pulling the atoms apart. Our tensile load is resisted by the forces of attraction between the atoms, at the same time our force acts with the forces of repulsion between the atoms. Therefore the atoms move apart until the applied force is balanced once more by the forces of attraction. If we release our force, the attractive forces pull the atoms back to the original point of balance, and our temporary elastic extension goes back to zero. The same thing happens in reverse if we apply a compressive load. In this case we work with the forces of attraction and against the forces of repulsion, and so we get a temporary elastic compression. When we release our applied load, the forces of repulsion push the atoms back to their original spacing.

(a) (b)

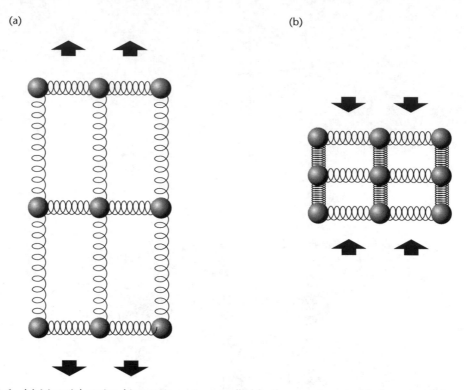

Figure 3.6   (a) Material strained in tension, atoms pulled further apart, material elongates; (b) material strained in compression, atoms pushed closer together, material becomes shorter

(After Gordon, 1971)

3.3.3 The elastic modulus or stiffness – a fundamental property

When we apply any load to any structure, it will deflect elastically. Even the weight of a couple of seagulls standing on a battleship gun-barrel will cause it to deflect elastically. In this case the deflection will be too small to measure with any extensometer or normal strain measurement device, but deflection will occur nevertheless. Finally, note that with elastic deformation the atoms stay in the same relative positions to each other; i.e. no atom shifts its position relative to its neighbours.

In construction, we require materials with a high stiffness for the structural or load-bearing elements of our buildings. By high stiffness we mean capable of load bearing with relatively small elastic deformation. For example, the Empire State Building in New York weighs something over 300,000 tonnes, and this enormous weight has to be carried by the structural frame of the building at ground level. As we move up the building, the load that has to be carried by the structure at that level decreases, of course. However, 300,000 tonnes is a very high load, and it is important that the material of the frame is both strong in compression and also of high stiffness. In fact, the Empire State Building is about 165 mm shorter than it should be in the unloaded condition, purely because of its high weight. The action of 300,000 tonnes acting in compression has shortened the building elastically by 165 mm. The frame of the Empire State Building is made from steel, which is both strong in compression and also of high stiffness.

The value of Young's modulus depends upon the strength of the bonds between the atoms in the crystal. This bond strength also plays an important part in determining how strong the metal is, and

Table 3.2 Values of strength, stiffness and melting temperature for a range of elements

Element	Yield strength (MPa)	Stiffness (E) (GPa)	Melting/softening temperature °C
Diamond	50,000	1,000	3,800
Tungsten	6,000	450–650	3,380
Steel, 0.4% C	400	200	1,450
Iron	50	196	1,537
Copper	60	124	1,083
Aluminium	40	69	660
Lead	11	14	327

since melting also involves breaking bonds, it will play a part in determining the melting temp-erature as well. Therefore we find that, as a general rule, elements having high strength and hardness also have high values of elastic modulus, *E*, and melting temperature. The data given in Table 3.2 illustrate this.

The data in Table 3.2 illustrate that diamond, which has the strongest bonding between its atoms, has the highest values of strength, stiffness and melting/softening temperature. Lead, with the weakest bonding of the materials shown, has the lowest values in each case.

To gain an idea of what is involved in elastic deformation, imagine a piece of material being loaded in tension. Looking at Figure 3.7, we can see that the load is carried by all the bonds in the cross–section of the material. The result, as we know, will be a slight increase in the length of the piece of material. If we release the tension, the piece of material will immediately spring back to its original length. This is *elastic* behaviour. Elastic strains are temporary, and are relaxed as soon as loading is removed. The same thing will be observed if we apply a compressive load.

Having examined elastic behaviour, now let us look at *plastic* behaviour.

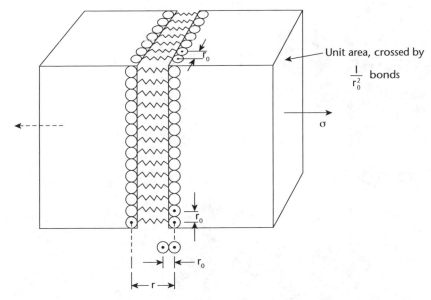

Figure 3.7 Showing the elastic straining of bonds in a material put into tension

(After Ashby & Jones, 1980)

3.3.4 Plastic behaviour

This is sometimes referred to as inelastic behaviour to distinguish it from elasticity. Plastic strains are *permanent* strains. There is no 'spring back'. Since there is a permanent shape change when plastic deformation occurs, there *must be some relative movement of the atoms* in the crystals. Furthermore, this movement must be able to occur without fracturing of the crystals. What mechanism allows this to happen?

In Figure 3.3, we saw that the initial portion of the graph was a straight line; this is the elastic part of the stress–strain behaviour. Within this region, if we double the load or stress, we double the extension or strain. This is known as Hook's Law, as it was first enunciated by Robert Hooke in the seventeenth century. Elastic behaviour is sometimes called linear behaviour because it is described mathematically by a straight line. However, the straight line reaches a peak, and then (in the case of steel) it falls, and thereafter the line is non-linear. This non-linear behaviour is called *plastic* deformation. The elastic or straight line behaviour is a fundamental property of the metal, and Young's modulus is a constant for the particular metal. The plastic behaviour is *not* fundamental, and the shape of the plastic flow curve can vary depending upon how fast we deform the metal, or upon the state of stress in the metal, or upon the temperature of the metal. Furthermore, the end of the plastic flow curve is the process of fracture, when the metal reaches the limit of its capacity to deform further, and it fails. What mechanism is required to accomplish this? The answer lies in a type of crystal defect called dislocations, and we shall examine these next.

3.4 Defects in crystals

The structure of a perfect crystal with a cubic lattice is shown in Figure 3.8. We can see that every atom is at a cube corner, i.e. they all occupy the correct places on the crystal lattice.

However, this situation will obtain at a temperature of absolute zero, but not at normal room temperature. In reality, metal crystals will contain crystal defects of various kinds, including vacancies, interstitialcies,

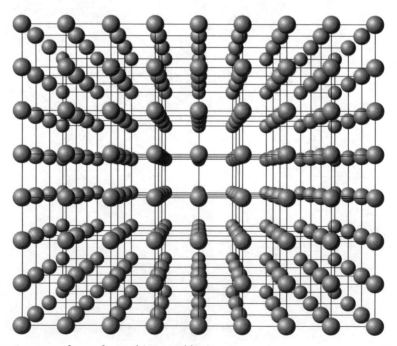

Figure 3.8 The structure of a perfect cubic crystal lattice

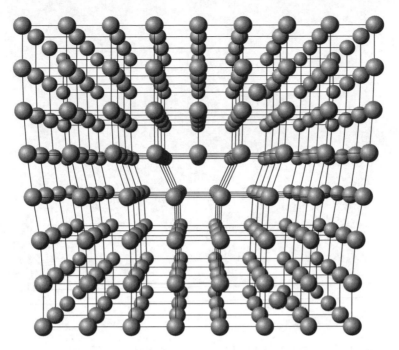

Figure 3.9 A dislocation in a cubic crystal

stacking faults and dislocations. We shall mainly consider dislocations here, as the others play a lesser role in plastic behaviour. What is a dislocation? A simple cubic crystal containing a dislocation is shown in Figure 3.9.

How does the dislocation make possible plastic deformation without failure? To answer this we need to consider what happens if a shear force is applied to a crystal containing a dislocation.

3.4.1 Crystal imperfections

The arrangements of atoms shown in Figure 3.8 in the last section are ideal or perfect arrangements. They would be seen as described at absolute zero, i.e. a temperature of −273 °C. However, at normal room temperature of, say, 20 °C, the crystals will not be perfect, but will contain imperfections or defects of various kinds: point, line and area defects.

- Point defects include:
    - vacancies: a vacancy is a site on a lattice not occupied by an atom;
    - solute atoms: this is a site on a lattice occupied by an atom of a different species;
    - interstitials: these are atoms forced into the spaces between other atoms on the lattice.
- Line defects: dislocations are easily the most important of this type.
- Area defects: stacking faults are examples of this type.

From the point of view of construction materials, easily the most important type of defect is the edge dislocation. This is because an understanding of the dislocation enables us to understand the plastic deformation and fracture of engineering materials. Figure 3.10 shows the various types of point defect that can exist in a metal crystal.

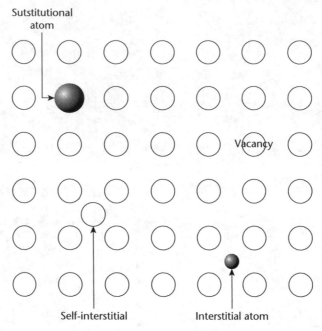

Figure 3.10   The various types of point defect that can exist in a metal crystal

(After Anderson et al., 1974)

3.4.2 Dislocations and plastic flow

Figure 3.9 shows a crystal containing a single edge dislocation, and the presence of such dislocations in metals was first proposed in 1934 to explain their plastic properties, although at that time there was no direct evidence for their existence. Experimental evidence for the actual existence of dislocations had to wait for 20 years; however, they are now very well understood. The movement of dislocations is the mechanism, at the microscopic level, by which the metal can be given a permanent change of shape. So when a metal is rolled into sheet or strip, or forged, the rolls or the forging dies are causing millions upon millions of dislocations to move within the crystals of which the metal is composed.

Figure 3.11 shows how the application of shear forces can cause a dislocation to move along a slip plane, resulting in the permanent movement of the block of atoms above the slip plane by one inter-atomic spacing relative to the lower block of atoms. Cleavage and fracture of the metal crystal does not occur. The passage of ten dislocations will result in shear movement of ten inter-atomic distances, and so on. Since hundreds of millions of dislocations will be moved in each crystal, when they are deformed we can easily see how measurable plastic strains are produced.

The dislocation density increases as a metal is cold worked. The number of dislocations in unit volume of crystal, *dislocation density N*, is defined as the total length of dislocations l per unit volume, $N = l / V$, normally quoted in units of cm^{-2}. For a well-annealed metal crystal, N is usually between 10^6 and 10^8 cm^{-2}, but it can be as low as 10^2 cm^{-2} with very careful preparation. For a heavily cold-rolled metal N can be around 5×10^{11} cm^{-2} (Hull, 1968).

3.4.3 Plastic deformation of metals

Plastic deformation is *permanent* deformation, as distinct from elastic deformation, which is *temporary*. We now know what a dislocation is, so we shall next examine what happens when a

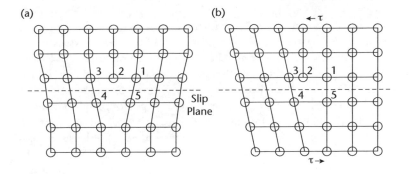

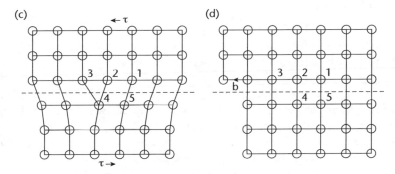

Figure 3.11 How a dislocation enables slip to occur in a metal crystal without cleavage fracture occurring

(After Anderson et al., 1974)

dislocation moves. We shall look at what happens when the dislocation moves by one inter-atomic spacing.

Deformation occurs essentially by a shearing process, i.e. a process where one plane of atoms slides or glides over the underlying layer. However, if the plane were to glide all at once, this would imply that all the bonds between the two planes of atoms would need to be broken at once. If this happened, the metal crystal would split and cleave into two halves. That this does not occur is proved by the excellent ductility that most pure metals and alloys normally exhibit. The exact way in which this shearing was accomplished puzzled metallurgists at first, until in 1934 three people independently published papers describing how it could happen. At the time they had no direct physical evidence for the existence of the crystal dislocations described above. They had to postulate their existence in metal crystals, and their predictions were eventually proved to be correct over 20 years later following the development of powerful electron microscopes.

3.4.4 Effects of temperature on plastic flow

One of the earliest references to the working of metals is found in the Bible, in the Book of Genesis. It refers to one Tubal Cain, a skilled worker of metals. From the earliest times, smiths have known that metals can be shaped by forging when they are hot. In the case of steel, this means when they are heated to at least red heat. When cold, metals are much harder and stronger, and more difficult to forge. Why is this?

Again, it is explicable in terms of the defect population, and in particular, the numbers of dislocations present in the crystals of which they are composed. The number of dislocations present in a metal crystal is strongly influenced by temperature. At absolute zero (–273 °C), metal crystals may be thought of as perfect for all intents and purposes. As they are heated up from absolute zero, dislocations appear in the microstructure. As heating continues, the multiplication of dislocations continues. So the structure of hot metal is said to be more disordered than that of cold metal. Dislocation multiplication continues until the metal reaches its melting temperature, when there is a state change from solid to liquid.

However, the multiplication of dislocations does not increase linearly with temperature, but rather follows an Arrhenius or exponential relationship. The number of dislocations N is proportional to $\exp-(Q/RT)$, where Q is an activation energy, R is the gas constant and T is the temperature. Because the index is negative, an increase in temperature produces a rapid non-linear increase in N.

3.4.5 Polycrystalline aggregates

In real metals as we encounter them in the construction industry, they do not consist of single crystals, but rather of polycrystalline aggregates. That is to say, a piece of steel will not consist of a single iron crystal but of thousands of iron crystals all joined together at their *grain boundaries*.

Each grain in Figure 3.12 represents a single crystal, i.e. the atomic arrangement in each grain is a regular crystal. The planes of atoms in the grain next door will also be regular, but will not be parallel to the planes of atoms in its neighbouring crystals. So each grain represents a very ordered structure, and the grain boundaries are relatively disordered. Such a structure is called a *polycrystalline aggregate*, because it is made up of many crystals. A steel I-beam in a building structure will be made up of billions of iron crystals containing typically 0.3–0.4 per cent carbon. Nearly all metallic engineering components are polycrystalline. Very occasionally, we deliberately make components out of a single crystal, one example being a turbine blade for the high-temperature portion of an aircraft engine.

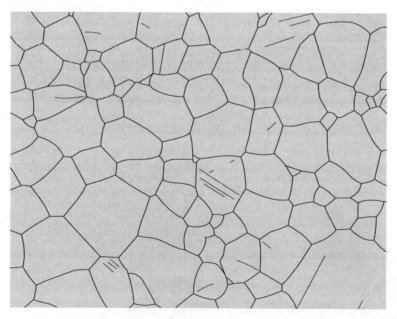

Figure 3.12 The microstructure of a pure metal, showing the grain (crystal) boundary arrangement
(After Tylecote, 1992)

3.4.6 Importance of ductility in structural steel

We know that for construction purposes, structural steels are often selected with a carbon content of 0.4 per cent, a composition offering a good combination of strength and ductility. The ductility is important, for reasons of safety. We can never predict every loading situation on a building during its design life, but the over-riding consideration is always life safety. While we can predict wind speeds and loadings, and all the likely effects of normal day-to-day operation of the building, there are possibilities that are not foreseeable. The impact of the B25 bomber that struck the Empire State Building in New York in July 1945 was not predictable, it was an accident. Since those days we have become familiar with urban terrorism, and the planting of bombs and their effects on buildings. The attack on the World Trade Center in New York in September 2001 was also an aircraft impact, and it was not accidental, but deliberately done.

While we cannot foresee these events, we still need to make our buildings as safe as possible. Iconic buildings of the past, such as the great Pharos Lighthouse in Alexandria in Egypt, or the great tomb of Mausolus at Harlicarnassus were two of the Wonders of the Ancient World, and lasted for many centuries. They were both destroyed by earthquakes in the thirteenth century, and so did not survive for us to see them. The reason they did not survive the earthquakes was that they were made of *brittle materials*.

When a brittle material is loaded, it deforms elastically until it reaches its elastic limit. However, when its capacity for elastic distortion is reached, it just fractures without warning. So if a building is made of natural stone, which is brittle, the kind of violent dynamic loading experienced during an earthquake would be likely to cause the stone to be loaded to its elastic limit. Portions of the structure loaded to the limit would then fail by brittle fracture, resulting in collapse of the whole edifice. This was the fate of the Pharos Lighthouse and King Mausolus' tomb. It is thought that the Pharos Lighthouse was perhaps close to 100 metres in height, and it was this great height that made it an ancient wonder. In our modern age we routinely construct buildings three or four times as high. This has been made possible by the availability of steel in large quantities and at low cost. Buildings this high have either a steel frame or a steel-reinforced concrete frame.

3.4.7 Strengthening mechanisms in metals

Since deforming metals plastically involves moving dislocations, anything that makes dislocations harder to move will have the effect of hardening and strengthening a metal. Now that we know about dislocation glide, we can appreciate that any impediment to the glide process, which either prevents it from occurring or which makes it more difficult, will cause the metal to become harder and stronger. By harder and stronger, we imply that the metal becomes more resistant to permanent deformation.

The things that will make dislocation movement more difficult include the presence of solute atoms in the crystal lattice, the presence of grain boundaries, because dislocation movement must stop when the line reaches the edge of the crystal and the effects of cold work. Let us examine, in simple terms, how these strengthening mechanisms work.

Solute hardening works because solute atoms will not be the same size as those of the parent lattice. They will therefore have a distorting effect on the parent lattice. To accommodate the larger (or smaller) diameter atom the lattice will be deformed in the zone around the solute atom. Therefore the crystal planes on which slip might occur are curved or bent instead of being straight, and this makes slip along that plane more difficult.

Grain boundaries represent areas of relative disorder, and since they also represent the place where the slip plane terminates, slip will stop at the grain boundary. This is shown in Figure 3.13.

The other phenomenon that will make slip more difficult is the presence of cold work. As plastic deformation occurs, the density of dislocations increases, and the crystals become distorted. Those

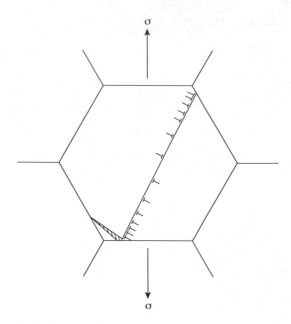

Figure 3.13 Edge dislocation 'pile up' at a grain boundary. If many dislocations of the same size pile up, then a micro-crack is nucleated

(After Cottrell, 1958)

dislocations which are most favourably oriented to slip do so, and when they are exhausted, slip can only continue by moving dislocations less favourably oriented; this effectively raises the yield stress. Moving dislocations through distorted crystals becomes progressively more difficult, and this phenomenon is known as cold work.

Eventually, we arrive at a situation where if we attempt to deform the metal any further, it becomes easier to enlarge and propagate cracks rather than move more dislocations, and at this point final fracture will occur. We shall examine the phenomenon of fracture in more detail in the next section.

3.5 Fracture behaviour in metals

Like all good scientific theories, once put forward, the theory of the dislocation enables us to understand a good deal more than just permanent, plastic deformation. Once the yield point has been exceeded, plastic deformation by dislocation slip occurs, with the easiest to move dislocations being moved first. When all these are exhausted, those that are slightly more difficult to move will move next until they are exhausted. Then those that are fractionally more difficult still will be the next to move, and so on. This explains why the plastic flow curve is a rising curve.

This process continues and the stress vs strain graph continues to show an increase, but with the rate of increase gradually slowing. If we were to arrest the deformation at some point on the rising curve, and completely relax the applied load, we should observe that our test piece was permanently elongated. If we replaced the test piece in the testing machine and re-applied the straining, we should observe that the test piece showed elastic behaviour up to the stress level at which we arrested the test. In other words, the yield point would be higher than it was when we began with unstrained metal. This apparent increase in strength is called *work hardening*. Work hardening can be removed by a heat treatment process called *annealing*, in which the cold-worked metal is heated to a temperature below its melting temperature, and the process of recrystallisation occurs.

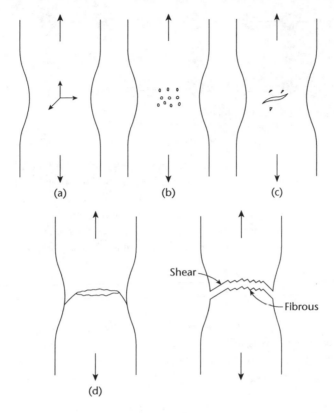

(a) (b) (c)

Shear

Fibrous

(d)

Figure 3.14   The stages involved in the familiar cup-and-cone fracture of a ductile metal

However, as straining proceeds, the metal continues to get stronger as the stress increases until the graph levels out and we reach a maximum stress. If straining is continued, the stress begins to decrease, slowly at first, and then at a faster rate. At some point past the peak stress, if we observe the deforming test piece, we shall see the formation of a waist or neck at some point along the gauge length. This neck will be the site of final fracture. Once the neck appears, deformation ceases in the metal except for the neck region. All further deformation is concentrated in the neck.

The stress level will decrease at an accelerating rate until final fracture occurs, very abruptly and with a bang. Figure 3.14 illustrates what happens inside the deforming metal during the final necking and fracture. This was first investigated in 1959 by Puttick at the Department of Metallurgy at Cambridge University.

When a metal test-piece is strained plastically in tension, the strain is initially uniform. The whole gauge increases in length and the diameter decreases to maintain constancy of volume. As straining proceeds and billions of dislocations are moved, we observe that the whole gauge length zone shows the appearance of micro-cracks. These micro-cracks grow in size up to and past the point of maximum load. At some point past this maximum load, inside the test-piece a few of the larger cracks coalesce to form a larger crack. This large crack weakens the material and this becomes the region of final failure. All deformation is now concentrated in the region of the internal crack, and this is the reason for the appearance of the 'neck', which is the site of final fracture. We can never predict where the neck will form, whether at the centre of the gauge length or nearer one end, and the reason is that we can never predict where the crack coalescence will occur within the deforming test-piece. By this point in a tension test, the load will be falling quite quickly, and the neck begins to reduce in diameter quickly, too. Final fracture is usually very abrupt, being marked by a loud crack, and the broken test-piece ends can be removed from the test for examination. The

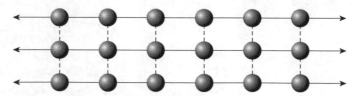

Figure 3.15   Array of atoms subject to a tensile load

(After Gordon, 1971)

fracture surface will be observed to be a dull grey in the centre, with a surrounding rim of shiny material at about 45° to the axis of tension. The grey region will be the inside surface of the large internal crack, and the shiny surfaces are the surfaces of the final, very sudden shear failure that concluded the test.

This appearance of the neck is a characteristic of ductile fracture; ductile failure always gives warning of its advent. The material always signals that it is approaching the point of failure. Brittle fracture, on the other hand, does not do this. Failure occurs very suddenly, usually without any warning and is frequently catastrophic.

3.5.1 Role of cracks in fracture

We have seen in the previous section that the fracture of a test piece in a mechanical tensile test involves the formation of cracks. Crack formation is always involved in material failures, in all materials including concrete and polymers – not just in metals. We have seen that micro-cracks are formed when dislocations of similar sign pile up at barriers such as grain boundaries and solute particles. The question that we need to answer is: why are cracks so dangerous?

Cracks are the precursors of failure. Often, the metal will continue to deform plastically even when cracks are present; sometimes, however, the metal will fail in a catastrophic manner, as happens when *fatigue* occurs. It is important to understand the principles involved in failure, and the role played by cracks. When a crack exists within a metal, the material on one side of the crack is disconnected from that on the other side. Consider a single array of metal atoms as shown in Figure 3.15.

If we apply a tensile force to the array it is carried equally by all the bonds between the vertical rows of atoms. However, if we imagine that all but one of the bonds are cut, we have the situation shown in Figure 3.16a. The vertical row of severed bonds effectively represents a crack in the structure. If we then apply a tensile force to this array, then no force will be transmitted across the crack because there are no linked atoms. All the force will be taken by the bond at the root of the crack (Figure 3.16b). The force acting on this bond at the root of the crack will be very high, many times higher than it would have been if the crack was not present. Therefore we can see that crack tips are regions of very high stress. This is why engineers refer to cracks as 'stress-raisers'. However, the crack needs to be normal to the direction of applied stress if it is to act as a stress-raiser. Cracks that are aligned with the stress direction have no effect. The crack has to be aligned so that it tends to open under load. This is why compressive stress has no effect; compressive loading will tend to close cracks rather than to open them. Concrete is full of tiny cracks and voids and is well known for its excellent load-bearing properties in compression. It is equally well-known for being very poor in tension, and this is the reason; it is the presence of thousands of tiny cracks that have a weakening effect.

We can see from the above figure that the tensile loading applied to the rows of separated bonds is carried by the first bond at the end of the crack. This means that the local loading on the first bond is very much greater than that carried by the others. This represents a very high local stress at the crack tip. In practice, engineers know that while the nominal stress on a piece of metal may be at a certain value, the actual stress at the tips of cracks present in the metal can be 10–20 times greater. Figure 3.17 shows the

(a)

(b)

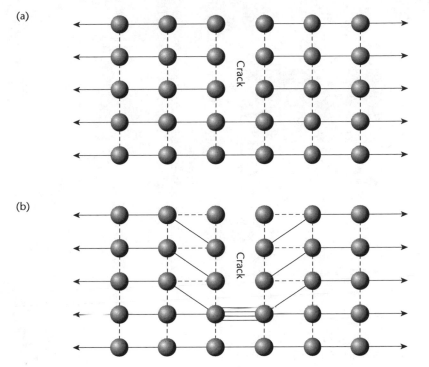

Figure 3.16 Array of atoms with crack, subject to a tensile load. (a) unloaded; (b) subject to load. All the load falls on the bond at the crack root

(After Gordon, 1971)

stress pattern at the tips of a crack normal to applied load direction in a piece of material. The diagram shows both a surface and an internal crack. In both cases the crack root radius is not very small. The smaller the tip radius (or the sharper the crack), the higher will be the stress intensity at the tip. Given the very high local stresses caused by the presence of a crack or cracks, the component will fail when it becomes easier to propagate the crack than to move more dislocations. So a ductile material will eventually fail because of this fact. The energy needed to move dislocations becomes greater than the energy needed to create more fracture surface.

3.5.2 Creep

Another phenomenon that can occur in metals is what is called creep. Creep occurs in materials under stress (stress less than the yield stress), when they suffer a small but measurable amount of plastic deformation. Clearly, at a stress below yield stress they should not deform plastically, but they do. In nearly all metals creep only occurs at elevated temperatures, and therefore creep is of little significance for civil engineers and construction managers. However, the metal lead (Pb) suffers from creep at room temperature. Why does lead creep at room temperature and steel and copper do not? The answer is found in how far above room temperature the melting temperature of the particular metal lies. As a general rule creep and other effects occur when the temperature of a metal exceeds about 0.5 of its melting temperature in degrees absolute (K). So in the case of lead, its melting temperature is 327 °C. This translates to a temperature of 600K (327° + 273° = 600K). Half of 600K is 300K, and if we turn this into degrees Celsius we get 300 °C − 273 °C = 27 °C, which is about room temperature. A similar calculation for steel gives a

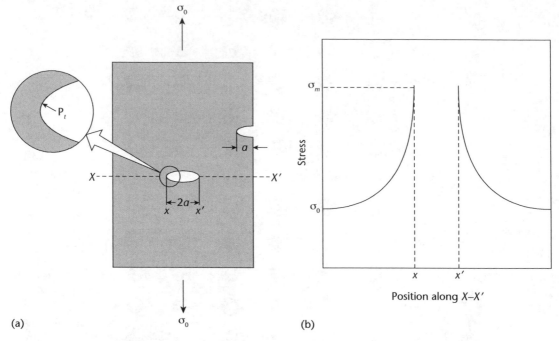

(a)

(b)

Figure 3.17 (a) Geometry of surface and internal cracks; (b) schematic stress profile along the line $X - X^1$ in (a)

temperature of about 590 °C, a temperature at which steel is red hot.

So at room temperature lead is in the same state as red-hot steel. This means that lead will not work harden at room temperature, and this is one reason why lead is such a good roofing material. It can be bent and rolled and tucked around roof details without suffering the low cycle fatigue and cracking that would occur if thin steel strip were used.

3.5.3 Fatigue

Fatigue is another mode of failure that can occur at stresses below the yield strength of a material. Fatigue occurs where we have repeated cycling of stress, i.e. where a material is repeatedly loaded and unloaded. Such situations occur in civil engineering structures such as bridges. The passage of traffic over the bridge causes the structure to be loaded and the component materials stressed, and after the passage of the vehicle the loading is removed and the stress is removed. When the next vehicle passes over the bridge the stress cycle is repeated. Fatigue failure is often characterised by sudden catastrophic failure. It is beyond the scope of this text to give a full treatment of the mechanisms involved in fatigue, but suffice to say that dislocation movements are involved together with crack growth processes; the process will be dealt with in these terms in Chapter 19.

3.6 Electrical and thermal properties of metals

We have looked in some detail at mechanical or strength properties because these are very important in building structures. However, the electrical and thermal properties of metals are also important, although there is no need to treat these in anything like the detail devoted to strength properties.

3.6.1 Electrical properties

Each atom in a crystal lattice will carry a positive charge, because it has donated electrons (negatively charged) to the inter–atomic bonding process. However, because all the atoms (positive charges) are neatly aligned in a very orderly way in three dimensions, the lattice will present a space characterised by very regular periodicity of electric charges. We have also seen that the metallic bond is characterised by the presence of a 'sea' of electrons permeating the metal lattice. The electrical periodicity will make the flow of electric current (flow of electrons) very easy. The electrons will flow with the application of only a small voltage (potential difference). Thus as a group, because metals are crystalline, they are all very good electrical conductors. Some metals are better than others; for example, pure copper and pure aluminium are excellent conductors.

Note that purity is important. If impurities are introduced into metals, the impurity atoms will interrupt the regular lattice periodicity, and increase the resistance to electron (current) flow. For the UK National Grid for the distribution of electricity, the overhead high voltage power lines use a composite; a pure aluminium conductor coiled round steel rope to bear the weight.

3.6.2 Thermal properties

As a group of materials, metals are generally excellent conductors of heat, and better than all the other construction materials. Again, the reasons for this are to be found in the regularity of their crystal structures. At absolute zero, the atoms in a metal crystal will be at rest, and we shall have a perfect crystal (as shown in Figure 3.8). As the temperature is increased, the atoms will begin to vibrate gently about their mean positions on the lattice. These vibrations will initially be of small amplitude. As the temperature is increased further, the amplitude of the vibrations will increase. As the amplitude of vibration increases, the atoms will begin to impinge on their neighbours, causing them to vibrate. Because all the atoms are neatly arranged in rows, the passing on of thermal vibrations is very easy and quick. If heat is applied to one side of a piece of metal, the heat is transmitted or conducted easily and quickly by lattice vibrations. If the amount of heat added reaches the upper limit, lattice vibration eventually becomes so violent that the atom acquires enough energy to break away from the lattice altogether, and this, of course, is the melting point.

This has been a purely qualitative treatment and a more quantitative discussion of the thermal properties of metals will be outlined in Chapter 6.

3.7 Critical thinking and concept review

1. How many nearest neighbour atoms are there for each atom in: (a) a BCC structure, (b) an FCC structure?
2. Why do metal crystals not cleave and split when slip occurs?
3. Why do metals become more ductile when they are heated?
4. Why is the metal lead so ductile at room temperature?
5. Explain why in a component under tensile stress, a crack normal to the stress direction puts the component at risk of failure whereas a crack in line with the stress direction does not.
6. Explain why the stress at the tip of a crack in a stressed material can be an order of magnitude higher than the nominal stress in the material.
7. The distance between the iron atoms in a piece of α-iron is 0.24824 nm, and is always the same, regardless of where on Earth the iron was made. Explain why this is so.
8. Explain in a simple qualitative way why metals as a group are very good conductors of: (a) heat and (b) electricity.

9. The inter-atomic distance in a crystal of pure copper is 0.2556 nm. If 1,000,000 edge dislocations move along the slip plane, by how much will the upper plane move with respect to the lower plane?
10. Explain why plastic deformation in metals leads to the formation of micro-cracks, and ultimately to fracture.
11. What will be the effect of fine grain size on the hardness and strength of metals?
12. Briefly outline the sequence of events inside a tension test-piece during the final stages of a test to failure.
13. Explain why lead suffers creep at room temperature while steel does not.
14. Why do we need to take account of creep effects in concrete?
15. Why will the presence of impurities in a metal reduce its electrical conductivity?

3.8 References and further reading

ANDERSON, J.C., LEAVER, K.D., ALEXANDER, J.M. & RAWLINGS, R.D. (1974), *Materials Science*, 2nd edition, Nelson, Sunbury on Thames.

ASHBY, M.F. & JONES, D.R.H. (1980), *Engineering Materials: An Introduction to their Properties and Applications*, Pergamon Press, Oxford.

CALLISTER, W.D. (1994), *Materials Science and Engineering: An Introduction*, John Wiley & Sons, New York.

COTTRELL, A.H. (1958), *Brittle Fracture in Steel and Other Materials*, trans. American Institute of Mechanical Engineers.

COTTRELL, A.H. (1964), *The Mechanical Properties of Matter*, John Wiley & Sons, New York and London.

DIETER, G.E. (1961), *Mechanical Metallurgy*, McGraw-Hill, New York.

GORDON, J.E. (1971), *The New Science of Strong Materials: Or Why You Don't Fall Through the Floor*, Penguin Books, Harmondsworth.

HULL, D. (1968), *Introduction to Dislocations*, Pergamon Press, Oxford.

SMALLMAN, R.E. & BISHOP, R.J. (1995), *Metals and Materials: Science, Processes Applications*, Butterworth-Heinemann, Oxford.

TABOR, D. (1979), *Gases, Liquids and Solids*, Cambridge University Press, Cambridge.

TYLECOTE, R.F. (1992), *A History of Metallurgy*, The Institute of Metals, London.

4

Mechanical properties of materials

Contents

4.1 Introduction

Materials are used because they have properties appropriate for the purpose to which they are put. For example, if we wish to make something water-proof, it is no use employing a material which is porous (i.e. which contains holes or voids). If we want to thermally insulate a building, on the other hand, the best material we can use would be a porous material, and not a solid one. The reason for this is that the holes or pores will contain pockets of air, which is a poor conductor of heat. Put another way, air is a good heat insulator. So some of our materials are used because they are good thermal insulators. However, we need other materials because of their strength. Buildings are structures, and some of them carry very heavy loads. The Petronas Towers in Kuala Lumpur, Malaysia, for example, weighs over 300,000 tonnes. The materials at the bottom of the building carry this load and therefore they must have high strength. Strong materials are usually not porous because holes and fresh air are not very strong!

Materials in construction, indeed in any application, are subjected to forces or loads, e.g. the aforementioned Petronas Towers. Why was concrete selected? We select a material by matching its mechanical properties to the service conditions required of the component. The initial process requires the engineer/materials scientist to determine the most important characteristics that the material must possess. Should it be strong, stiff or ductile? Or a combination of all three properties? Will it be subjected to repeated application of high forces, a sudden force, high stresses at high temperatures or abrasive conditions? Once we have determined this, we can select the appropriate material with the requisite properties. However, we must know how the properties have been determined and, more importantly, what the properties actually mean.

In this chapter we take an in-depth look into what exactly the properties mean and how they are determined. We use a very wide range of materials in building construction, including steel and other metals, bricks, concrete, timber, glass, plaster, polymers, bitumens, etc. In each case, these materials are used because of particular physical and mechanical properties that they possess, such as strength, hardness, stiffness, elasticity, durability, corrosion resistance, and so on.

4.2 Stress and strain

Strength is probably the most important mechanical property of a material. Strength is measured using stress. If a load is static or changes slowly with time and is applied uniformly over a cross-section or surface of a member, the mechanical behaviour may be determined by a simple stress–strain test; these are most commonly conducted for metals at ambient (room temperature). The two most common ways in which a load may be applied are in tension or compression – the difference is illustrated in Figures 4.1 and 4.2. When an object is subjected to a tensile force it is in effect being pulled apart like a rope in a tug-of-war contest. In contrast, when an object is subjected to a compressive force it is being pushed together or squeezed, as depicted in Figure 4.2.

4.2.1 The tensile test for determining the tensile strength

This is one of the most common mechanical stress–strain tests performed in tension. The test measures the resistance of a material to a static or slowly applied force. The specimen is placed in a testing machine and a force F, called the load, is applied as illustrated in Figure 4.1. The amount the specimen stretches is measured. The tensile testing machine is designed to elongate the specimen at a constant rate, and to continuously and simultaneously measure the instantaneous applied load and the resulting elongations. A stress–strain test typically takes 5–50 minutes, depending on the loading rate, and is destructive; that is, the test specimen is permanently deformed and usually fractured.

The output of a tensile test is recorded as load (or force) versus elongation. These load deformation characteristics are dependent on the specimen size, i.e. if the cross-sectional area is doubled it will take twice the load to produce the same elongation. To minimise the geometrical factors, load and elongation are normalised to the respective parameters of engineering stress and engineering strain.

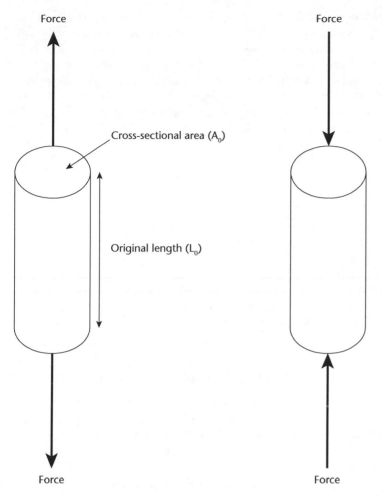

Force

Cross-sectional area (A_0)

Original length (L_0)

Force

Force

Force

Figure 4.1 Schematic illustration of an object subjected to a tensile force or load

Figure 4.2 Schematic illustration of an object subjected to a compressive force or load

(Engineering) stress (σ) is a measure of the average amount of force exerted per unit area. It is a measure of the intensity of the total internal forces acting within a body across imaginary internal surfaces, as a reaction to external applied forces and body forces. Stress is a concept that is based on the concept of continuum. In general, stress is expressed as:

$$\sigma = \frac{\text{Load or Force (F)}}{\text{Area (A)}}$$

where

σ is the average stress, also called engineering or nominal stress
F   is the force acting over the area A
A   is the cross–sectional area.

(Engineering) strain is the unit of change in the size or shape of a body due to force; a dimensionless number that characterises the change in dimensions of an object during a deformation or flow process. Strain is the

geometrical expression of deformation caused by the action of stress on a physical body. Strain is calculated by first assuming a change between two body states: the beginning state and the final state. Then the difference in placement of two points in this body in those two states expresses the numerical value of strain. Strain therefore expresses itself as a change in size and/or shape. Given that strain results in the deformation of a body, it can be measured by calculating the change in length of a line or by the change in angle between two lines (where these lines are theoretical constructs within the deformed body). The change in length of a line is termed the stretch, absolute strain or extension, and may be written as ΔL.

Then the (relative) strain, ε, (or engineering strain) is given by:

$$\varepsilon = \frac{\Delta L}{L}$$

where

ε is strain in measured direction
ΔL is the original length of the material
L is the current length of the material.

The extension is positive if the material has gained length (in tension) and negative if it has reduced length (in compression). Because L is always positive, the sign of the strain is always the same as the sign of the extension. Strain is a dimensionless quantity. It has no units of measurement because in the formula the units of length cancel out.

4.2.2 The units for measuring force (load), stresses and strengths

The SI unit for stress is the pascal (Pa), which is a shorthand name for one newton (force) per square metre (unit area). The unit for stress is the same as that of pressure, which is also a measure of force per unit area. Engineering quantities are usually measured in megapascals (MPa), newtons per squared millimetres (N mm^{-2}), or gigapascals (GPa). In imperial units, stress is expressed in pounds-force per square inch (psi) or kilopounds-force per square inch (ksi).

Most materials science texts will measure stresses or strengths in MPa; most engineering texts, on the other hand, usually use N mm^{-2}. North American texts tend to use psi. In this text we will use N mm^{-2}, note 1 N mm^{-2} = 1 MPa

Strain is often expressed in dimensions of millimetres/millimetre or inches/inch anyway, as a reminder that the number represents a change of length. But the units of length are redundant in such expressions, because they cancel out. When the units of length are left off, strain is seen to be a pure number, which can be expressed as a decimal fraction, a percentage or in parts-per notation. In common solid materials, the change in length is generally a very small fraction of the length, so strain tends to be a very small number.

Stress–strain data are very useful as they can then be applied to any size of component or structure. The stress–strain graph can tell us a great deal about the properties and behaviour of materials and their response to loading. Lets commence with Young's modulus.

4.3 Young's modulus/modulus of elasticity

Figure 4.3 shows a stress–strain plot for a typical brittle material; we can see a straight (linear) line. The slope or gradient of this line indicates how easy or how difficult it is to produce elastic distortion. In other words, how flexible or stiff the material is, i.e. how easily it can bend. This property is given by the Young's modulus or modulus of elasticity, E, thus

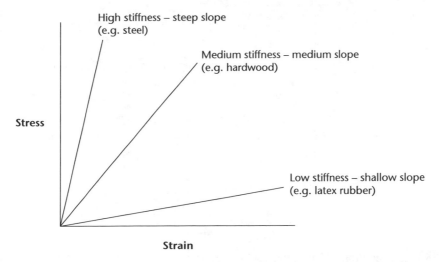

Figure 4.3 A stress–strain graph of a brittle material; the gradient of the slope determines the Young's modulus of the material

$$E = \frac{\sigma}{\varepsilon}$$

Units of E are usually given as GPa. Please note 1 GPa = 1,000 N mm^2

Therefore, a shallow slope indicates a low stiffness and a steep line indicates a high stiffness; the value of stiffness is given by Young's modulus of elasticity (E). All materials are elastic. We know that rubber is elastic – it is easy to stretch a rubber band. It is less easy to flex a wooden ruler, and even more difficult to compress a steel valve spring. However, they are all elastic (rubber, wood and steel). By elastic we mean that when they are put under load they deform or change shape, when unloaded they spring back immediately to their original shape. How easy they are to deform elastically is indicated by their stiffness. Stiffness is a measure of how easy it is to elastically deform a material. So it describes the nature of a material's elasticity. Stiffness is the value of the elastic modulus (Young's modulus). For example, rubber has a low modulus, it is easy to deform. Timber has a higher modulus and steel has a much higher modulus, being the hardest of the three to deform elastically, as illustrated in Figure 4.4.

Apart from Young's modulus, a stress–strain plot tells us the strength of a material, which is arguably the most important property. The strength of a material is a measure of how much resistance it can offer to the processes of deformation (shape change when put under load) and fracture (ultimate rupture into two or more pieces). We can identify a number of strength properties, including:

- yield strength
- tensile strength
- compressive strength.

For any type of strength:

$$\text{Strength or Stress (N/mm}^2) = \frac{\text{Load or Force (N)}}{\text{Cross-sectional area (mm}^2)}$$

Note: units of strength or stress are MPa or N mm^{-2}.

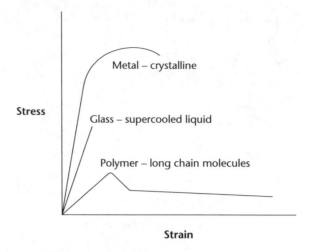

Figure 4.4   Stress–strain graphs for materials of various stiffness

4.4 Yield strength

All materials behave elastically up to a point. Let's take a rubber band for example – if we pull it with a small force and let go, it will return to its original length/size. However, if we continue to increase the load there will come a point when it will permanently increase in length. At this point the rubber band has permanently deformed – it has reached the yield point. When a material has permanently deformed it is in the 'plastic region'.

Yield strength is the stress required to produce a very slight yet specified amount of plastic strain; this strength is normally associated with the onset of plastic deformation. Once the yield strength is exceeded the material is permanently deformed; this is an irreversible process. Illustrated in Figure 4.5 is a typical stress–strain plot for a simple ductile metal/material. If the metal is subject to a uniaxial tensile load (as in Figure 4.1), the initial response to loading will be elastic. The stress–strain graph is linear in this region – up to point 3 on the graph; the slope of this line indicates how easy or how difficult it is to produce elastic distortion. Eventually, the material reaches its elastic limit or yield strength (point 3 on the graph) and permanent (non-reversible) elongation occurs. We are now in the region of plastic deformation and the stress–strain line is no longer linear. In the case of ductile metals the graph will continue to rise until a maximum is reached (point 1) – this is the maximum stress or UTS (ultimate tensile strength) or tensile strength as it is commonly known. The stress begins to fall and the negative slope is observed until final fracture occurs (point 4). Brittle materials will fail between points 2 and 3, as depicted in Figure 4.3.

In some cases the yield point may not be easily discernible from the stress–strain, thus, as a consequence, a convention has been established wherein a straight line is constructed parallel to the elastic portion of the stress–strain curve at some specified strain offset, usually 0.002, as illustrated in Figure 4.6. The stress corresponding to the intersection of this line and the stress–strain curve as it bends over in the plastic region (where line A–B intersects the stress–strain curve) will determine the yield strength. This is usually the common procedure for determining the yield strength.

Thus the yield point is simply the stress that must be exceeded to cause plastic (permanent) deformation. In general, ductile materials show a yield point, whereas brittle materials do not. Steel is unusual in showing a very marked yield point. So yield strength will be calculated as:

$$\text{Yield Strength (N/mm}^2) = \frac{\text{Load to cause Yielding (N)}}{\text{Cross-sectional area (mm}^2)}$$

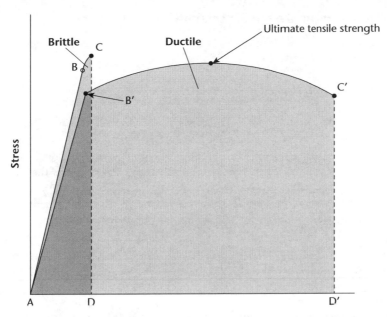

Figure 4.5 Typical stress–strain behaviour for brittle (ABCD) and ductile (AB'C'D') materials loaded (tensile) to fracture. In both cases B and B' represent the yield points

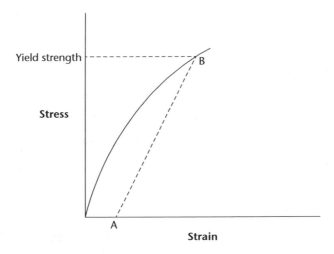

Figure 4.6 Typical stress–strain behaviour for a metal showing elastic and plastic deformations. As the precise yield point is not easily discernible a straight line is drawn (AB); the point of intersection represents the yield point. This is known as the 0.002 strain offset method

The yield strength is often referred to as the 'danger level' for a material, as once this is reached it is only a matter of time before the material will fail (break). As a result engineers ensure that for any material in any application the maximum stress the material is subjected to is no more than 50% of the yield strength, i.e. 0.5 × yield strength. In high-risk applications such as aeroplanes or cars, normally it is about 0.2–0.3 × yield strength.

4.5 Tensile strength (ultimate tensile strength (UTS))

If a material is put into tension, this is the stress that must be exceeded to cause tensile fracture. Tensile strength is calculated as:

$$\text{Tensile Strength (UTS) (N/mm}^2) = \frac{\text{Tensile Load to cause Fracture (N)}}{\text{Cross-sectional area (mm}^2)}$$

The UTS is simply the maximum stress any material can endure in tension. Thus, as illustrated in Figures 4.5 and 4.6, the UTS is the maximum value from a stress–strain plot for a material when subjected to a tensile load. The tensile strength can thus be defined as the maximum engineering stress, in tension, that may be sustained without fracture. If a material is put into tension, this is the stress that must be exceeded to cause tensile fracture.

Figures 4.5–4.7 are stress–strain plots for a ductile material; up to this point all deformation is uniform. However, at this maximum stress a small neck (constriction) begins to form and all subsequent deformation is confined to this neck. This phenomenon is known as 'necking', and is defined as the reduction of the cross-sectional area of a material in a localised area caused by uniaxial tension. During necking, the material can no longer withstand the maximum stress and the strain in the specimen rapidly increases. Fracture ultimately occurs at the neck, as illustrated in Figure 4.8. The fracture strength corresponds to the stress at fracture (σ_f in Figure 4.7).

Tensile strengths can vary from 50 N mm^{-2} for aluminium to 3000 N mm^{-2} for certain steels. Usually, when the strength of a material is cited for design purposes, the yield strength is used. This is because by the time the UTS has been reached, often the material or structure has experienced so much plastic deformation that it is rendered useless. Furthermore, fracture strengths are not normally specified for engineering design purposes.

4.6 Brittleness and ductility

Before we define brittle and ductile fracture it is imperative to grasp the concept of deformation.

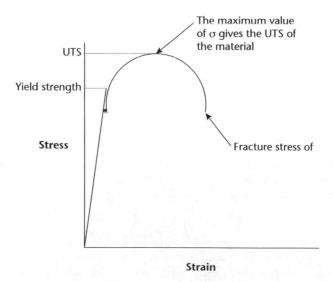

Figure 4.7 A stress–strain graph for a ductile material showing the ultimate tensile strength (UTS) and fracture stress

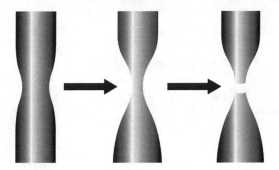

Figure 4.8   Necking in a ductile material under a tensile load ultimately leading to fracture

4.6.1 Elastic and plastic deformation

Elastic deformation is non-permanent deformation, that is, totally recovered upon release of an applied load. This type of deformation occurs below the yield strength. Plastic deformation is permanent or non-recoverable deformation after release of the applied load. To effect plastic deformation the material must be subjected to a force higher than the yield stress. However, an object in the plastic deformation range will first have undergone elastic deformation, which is reversible, so the object will return part way to its original shape. Plastic deformation ends with the fracture of the material.

Therefore, brittle materials experience little or no plastic deformation prior to fracture; brittleness is the tendency of a material to fracture with little or no plastic deformation. Brittle materials have little or insignificant ductility.

Ductile metals experience observable or gross plastic deformation prior to fracture; ductility is a measure of a material's ability to undergo appreciable plastic deformation before fracture or how easy it is to permanently deform a material. Lead is an example of a ductile metal. Grey cast iron is not ductile – it will break before it bends. Lead is also a soft metal, grey cast iron is a hard metal. So we frequently find that materials that are hard and strong have low ductility, whereas materials which are weak and soft are very ductile.

Note: ductility is *not* the same thing as elasticity. Elasticity is non-permanent shape change, whereas ductility involves exceeding the elastic limit and causing permanent shape change. Ductility is usually measured as a strain to failure, and strain is defined thus:

$$\text{Strain} = \frac{\text{Change in dimension (mm)}}{\text{Original dimension (mm)}} \times 100\%$$

4.6.2 Comparison between different types of materials

Beside metals, ceramics and polymers – both natural and man-made – are used in building. In terms of stress–strain diagrams, Figure 4.9 shows a family of curves for metals, ceramics and polymers drawn on the same axes for comparison. It can immediately be seen that polymers are much weaker and have very much lower stiffness values than metals. They are ductile, and they do possess tensile strength but as a class they are weak and soft. Ceramics, on the other hand, can posses great stiffness and good strength in compression, but they are not ductile and fail at low strains. So in terms of the work done to failure, metals are capable of absorbing more energy than any other class of material.

4.7 Compressive strength

Another type of strength is compressive strength. When a material is loaded in compression, as illustrated in Figure 4.2, this is the stress level that must be exceeded to cause compressive failure (inability to

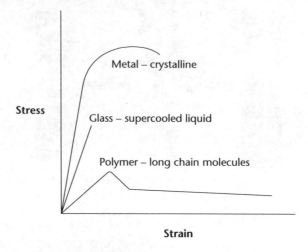

Figure 4.9 Stress–strain graphs for steel and various non-metallic materials

continue carrying the compressive load). Concrete cubes are tested to failure in compression, and strength values for concrete are compressive strength values.

$$\text{Compressive Strength}\,(\text{N}/\text{mm}^2) = \frac{\text{Compressive load to cause failure (N)}}{\text{Cross-sectional area (mm}^2)}$$

4.8 Shear strength

For tensile tests, a material is subjected to a uniaxial tensile force; however, if the material undergoes a non-uniaxial tensile load it experiences a shear stress as illustrated in Figure 4.10. Like other strength properties the unit of shear strength is N mm^{-2} or MPa.

4.9 Poisson's ratio

The Poisson's ratio (v) is taken to be the ratio of the strains a material experiences when subjected to a tensile force; the strains are in the x, y and z directions as illustrated in Figure 4.11.

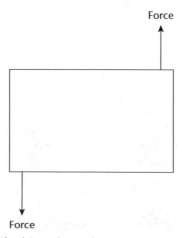

Figure 4.10 Illustration of a material subjected to a shear force

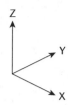

Figure 4.11   The directions in which a material experiences strain movement as a result of a uniaxial tensile force

It is unlikely, with the exception of some civil engineers, that students of a construction/architecture discipline will encounter Poisson's ratio; however, it is useful to bear in mind most metals have a Poisson's ratio between 0.25 and 0.35.

4.10 Worked examples

4.10.1 Strength calculation

A concrete cube is tested for strength in a laboratory test. The cube dimensions are 100 mm × 100 mm × 100 mm, and the cube fails under a compressive load of 600 kN. What is the compressive strength of the concrete?

Solution

We use the simple formula:

$$\text{Stress or strength} = \frac{\text{Load (N)}}{\text{Cross-sectional area (mm}^2)}$$

To calculate the strength we need to know two things:

1.   the load
2.   the cross-sectional area

We are told that the failure load is 600 kN. This is 600,000 N (kilo means 1,000). So the load is 600,000 newtons.

The cube is a 100 mm cube. Therefore the cross-sectional area will be 100 mm × 100 mm, i.e. 10,000 mm².

So the strength will be:

$$\text{Strength} = \frac{600,000 \text{ (N)}}{10,000 \text{ mm}^2}$$

So the strength of this concrete sample is 60 N mm⁻² .

4.10.2 Calculation of load-carrying ability: component or element

Consider a steel cable on a suspension bridge. Each cable is made of steel with a yield strength of 400 N mm⁻², and each cable carries a load of 4 kN. If each cable has a cross-sectional area of 100 mm², are these cables strong enough to carry the load?

Solution

We use the simple formula:

$$\text{Stress or strength} = \frac{\text{Load (N)}}{\text{Cross-sectional area (mm}^2)}$$

The load on each cable is 4 kN, i.e. 4,000 N. The cross-sectional area of each cable is 100 mm², and the yield strength of the cable steel is 400 N mm⁻². So the stress in each cable will be:

$$= \frac{4,000 \text{ N}}{100 \text{ mm}^2} = 40 \text{ N/mm}^2$$

The stress is well below the yield stress for the steel of 400 N mm⁻². These cables would easily be able to carry the load on them.

4.10.3 Calculation of load-carrying ability: building design

We are putting up a very tall building which weighs 200,000 tonnes. This weight is to be carried on 500 steel I-beams. We are using normal structural steel, with a yield strength of 300 N mm⁻². Each beam has a cross-sectional area of 10,000 mm². Ignoring the possibility of other failure mechanisms, will these 500 steel columns carry the building's weight?

Solution

We use the simple formula:

$$\text{Stress or strength} = \frac{\text{Load (N)}}{\text{Cross-sectional area (mm}^2)}$$

What is the load? The building weighs 200,000 tonnes (1 tonne = 1,000 kg). We need to convert tonnes to newtons: 1 tonne = 10,000 N.

Therefore, the load will be:

$$200,000 \text{ tonnes} \times 10,000 \text{ N tonne}^{-1} = 2.0 \times 10^5 \text{ tonnes} \times 10^4 \text{ N tonne}^{-1}$$
$$= 2.0 \times 10^9 \text{ N}$$

We also need to know the total cross-sectional area of all the steel columns carrying the weight. We have 500 columns, each of cross-sectional area 10,000 mm². Therefore, total cross-sectional area:

$$= 500 \times 10,000 \text{ mm}^2$$
$$= 5,000,000 \text{ mm}^2$$
$$= 5 \times 10^6 \text{ mm}$$

So the stress in the steel column can be found using our simple formula:

$$\text{Stress} = \frac{\text{Load (N)}}{\text{cross-sectional area (mm}^2)}$$
$$\text{Stress} = \frac{2 \times 10^9 \text{ (N)}}{5 \times 10^6 \text{ (mm}^2)}$$
$$= 0.4 \times 10^3 \text{ N/mm}^2 \text{ or}$$
$$= 400 \text{ N/mm}^2$$

Now we know that the yield strength of our structural steel is only 300 N mm^{-2}, and yet the actual stress in the columns will be 400 N mm^{-2}. This means either:

1.   that the steel we are using is not strong enough, or
2.   we do not have enough columns to carry the weight.

4.10.4 Elastic deformation of metals

A piece of copper originally 300 mm long is pulled in tension with a stress 276 N mm^{-2}. If the deformation is entirely elastic, what will be the resultant elongation? The Young's modulus of copper is 110 kN mm^{-2} (or 110 GPa).

Solution

Firstly lets visualise the problem:

We have a piece of copper which is 300mm long:

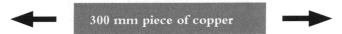

300 mm piece of copper

Now imagine the copper piece is pulled from both directions:

← **300 mm piece of copper** →

As a result the copper piece has expanded:

$$\Delta L$$
←→

300 mm piece of copper

This expansion is also known as change in length or ΔL
   In this problem we need to work out ΔL
   Since the deformation is elastic, strain is dependent on stress according to the equation:

$$E = \frac{\sigma}{\varepsilon}$$ Equation [1]

where E = Young's Modulus

$$\sigma = \text{stress} = \frac{\text{force}}{\text{area}}$$ Equation [2]

$$\varepsilon = \text{strain} = \frac{\text{change in length } (\Delta L)}{\text{original length } (L_0)}$$ Equation [3]

therefore, if we rearrange Equation [1] above:
   we get

$$\varepsilon = \frac{\sigma}{E}$$

Now from Equation [3] we know

$$\varepsilon = \frac{\Delta L}{L_0}$$

If we now combine the two equations above we have:

$$\frac{\Delta L}{L_0} = \frac{\sigma}{E}$$

We know that in this problem we have to work out Δl, therefore we must rearrange the equation above to get Δl on its own:

$$\Delta l = \frac{\sigma L_0}{E}$$

The values of σ and L_0 are given as 276 N mm^{-2} and 300 mm, respectively, and the magnitude of E for copper is 110 kN mm^{-2}. Elongation is obtained by substitution into the expression above as:

$$\Delta l = \frac{(276 \text{ N/mm}^2)(300\text{mm})}{110 \times 10^3 \text{ N/mm}^2} = 0.75 \text{ mm}$$

4.11 Hardness

Hardness is linked to strength – high-strength materials are nearly always very hard, and weak materials are usually relatively soft. For example, diamond is very hard but brittle, and the metal lead is very soft and very ductile. Hardness is the measure of a material's resistance to deformation by surface indentation, e.g. scratches. The hardness of a material is generally directly proportional to strength and brittleness, i.e. the harder the material the lower the ductility. There are different types of hardness tests to determine the hardness of the material.

4.11.1 Vickers hardness test

Sometimes known as the diamond pyramid hardness test, this is a popular method of determining the hardness value of materials, especially metals. For each test a very small diamond indenter having pyramidal geometry is forced into the surface of the specimen. Applied loads range from 1 g to 1,000g. The resulting impression is observed under a microscope and measured; the measurement is then converted into a hardness number.

4.11.2 Brinell hardness test

This is a test for determining the hardness of a material by forcing a hard steel or carbide ball of specified diameter into it under a specified load; the result is expressed as the Brinell hardness number.

For both Brinell and Vickers tests, as a general rule of thumb for most steels and some other metals:

tensile strength = ~3 × hardness value

4.11.3 Knoop hardness test

This is an indentation hardness test using calibrated machines to force a rhombic-based pyramidal diamond indenter having specified edge angles, under specified conditions, into the surface of the test material and to measure the long diagonal after load removal.

4.11.4 Rockwell hardness test

This is a test method consisting of indenting the test material with a diamond cone or hardened steel ball indenter. The indenter is forced into the test material under a preliminary minor load, usually 10 kgf. When equilibrium has been reached, an indicating device, which follows the movements of the indenter and so responds to changes in depth of penetration of the indenter, is set to a datum position. While the preliminary minor load is still applied, an additional major load is applied with resulting increase in penetration. When equilibrium has again been reached, the additional major load is removed but the preliminary minor load is still maintained. Removal of the additional major load allows a partial recovery, so reducing the depth of penetration. The permanent increase in depth of penetration, resulting from the application and removal of the additional major load, is used to calculate the Rockwell hardness number.

4.12 Durability

As buildings last, on average, much longer than most manufactured goods, durability is very important. Durability is the way we describe the property of lasting a long time. Durability is the ability of a material to resist wear, and loss of material through continual use. A material that is durable will last a long time. Low durability means that the material will not survive for long. Flint and granite are both used in building, and both of these will last for centuries, resisting the effects of weather, rain and frost very well. Plaster, on the other hand, if exposed to the weather, will not last very long at all. So we say that flint and granite are durable, and plaster is not.

4.13 Toughness

As discussed in earlier sections, the stress–strain diagram can tell us a lot more about the material. The area under the stress–strain diagram is a measure of how much energy or work has to be expended to strain the material to fracture. So anything which reduces the strain to failure in a material will reduce the energy required to break it. In other words, it will reduce the toughness of the material. We can compare the relative toughness levels for the various classes of materials in Figure 4.12, and we can see why metals are used for so many load-bearing situations in engineering and construction.

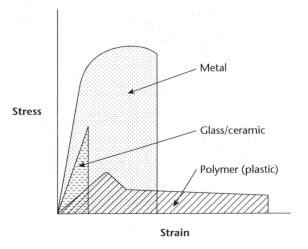

Figure 4.12 A stress–strain plot for typical metals, ceramics and plastics, illustrating the relative toughness of each material

Therefore, toughness is equal to the area under the tensile stress–strain curve. Toughness is the ability of a material to deform plastically and to absorb energy in the process before fracture. Ductility is a measure of how much a material deforms plastically before fracture – a ductile material is not necessarily tough. The key to toughness is a good combination of strength and ductility. A material with high strength and high ductility will have more toughness than a material with low strength and high ductility. Therefore, one way to measure toughness is by calculating the area under the stress–strain curve from a tensile test. This value is simply called material toughness and it has units of energy per volume. Material toughness equates to a slow absorption of energy by the material.

4.14 Critical thinking and concept review

1. How do we determine the mechanical properties of materials?
2. What are stress and strain? What are the units?
3. What does Young's modulus tell us?
4. Why is the yield strength very important for an engineer?
5. What is the difference between brittle and ductile materials? Which one is more desirable?
6. What is the UTS?
7. What is necking?
8. What is the difference between elastic and plastic deformation?
9. What is hardness?
10. What is toughness and how can this be deduced?

Microstructure and phase transformations in alloys

Contents

5.1 Introduction

The understanding of phase diagrams in metallurgy is hugely important as the microstructure of a metal directly determines its mechanical properties, i.e. tensile strength, ductility, etc., as discussed in Chapter 3. Additionally, phase diagrams also give us valuable information about the melting points of each alloying constituent and other important chemical properties.

While this area is of paramount importance to students of materials science, metallurgy and other physical sciences disciplines, those studying civil engineering and architecture-related disciplines will not need to know much about this area; therefore, accordingly this chapter will touch upon the main fundamentals of microstructure and phase transformations.

5.2 Basic concepts

5.2.1 Solubility limit

Most metals utilised in any industry, particularly construction, are alloys (a metallic material composed of more than two constituents, e.g. steel which is made from iron, carbon and other alloying elements); thus, a materials scientist needs to understand how the parent metal and its alloying element behave chemically.

In chemistry terms we need to know how much solute may dissolve in the solvent to form a solid solution, e.g. in steel the iron is the solvent and carbon is the solute, the two combined form the steel solid solution. Whenever we add a solute to a solvent there is a maximum concentration – for example, if we consider a glass of water as the solvent and sugar as the solute there will be a limit to the amount of sugar the glass of water will dissolve, this is known as the solubility limit. If we assume a glass of water dissolves five teaspoons of sugar but not six, then the solubility limit of sugar in a glass of water is five teaspoons. The important thing to remember is temperature, the aforementioned solubility limit is at room temperature; however, if we increase the temperature we will find that the solubility limit will increase, i.e. at 90 °C the solubility limit of sugar in a glass of hot water will increase to about nine teaspoons.

5.2.2 Phase

Another fundamental term is phase, which is a homogeneous part of a system that has the same chemical and physical properties. All metal alloy systems have at least one solid phase at any given chemical composition; as we shall see later, steel has many phases at different temperatures and chemical compositions which determines the mechanical properties of the material.

5.2.3 Microstructure

The properties of an alloy do not depend only on concentration of the phases, but how they are arranged structurally at the microscopic level. Thus, the microstructure is specified by the number of phases, their proportions and their arrangement in space.

A binary alloy may be:

- a single solid solution;
- two separated, essentially pure components;
- two separated solid solutions;
- a chemical compound, together with a solid solution.

The way to tell is to cut the material, polish it to a mirror finish, etch it with a weak acid (components etch at a different rate) and observe the surface under a microscope.

5.2.4 Phase equilibria

Equilibrium is the state of minimum energy. It is achieved given sufficient time. But the time to achieve equilibrium may be so long (the kinetics are so slow) that a state that is not at an energy minimum may have a long life and appear to be stable. This is called a metastable state.

A less strict, operational, definition of equilibrium is that of a system that does not change with time during observation.

Equilibrium phase diagrams

Equilibrium phase diagrams give the relationship of composition of a solution as a function of temperatures and the quantities of phases in equilibrium. These diagrams do not indicate the dynamics when one phase transforms into another. Sometimes diagrams are given with pressure as one of the variables. In the phase diagrams we will discuss, pressure is assumed to be constant at one atmosphere.

5.2.5 Binary isomorphous systems

This very simple case is one complete liquid and solid solubility, an isomorphous system. The example is the Cu–Ni (Cu is the symbol for copper and Ni is the symbol for nickel) alloy of Figure 5.1. Temperature is plotted on the y-axis and the alloy composition is plotted on the x-axis. The composition ranges from 0 wt% Ni (100 wt% Cu) on the left side of the x-axis to 100 wt% Ni (0 wt% Cu) on the right. There are three different phases present on the phase diagram, an alpha (α) field, a liquid (L) field and a two–phase $\alpha + L$ field. Each individual region is defined by the phase/phases that exist over the temperature range defined by the phase boundary lines.

The liquid L is a homogeneous liquid solution composed of both copper and nickel. The α phase is a solid solution consisting of both Cu and Ni atoms and has an FCC crystal structure. Below approximately 1080 °C the copper and nickel are completely soluble in each other; the complete solubility occurs because both Cu and Ni have the same crystal structure (FCC), near the same radii, electronegativity and valence. Due to this complete solubility between each other the copper–nickel system is called isomorphous.

With most binary phase diagrams for metallic alloys the solid solutions are usually designated by Greek letters e.g. α, β, γ, etc. The liquidus line separates the liquid phase from solid or solid + liquid phases. That is, the solution is liquid above the liquidus line. The solidus line is that below which the solution is completely solid (does not contain a liquid phase). In Figure 5.2, the solidus and liquidus lines intersect at

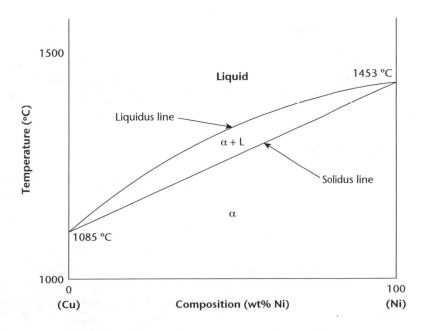

Figure 5.1 The copper–nickel phase diagram

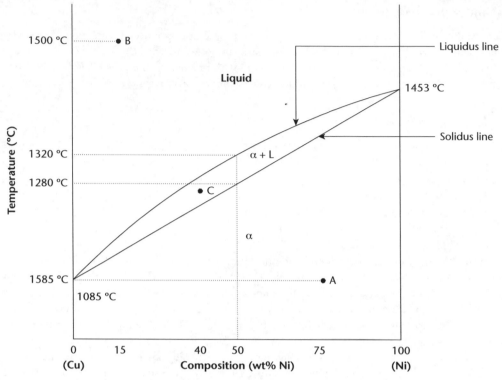

Figure 5.2 The copper–nickel phase diagram showing the melting point for a 50 wt% Cu–50 wt% Ni composition

both ends of the *x*-axis – this corresponds to the melting points for both metals, i.e. pure copper melts at 1085 °C and pure nickel has a melting point of 1453 °C. However, if we take a composition which is a mixture of copper and nickel, melting would take place between the solidus and liquidus line. For example, if we have an alloy containing 50 wt% copper and 50 wt% nickel (Figure 5.2), then melting starts at about 1280 °C, when it hits the solidus line; the melting process continues with increasing temperature until the vertical line touches the liquidus line (about 1320 °C); at this point the alloy is completely liquid.

5.3 Analysing phase diagrams

Binary phase diagrams are extremely useful to metallurgists. In general, three important types of information can be extrapolated from phase diagrams: the phases present, the concentration/composition of each phase and fractions of the phases.

5.3.1 Phases

This is relatively simple to deduce; all that is required is to locate the temperature-composition point on the phase diagram and note which phase(s) is/are present. For example, in Figure 5.2 an alloy composition of 75 wt% Ni–25 wt% Cu will be located at point A; as this point is within the α region therefore only the α phase would be present. If we take another alloy composition of 15 wt% Ni–85 wt% Cu (point B on the diagram), this will consist of liquid phase only.

5.3.2 Concentrations

To determine the phase compositions or concentrations the tie-line method is used. First, the composition and temperature on the diagram are located. The methods are different depending on whether we are dealing with a single- or two-phase region. It is very simple for a single-phase alloy. For example, if we take point A from Figure 5.2, the 75 wt% Ni–25 wt% Cu at 1,085 °C. At this temperature and composition, only the α phase is present, having a composition of 75 wt% Ni–40 wt% Cu.

However, for an alloy having temperature and composition located within a two-phase region, it becomes a little more complicated. With all two-phase regions a series of horizontal lines exist at each temperature; this line is called a tie line. The tie lines exist between the phase boundary lines within the two-phase region. Thus, to determine the equilibrium concentrations of two phases, the following procedure is utilised:

1. In the two-phase region draw the tie line or isotherm.
2. Note the intersection with phase boundaries. Read compositions.

For example, consider point C (40 wt% Ni–60 wt% Cu) from Figure 5.2 (a magnified version is shown in Figure 5.3). From the diagram it can be seen that point C lies within the $\alpha + L$ region/phase. We need to know the wt% Ni and Cu for both the α and L phases. The tie line is illustrated in the figure; the horizontal tie line intersects both the liquidus and solidus lines given by the compositions CL and Cα, respectively. The value of CL as shown in Figure 5.3 is 32 wt% Ni–68 wt% Cu; therefore this is the composition of the liquid phase. Similarly, the value of Cα is 46 wt% Ni–54 wt% Cu.

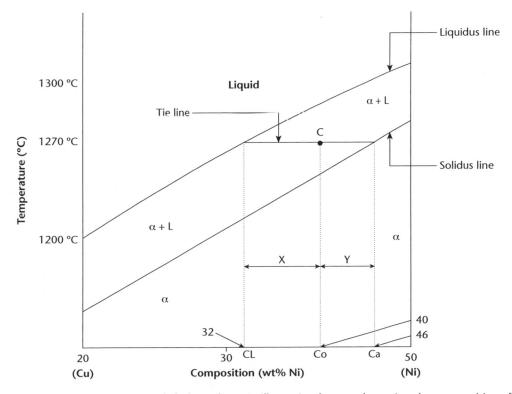

Figure 5.3 Part of the copper–nickel phase diagram illustrating how to determine the compositions for an alloy (point C from Figure 5.2)

5.3.3 Fractions

The fractions of phases can also be determined by using phase diagrams. As before, single and two-phase scenarios must be treated accordingly. Single phases are very simple as only one phase is present. For one-phase alloys, due to the presence of a single phase only, the phase fraction is 1.0 or expressed as a percentage. From Figure 5.2, for the 75 wt% Ni–25 wt% Cu alloy at 1085 °C at point A, the alloy is 100 per cent α.

If we now consider a scenario in which the temperature and composition is located within a two-phase region, then it is rather more complicated. As previously, the tie line is utilised with a procedure called the lever rule (or inverse lever rule).

The lever rule

The lever rule is a way in which to calculate the proportions of each phase present on a phase diagram in a two-phase field at a given temperature and composition. This rule is applied as follows:

1. Construct the tie line.
2. Obtain ratios of line segments lengths.
3. Calculate the fraction of one phase by taking the length of the tie line from the overall alloy composition to the phase boundary for the other phase; then divide by the total tie line length. *Note*: the fractions are inversely proportional to the length to the boundary for the particular phase. If the point in the diagram is close to the phase line, the fraction of that phase is large.
4. Express the values as a fraction or percentage.

If we consider the example from Figure 5.3, for which at 1270 °C both the α and liquid phases are present for a 40 wt% Ni–60 wt% Cu, we need to know the fraction of the α and liquid phases. Earlier we used the tie line to calculate the compositions of the α and L phases. As specified on the diagram, the alloy composition is Co, the fractions of the phases are $W\alpha$ and WL, respectively. Using the lever rule, we can calculate WL as:

$$WL = \frac{Y}{X + Y}$$

thus,

$$WL = \frac{C\alpha - Co}{C\alpha - CL}$$

From Figure 5.3 the values of wt% Ni are used, therefore, Co = 40 wt% Ni, $C\alpha$ = 46 wt% Ni and CL = 32 wt% Ni. If we substitute these values into the above equation:

$$WL = \frac{46 - 40}{46 - 32}$$

$$= \frac{6}{14}$$

$$= 0.43 \text{ (or 43\%)}$$

Conversely for the α phase:

$$W\alpha = \frac{X}{X + Y}$$

$$= \frac{Co - CL}{C\alpha - CL}$$

$$= \frac{40 - 32}{46 - 32}$$

$$= \frac{8}{14} = 0.57 \, (\text{or } 57\%)$$

Obviously, identical answers would be obtained if the compositions were expressed in weight percentage copper instead of nickel. Another important point is that both fractions should always add up to *1.0* (0.43 + 0.57) or *100 per cent* (43 + 57).

The lever rule may at first seem complex, but in practice it is quite simple – we shall see another example in the next section. In general it must be borne in mind that when a single phase exists the alloy is completely that phase (there is no need to use the lever rule). For a two-phase alloy the lever rule is used where a ratio of the tie line segment lengths is taken.

5.3.4 Binary eutectic systems

The most common form of metallic phase diagrams used in metallurgy are binary eutectic systems. Numerous binary phase diagrams exist; one of the most common and simple is the lead–tin phase diagram illustrated in Figure 5.4. It is not expected that students from a civil engineering/architectural background will be too familiar with phase diagrams, but it is important to appreciate the main concepts and features (it is possible some civil engineering courses will include binary phase diagrams as part of a materials module).

A number of features of this phase diagram are important and worth noting. First, there are three single-phase regions found on the diagram – α, β and liquid. The α phase is a solid solution rich in lead. It is worth noting at this point that for any binary phase diagram where α and β are used, the metal/material on the left-hand side of the x-axis is α and, conversely, the material on the right-hand side of the x-axis is β, thus in this case α is lead and β is tin.

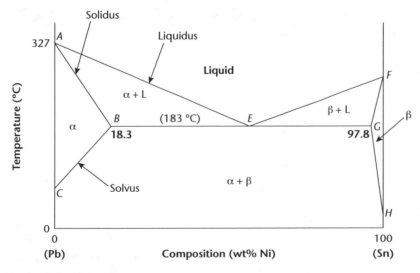

Figure 5.4 The lead–tin phase diagram

The solubility in each of these solid phases is limited, thus at any temperature below the line BEG only a limited concentration of tin will dissolve in lead (for the α phase) and the same for lead in tin (for the β phase). The solubility limit for the α phase is given by the boundary line CBA on the left-hand side of the diagram, labelled α; it increases to a maximum of 18.3 per cent tin (Sn) at 183 °C at point B and decreases back to zero at A (327 °C). At temperatures below 183 °C, the solubility limit line separating the α and $\alpha + \beta$ phase regions is called the solvus line, as illustrated; the solvus line separates a homogeneous solid solution from a field of several phases which may form by melting. The boundary line AB is the solidus line. The solidus line is that below which the solution is completely solid (does not contain a liquid phase). The solvus and solidus lines also exist for the β phase, HG and GF, respectively. The maximum solubility of lead (Pb) in the β phase at point G is 2.2 wt% (100 − 97.8, as shown on the right-hand side of the diagram). The horizontal line BEG is also a solidus line as it represents the lowest temperature at which a liquid phase may exist for any stable lead–tin alloy.

The lead–tin alloy system is very useful, especially as a solder, and is used extensively in many industries, especially construction. This low melting temperature alloy is usually prepared at near eutectic composition, e.g. from Figure 5.4 for the lead–tin system the eutectic composition 61.9 per cent tin (Sn) and 38.1 per cent lead (Pb) has the lowest melting point; therefore a familiar example is the 60–40 solder, containing 60 wt% Sn and 40 wt% Pb. The eutectic lead–tin alloy is completely molten by 185 °C, which makes this material particularly useful as a low temperature solder as it is easily melted.

Calculation of fractions using the lever rule

What composition and mass fraction of the Pb and Sn phases are present at 150 °C at the eutectic composition?

From Figure 5.5 we can see the eutectic composition is 61.9 wt% Sn–38.1 wt% Pb. First, we need to locate the temperature–composition point on the phase diagram (point A); as this lies in the $\alpha + \beta$ region we can expect both the α and β phases will be present.

As two phases will be present we therefore need to draw a tie line in the $\alpha + \beta$ phase at 150 °C, as indicated in Figure 5.5. The composition for the α phase is denoted by $C\alpha$ and equates to approximately 10 wt% Sn–90 wt% Pb. Similarly, the composition for the β phase is denoted by $C\beta$ and equates to approximately 98 wt% Sn–2 wt% Pb.

To calculate the mass fraction we need to utilise the lever rule. From Figure 5.5 and Figure 5.3 (part of the copper–nickel phase diagram illustrating how to determine the compositions for an alloy) we have taken CA to be the overall alloy composition, thus the mass fraction may be calculated as follows:

$$W\alpha = \frac{C\beta - CA}{C\beta - C\alpha}$$

$$= \frac{98 - 61.9}{98 - 10} = 0.41 \text{ or } 41\%$$

$$W\beta = \frac{C\alpha - C\alpha}{C\beta - C\alpha}$$

$$= \frac{61.9 - 10}{98 - 10} = 0.59 \text{ or } 59\%$$

Just to double check $W\alpha + W\beta = 1.0$ or 100 per cent. Therefore, at the eutectic composition at 150 °C, the alloy contains 41 per cent Pb and 59 per cent Sn.

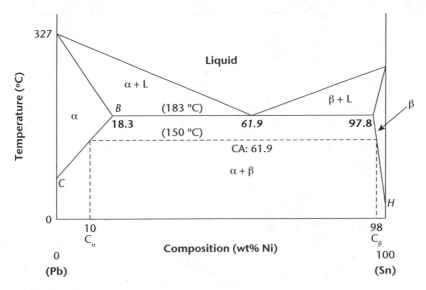

Figure 5.5 The lead–tin phase diagram showing the tie line which corresponds to the eutectic composition at 150 °C

The above example is at the eutectic composition; however, in practice other compositions are frequently used. The eutectic composition for a Pb–Sn alloy is 61.9 wt% Sn–38.1 wt% Pb. If an alloy contains less than 61.9 wt% Sn then the alloy is said to be a *hypoeutectic alloy*; if an alloy contains more than 61.9 wt% Sn then it is a *hypereutectic alloy*.

5.4 The iron–iron carbide (Fe–Fe₃C) phase diagram: steels and cast iron

This is without doubt the most important alloy system for most industrial applications worldwide. The diagram Fe–C is simplified at low carbon concentrations by assuming it is the Fe–Fe₃C diagram and is illustrated in Figure 5.6. Concentrations are usually given in weight percent. The possible phases are:

- α-ferrite (BCC) Fe–C solution
- γ-austenite (FCC) Fe–C solution
- δ-ferrite (BCC) Fe–C solution
- liquid Fe–C solution
- Fe_3C (iron carbide) or cementite, an intermetallic compound.

The maximum solubility of C in α-ferrite is 0.022 wt%. δ-ferrite is only stable at high temperatures. It is not important in practice. Austenite has a maximum C concentration of 2.14 wt%. It is not stable below the eutectic temperature (727 °C) unless cooled rapidly (see Chapter 10). Cementite is in reality metastable, decomposing into α-Fe and C when heated for several years between 650 and 770 °C.

For their role in mechanical properties of the alloy, it is important to note that ferrite is soft and ductile and cementite is hard and brittle. Thus, combining these two phases in solution an alloy can be obtained with intermediate properties. (Mechanical properties also depend on the microstructure, that is, how ferrite and cementite are mixed.)

75

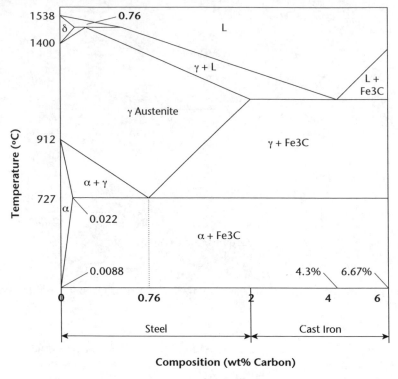

Figure 5.6   The iron–iron carbide phase diagram showing the different compositions and phases in steel and cast iron

5.4.1 Development of microstructures in iron–carbon alloys

The eutectoid composition of austenite is 0.76 wt%. When it cools slowly it forms perlite, a lamellar or layered structure of two phases: α-ferrite and cementite (Fe$_3$C). Hypoeutectoid alloys contain proeutectoid ferrite plus the eutectoid perlite. Hypereutectoid alloys contain proeutectoid cementite plus perlite.

Since reactions below the eutectoid temperature are in the solid phase, the equilibrium is not achieved by usual cooling from austenite.

The influence of other alloying elements

As mentioned in Chapter 3, alloying strengthens metals by hindering the motion of dislocations. Thus, the strength of Fe–C alloys increase with C content and also with the addition of other elements.

5.5 Critical thinking and concept review

1.   What is a solubility limit?
2.   What is a binary alloy?
3.   What is a microstructure?
4.   What is an equilibrium phase diagram?
5.   What are binary isomorphous systems?
6.   How are the proportions of phases present in a binary phase diagram calculated?
7.   How does the Lever Rule function?

8. What is a eutectic system?
9. What is the eutectic temperature? What happens at this temperature?
10. Which range of metal alloys does the $Fe-Fe_3C$ diagram represent?
11. What is the austenite phase?
12. How many types of ferrite phases exist?
13. What is the cementite phase?
14. How do the different phases in the $Fe-Fe_3C$ diagram contribute to the mechanical properties of steel?
15. How do these different phases in the $Fe-Fe_3C$ diagram develop with a change in temperature and carbon content?

This chapter deals in detail with the thermal properties of materials. When put into service, buildings are exposed to quite large variations in temperature day by day and year by year. These temperature variations and their effects, if not allowed for, can cause serious problems in buildings, leading to premature failure and the need for expensive repair and remedial work. The whole area of relative movements between building materials is dealt with here.

We need to design our buildings to conserve heat, and to be comfortable in both summer and winter; to do this we require a good understanding of the thermal properties of materials. The incidence of fire can also cause buildings to be subjected to extreme heating conditions, and this can lead to severe damage to, or destruction of, the building fabric. The importance of the thermal properties of materials (thermal diffusivity and thermal inertia) in determining their response to fire is discussed in this chapter.

Contents

6.1 Introduction

The nature of heat has not always been understood; until about 1800 it was thought to be a substance called 'caloric' that flowed into any material that was exposed to flame. Only later was it found that heat is in fact atoms and molecules in motion. We now have the kinetic theory to explain heat in terms of these atomic and molecular motions in solids, liquids and gases. We live in a world where the ambient temperature can vary by a large margin, and so it is important that we understand the response of materials to heat. When used in buildings, materials may be subject to heating effects in a variety of ways:

- the normal daily heating and cooling cycle due to the sun;
- daily internal heating system cycle;
- heating due to repair and maintenance operations;
- heating caused by a building fire.

The heat variations due to the first two of these causes will occur on a daily basis, and will be largely predictable and fairly modest in extent. Inside a building, temperature variations of no more than 20 °C would be likely. Variations diurnally and nocturnally could be as high as 30 °C or, exceptionally, up to 50 °C. Heat variations due to repair and maintenance operations will occur infrequently, and at irregular intervals, and may be more severe. Heating due to fire may never happen, but if it does, temperatures of several hundred degrees Celsius may be achieved, and could cause serious damage to, or the destruction of the materials of the building.

We know that temperatures inside and outside a building can vary over the usual 24-hour period. So we need to bear in mind the *time dimension*. In other words, temperatures are not static but rather their behaviour is *dynamic*. This is even more the case with building fires, where the temperatures can rise by hundreds of degrees in a matter of minutes. In this circumstance, the time dimension becomes critically important, as it plays a vital role in life safety because people need time to escape from the burning building. In this chapter we shall deal with normal heating cycles and the thermal behaviour of materials, and also aspects of thermal behaviour in the extreme heating conditions brought about by a fire.

It is very important to minimise heat losses from buildings in order to reduce energy consumption and the associated carbon footprints. To do this we need to understand how to make appropriate use of the thermal properties of materials. Another concept of importance in an energy-conscious world is that of *thermal mass*. Materials are being used to improve the thermal performance of buildings because of their ability to store thermal energy while the heating is turned on and release it later when the heating is switched off, thereby reducing the thermal load on heating systems. We shall explore these concepts later in this chapter.

6.2 Nature of thermal energy

Before we proceed, it will be useful to briefly discuss the nature of heat, or thermal energy as it is also called. Like all forms of energy, heat is measured in joules (J), and it can be converted into other forms of energy (1 joule = 1 Nm).

The science of *thermodynamics* deals with the nature of heat and its relationships with other forms of energy, such as mechanical, electrical and chemical. It deals with the relationships and inter-convertibility of all forms of energy. This is important because of the current concern over global warming, and the recognition that consumption of energy is usually accompanied by the release of carbon dioxide gas into the Earth's atmosphere when the source of that energy is fossil fuel (coal, oil, natural gas, peat, etc.). We do not go into the mathematics of thermodynamics here, but it will be useful to cite the first two laws:

1. The first law of thermodynamics is the principle of the conservation of energy. Energy cannot be created or destroyed, but only converted from one form into another. Heat and work are inter-convertible.
2. The second law of thermodynamics states that heat (or energy) transfer can take place spontaneously in one direction only: from a hotter to a cooler body or, generally, from a higher energy state to a lower energy state.

We must also note that whenever we transform one form of energy into another, *there will always be a loss of energy in the transformation*. In other words, *it is impossible to convert one form of energy into another with 100 per cent efficiency*. We have discussed the nature of heat energy; it will be useful to define what we mean by temperature.

6.2.1 Significance of temperature

Temperature can be considered as a symptom of the presence of heat in a material; it is a measure or index of the thermal state of that material. The higher the temperature of a material, the more thermal energy (heat) it contains. *Heat is a form of energy that flows in response to a temperature difference*, and it flows from a region of higher temperature to a region of lower temperature. This behaviour is fundamental in the same way that bodies do not roll uphill. Both of these observations stem from our knowledge that flows always occur from zones of high energy to zones of lower energy. At its deepest level, this is how the universe works. It has *never* been observed to do otherwise.

We are familiar with a world in which there are solids, liquids and gases. On planet Earth we enjoy an ambient temperature of around 10–20 °C for most of the year in the UK. The things that are solid, liquid or gas are in these states at 10–20 °C. However, there is no solid known to science that cannot be turned into a vapour, by heating to a sufficiently high temperature, nor any gas that cannot be solidified by cooling it to a sufficiently low temperature under a high pressure. In other words, we can vaporise steel and solidify air by adjusting their heat content. Heat is therefore a form of energy of fundamental importance. The amount of heat contained by a body will determine its temperature.

When considering buildings and civil engineering structures, it is obvious that mechanical/strength properties are very important, and we have already examined these. In the energy-conscious twenty-first century, and with sustainability being the current preoccupation, the thermal properties of materials are also of equal importance. There are a number of important reasons for this; first, we require our buildings to conserve energy; second, in a fire the thermal properties will determine how well the building structure performs; and third, with all the different materials in a building expanding and contracting by different amounts as the temperature varies, only a well-designed building will avoid problems caused by differential movement. Having materials with the correct thermal properties can make a major contribution to fire safety. The reverse is also true; incorrect selection of thermal properties can increase the fire hazard and the danger to life. Similarly, only the right materials and designs will prevent cracks from opening up and allowing the ingress of water.

We require our buildings to be comfortable at all times, and in winter especially we need to heat buildings. We have seen that heat is a form of energy that flows from regions at higher temperatures to those at lower temperatures, with the speed of heat flow being proportional to the temperature difference. Heat can be transferred in various ways, but in escaping from our buildings, the materials of which they are made will exert a great influence via their thermal properties.

Similarly, in the unfortunate event of a building fire, the speed with which materials heat up, and the speed with which they transfer heat (and perhaps help spread the fire) will depend upon their thermal properties. We need to look next at

- the ways in which heat energy can be transferred; and
- the effects of heating on the various materials of construction.

6.3 Nature of heat transfer

Heat transfer is the science that seeks to predict the energy transfer that occurs between bodies as a result of temperature differences. The science of thermodynamics teaches us that this energy transfer is defined as heat. The science of heat transfer is concerned with how this transfer occurs and, just as importantly, the rate at which it occurs. Thermodynamics can tell us nothing about transfer rates, being only about the energies of systems that are in equilibrium. We can use thermodynamics to predict how much heat is to be transferred, but not how quickly this will occur. As we shall see, some materials will conduct heat readily and quickly, while others will not. Heat will soak rapidly through some materials but not through others, depending upon their thermal properties. We need to understand the thermal behaviour of materials if we are to design and build energy-efficient buildings.

6.3.1 Mechanisms of heat transfer

When different parts of a body (of solid, liquid or gas) are at different temperatures, heat flows from the hotter regions to the cooler. When heat transference takes place, there are three distinct mechanisms by which it can occur:

1. *Conduction*, where the heat (thermal energy) passes through the body itself; this is essentially something that occurs in solids.
2. *Convection*, where heat is transferred by relative motion of portions of the body, and so convection can only occur in fluids (liquids and gases).
3. *Radiation*, in which heat is transferred directly from one body to another, more distant body by electromagnetic radiation. No contact or intervening conductive medium is necessary; radiation can occur in vacuum.

These three mechanisms of heat transfer are shown in Figure 6.1.

In liquids and gases, convection is of paramount importance, but in solids convection is completely absent and radiation is usually negligible. So in our consideration of solid materials, the thermal conductivity will clearly be important. The thermal performance of buildings is determined to a large degree by heat transfer through the building envelope, i.e. the outer walls and surfaces. We shall look first at the thermal conductivity of materials before moving on later to consider the heat losses through the outer building envelope.

6.4 Effects of heat on materials

When materials are heated they can respond in many ways, depending on whether they are solid, liquid or gaseous. In buildings, heating can occur in two main ways; we have the normal cycle of heating and cooling, and we can also have heating that occurs as a result of fire.

Considering solids alone, the response to heat can take one of many forms. For example, the material can melt, or soften; it can ignite and start to burn; it can crack and suffer from thermal

(a)

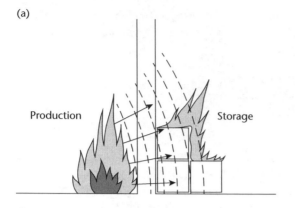

Production   Storage

Heat is conducted through
solid wall to the material
on the other side

(b)

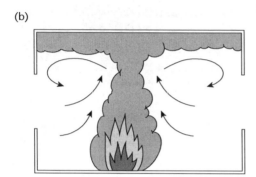

Buoyant plume of hot gas
rises above the fire drawing
cold air into the fire and
setting up a convection
current of air.

(c)

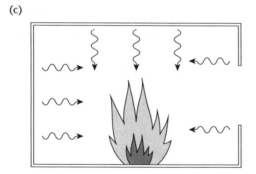

Heated walls of room radiate
heat back towards the fire.
No transfer medium is required
radiation can operate in vacuo.

Figure 6.1   The three modes of heat transfer: (a) conduction through a solid; (b) convection currents in a fluid; and (c) radiative heat transfer

(After Shields & Silcock, 1987)

shock; it can show a phase change (see Chapter 5). These effects can either be reversible or non-reversible.

These effects are all extreme effects, with temperature changes of hundreds of degrees Celsius, and they are only observed during an uncontrolled heating event such as would occur during a building fire. Since we are interested in the effects of fire on buildings, these effects will be considered later, in Chapter 20. In normal building use we shall only consider relatively small effects such as diurnal temperature variations, and the heating effects brought about by the operation of building heating systems. We can therefore divide this section into two parts:

1. Normal diurnal and heating system temperature variations.
2. Extreme and uncontrolled heating, building fires, etc.

Therefore it is important to gain an appreciation and awareness of the effects of heat on construction materials. When materials are heated either intentionally or inadvertently, they respond in *three* important ways:

1. They *absorb heat*; some materials absorb only a small amount for each degree of temperature rise, and some a great deal. Heat absorption capacity is measured by *specific heat capacity* (C_p).
2. They *conduct heat* away from the heat source, some very poorly and some extremely well. Their ability to do this is measured by their *thermal conductivity* (k).
3. They *expand*, some by only a small amount and some by much more.
   This expansion is measured by their *coefficient of thermal expansion* (α) (likewise, when materials cool, they contract).

These are all fundamental thermal properties of materials and are important in building construction, and we shall need to define all three in turn.

6.4.1 Thermal conductivity of a material

As we have seen, heat energy flows in response to a temperature difference, in the same way that an electric current flows in response to a potential (voltage) difference. The bigger the difference in temperature, the faster the heat will flow. Heat flow will also be determined by the *thickness* of the material through which it flows. So the units of thermal conductivity will be expressed as a rate of heat flow per unit of thickness per degree of temperature difference. Values of thermal conductivity for a range of building materials are given in Table 6.1.

Table 6.1 Thermal properties of some common building materials

Material	Thermal conductivity (k) $W\ m^{-1}\ K^{-1}$	Density (ρ) kg m^{-3}	Specific heat capacity (C_p) J. $kg^{-1}\ K^{-1}$
Aluminium	180.00	2,800	880
Copper	398.00	8,930	385
Steel	60.00	7,800	482
Brick	0.80	2,600	800
Concrete	1.40	2,400	880
Glass	1.00	2,600	670
Plaster finish	0.50	1,300	1,000
Plasterboard	0.16	950	840
Timber	0.11	545	2,720
Fibrous insulation	0.05	240	1,250
Cellular insulation	0.03	20	1,500
Rubber (polyisoprene)	0.15	910	2,500
Polypropylene (high density)	0.52	970	2,100

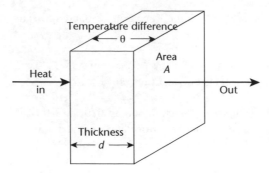

Figure 6.2   Unit cube of solid material with heat flux passing through it

The derivation is quite simply as follows. Consider a cube of solid material of side 1 metre, as shown in Figure 6.2.

Heat passes into the block from face 1 to face 2 without any transverse (sideways) flow, all the heat entering face 1 emerges at face 2. If the heat H joules flows in time t (seconds), the rate of passage of heat is:

$$\frac{H}{t} = \text{watts}$$

H is proportional to the block face area A, and also to the temperature difference θ, but inversely proportional to the block thickness d. If we insert a constant of proportionality k, the thermal conductivity, we can write an equation as follows:

$$\frac{H}{t} = \frac{k \times A \times \theta}{d} \text{ watts}$$

Re-arranging for k, we have:

$$k = \frac{\text{watts} \times d}{A \times \theta}$$

Dimensionally:

$$k = \frac{\text{watts} \times \text{length}}{\text{length}^2 \times {}^\circ\text{C}} = \text{W/m}^\circ\text{C or W/m K}$$

Units: Watts/m/°K (W m^{-1} K^{-1})

The reciprocal of thermal conductivity k is the *thermal resistivity* of a material, and this is a useful property in determining resistance to heat loss through building elements such as walls, ceilings, windows, etc. The units of thermal resistivity will therefore be m K W^{-1}, and we shall return to this later.

6.4.2 Specific heat capacity

In situations where the temperatures on either side of a wall or barrier are constant, the thermal conductivity value will be all we need to take into account. However, in dynamic situations where temperatures are changing with time, the conductivity will not be sufficient to explain and predict behaviour. In these situations we shall need to know the specific heat capacity (C_p) of the material. An example of a dynamic temperature situation could be very early in the morning when the heating system in a building is turned on. The walls will initially be cold right through. As time passes, the interior surfaces of the walls become heated, and the exterior surfaces remain cold. Eventually, a steady temperature gradient will be reached through the wall from inside to out, with temperatures inside and outside being constant. So we shall have a dynamic situation when heating is switched on, and again when it is turned off and the building cools.

If we have unit quantity (1 kg) of a material at temperature θ °C, the *specific heat capacity is the amount of heat energy in joules required to raise its temperature by an increment $\delta\theta$ °C (usually 1 °C)*. So heat capacity will be expressed in joules per kilogram per degree K.

Units: joules/kg/°K (J kg^{-1} K^{-1})

We have defined the two basic properties – namely, thermal conductivity and specific heat capacity. However, in the real world we shall rarely use these two properties as such. In practice we shall be interested in the bulk properties deriving from conductivity and heat capacity. For example, in a fire situation we shall need to know how quickly heat can diffuse into and through a structure, or how quickly a surface will become heated to the point where it begins to radiate heat. In order to determine these things, we need to take into account the conductivity, heat capacity and density of our materials.

6.4.3 Thermal mass

The thermal mass of a building element, e.g. a wall, is the amount of heat energy it can hold when it has reached its equilibrium temperature. It is possible in some situations to design a building with some materials and make use of the fact that the structure will give up its heat and keep the interior warm for some time after the heating has been switched off. The thermal mass is also the amount of heat that must be put into the structure to warm it to its equilibrium temperature; a high thermal mass means that the structure will probably take longer to heat up. To calculate the thermal mass of an element we need to know its volume, density and its specific heat capacity.

Specimen calculation

Calculate the thermal mass of a brick wall 4 m long by 2.5 m high and 100 mm thick, if the bricks have a density of 2,600 kg/m^3. The specific heat capacity of brick is 800 J/kg K.

$$
\begin{aligned}
\text{Thermal mass} \ &= \text{volume (m}^3) \times \text{density (kg/m}^3) \times \text{specific heat capacity (J/kg K)} \\
&= 4 \text{ m} \times 2.5 \text{ m} \times 0.10 \text{ m} \times 2,600 \text{ kg/m}^3 \times 800 \text{ J/kg K} \\
&= 1 \text{ m}^3 \times 2,600 \text{ kg/m}^3 \times 800 \text{ J/kg K} \\
&= 2,080,000 \text{ J/K}
\end{aligned}
$$

Therefore the wall will hold 2.08 MJ of heat for each 1 K above ambient temperature.

6.4.4 Thermal inertia and thermal diffusivity

Imagine heat energy is impinging on the surface of a solid body, as shown in Figure 6.3. We shall also imagine that our body is internally subdivided into small, cubic elements as shown. We shall want to know how quickly the surface of the body heats up, and also how quickly the heat will diffuse into the interior of the solid. To determine these we need to know the fundamental thermal properties of the material of which the solid is made, i.e. the thermal conductivity (k), the specific heat capacity (C_p) and the density (ρ). Let us first consider surface heating.

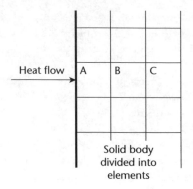

Figure 6.3   Solid body composed of small cubic elements being heated on its surface

The heat energy will first impinge on element A in the body. We first need to calculate how much heat will be required to heat up the unit volume of the material, and to do this we need values for specific heat capacity (C_p) and density (ρ). The specific heat capacity tells us how much heat is required to heat up *unit mass* of material (J/kg/K), and if we multiply this value by the density (kg/m³), we shall then obtain the amount of heat needed to heat up *unit volume* of the material (J/m³/K). If this amount of heat is small, then our element A will heat up quickly; if the amount of heat is large, then heating will take longer.

However, this is not the only factor in play. The thermal conductivity of the material (and so of element A) will also be an important factor. If the thermal conductivity (k) is high, once element A is hotter than element B, the heat arriving in element A will soon be conducted inwards to element B. This will make it more difficult to significantly raise the temperature of element A. If, on the other hand, the thermal conductivity of the material is low, then heat will not be rapidly conducted inwards to element B.

The speed with which the surface layer of a solid is heated is therefore governed by its *thermal inertia*, defined as follows:

$$\text{Thermal inertia} = \text{thermal conductivity} \times \text{specific heat capacity} \times \text{density}$$
$$(k) \qquad\qquad (C_p) \qquad\qquad (\rho)$$

A high value of thermal inertia will mean slow surface heating, whereas a low value will give rapid surface heating.

Now let us consider the diffusion of heat into the interior of the solid. Again, we shall need to know the values of k, C_p and ρ for the material. The heat impinging on the surface of the solid will first strike element A, as before. If the material of the solid has a high value of thermal conductivity, then heat in element A will be quickly conducted through to element B. However, we also need to know how much heat energy is required to heat up element A, and we know that this value will be determined by the product of specific heat capacity and density. If it takes a lot of heat energy to heat element A, then this will slow down the rate at which heat is conducted inwards. Therefore, the speed with which heat will diffuse into a solid is governed by its *thermal diffusivity*, defined as follows:

$$\text{Thermal diffusivity} = \frac{\text{Thermal conductivity } (k)}{\text{Specific heat capacity } (C_p) \times \text{Density } (\rho)}$$

Table 6.2 Thermal diffusivity and thermal inertia data for some common building materials

Material	Thermal diffusivity m² s⁻¹	Thermal Inertia W m² K⁻¹
Aluminium	73.050×10^{-6}	443.520×10^6
Brick	0.385×10^{-6}	1.664×10^6
Copper	111.980×10^{-6}	$1{,}323.649 \times 10^6$
Plasterboard	0.198×10^{-6}	0.127×10^6
Plaster finish	0.385×10^{-6}	0.650×10^6
Cellular insulation	1.000×10^{-6}	0.0009×10^6
Fibrous insulation	0.167×10^{-6}	0.015×10^6
Steel	15.959×10^{-6}	225.770×10^6
Concrete	0.663×10^{-6}	2.957×10^6
Timber	0.074×10^{-6}	0.163×10^6
Glass	0.574×10^{-6}	1.742×10^6
Rubber (polyisoprene)	0.066×10^{-6}	0.341×10^6
Polyethylene (high density)	0.255×10^{-6}	1.059×10^6

A high value of thermal diffusivity will result from high conductivity and low product of $C_p \times \rho$. Such a material will diffuse heat rapidly into the interior when heated. Metals as a group are good conductors of heat, and they also have high values of thermal diffusivity. Note that materials with high thermal inertia values are also rapid heat diffusers. Values of thermal inertia and thermal diffusivity are given in Table 6.2. In practice, we can see the effect of thermal diffusivity in Figure 6.4. On the left is a material with high thermal diffusivity, and on the right is a material with low thermal diffusivity.

When heat is applied to materials they can respond in a variety of ways. They can *melt*, they can *soften*, they can *expand*, they can *ignite and burn*, they can undergo a *phase change* and, in extreme cases, they can *vaporise*.

6.5 Thermal resistance, U-value calculations, etc.

We have already seen in Section 6.4.1 that thermal resistivity is the reciprocal of thermal conductivity, with the units of m K/W. This concept has great importance in building design, because we are concerned to ensure that our buildings conserve heat so that they use as little energy as possible. To calculate a U-value, or to calculate how much extra insulation will be required to make a building conform to current Building Regulations, we must be able to calculate the *thermal resistance (R)* of materials. This is a simple calculation; the thermal resistance of a layer of material is the product of the thermal resistivity of the material and the thickness of the layer, i.e.

$$\text{Thermal Resistance } R = \frac{1}{k} \times \text{Thickness (m)}$$

$$\textit{Units: } \frac{\text{m K}}{\text{W}} \times \text{m} = \frac{\text{m}^2\,\text{K}}{\text{W}}$$

Thermal resistance calculation. We may wish to improve the thermal insulation in the external walls, roof or other element of a building. To determine the correct thickness of insulating material we must be able to determine its thermal resistance.

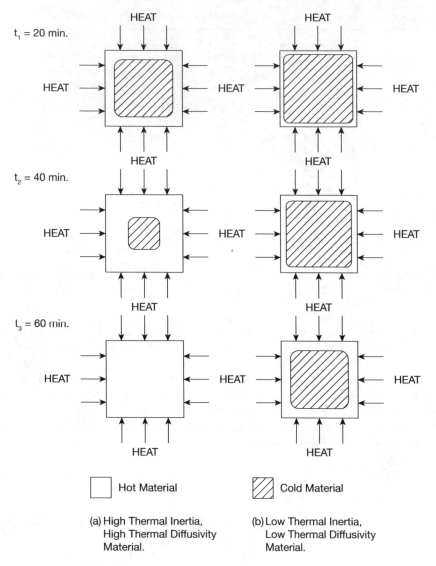

Figure 6.4 Blocks of (a) high thermal diffusivity material, and (b) low thermal diffusivity material heated at the same rate. Heat flows into the high diffusivity material (a) much more rapidly

Specimen calculation

Calculate the thermal resistance of a layer of cellular insulation 50 mm thick if the material has a thermal conductivity value of 0.03 W/m K.

$$\text{Resistance} = \frac{1 \times 0.05 \text{m}}{0.03 \text{ W/m K}} = 1.667 \text{ m}^2 \text{ K/W}$$

We can see that this layer has a high value of thermal resistance.

Table 6.3 Thickness, thermal conductivity and thermal resistances of materials in a typical building wall

Element	Thickness (mm)	Thermal conductivity (W m^{-2} K^{-1})	Thermal resistance (m^2 K W^{-1})
Outer surface	–	–	0.06†
External brickwork	105	0.77	0.136
Cavity insulation	70	0.04	1.75
Blockwork	100	0.19	0.526
Plaster	13	0.16	0.081
Inner surface	–	–	0.12†
			$\Sigma R = 2.673$

† 'Traditional' values used in the UK.

U-Value (thermal transmittance) calculation. In heat loss rate calculations, we use U-values (thermal transmittance values). The units of U-value are W/m^2 K. To determine a U-value, we must first calculate the total thermal resistance of all the elements (layers) in a wall, roof, window or other building element. These calculations are simple as they involve essentially plane, homogeneous layers of materials such as bricks, blocks, plaster, etc., as well as the boundary layers of air attached to the inside and outside surfaces.

For the example of the wall structure given in Table 6.3 (CIBSE Guide A, 1999), the U-value will be:

$$U = \frac{1}{\Sigma R} = \frac{1}{2.673} = 0.39 \text{ W/m}^2 \text{ K}$$

Worked example

Concrete blocks 100 mm thick are to be replaced by concrete/polymer foam blocks also 100 mm thick, to increase the insulation to a building. The composite will consist of 60 mm of concrete and 40 mm of foam. Calculate: (a) the thermal resistance of the concrete blocks, (b) the thermal resistance of the composite blocks.
   Thermal conductivities of concrete and foam are 1.4 and 0.06 W/m K, respectively.

(a)   Resistance of concrete $= \dfrac{1 \times 0.10 \text{ m}}{1.4} = 0.071 \text{ m}^2 \text{ K/W}$

(b)   Resistance of composite. First, resistance of concrete:

$$= \frac{1 \times 0.06 \text{ m}}{1.4} = 0.043 \text{ m}^2 \text{ K/W}$$

Second, resistance of foam:

$$= \frac{1 \times 0.04 \text{ m}}{0.06} = 0.667 \text{ m}^2 \text{ K/W}$$

Total resistance of composite:

$$= 0.043 + 0.667 = 0.710 \text{ m}^2 \text{ K/W}$$

So, the foam layer has increased the thermal resistance of the block from 0.071 to 0.710 m^2 K/W, i.e. the resistance is increased by a factor of ten times.

This U-value is too high to comply with the current UK Building Regulations, and is included merely as an illustration of how the calculation is performed.

Application to thermal composites. Another possible scenario is that we may consider using a composite material for, say, the inner leaf of an exterior wall, and we wish to determine the overall thermal resistance of our composite blocks or panels. Composite materials made for reasons of thermal insulation are dealt with in Chapter 17. The same calculation method outlined above is employed.

6.6 Thermal expansion effects in materials

Besides heat transfer and heat absorption, the other important response to heat is the thermal expansion of materials. Materials, whether solid, liquid or gaseous, tend to expand when heated. We must take account of this expansion if we are to properly understand the behaviour of materials.

The materials used in building vary widely in their thermal expansion coefficients. Clay bricks expand the least, whereas polymeric materials have the highest expansion coefficients.

6.7 Relative movement effects in building materials

By relative movement is meant the differential expansion and contraction effects that occur between differing materials in a building. These effects are very important, because they can cause gaps and cracks to open up which can lead to leaks and the ingress of water in situations where the building is supposed to be water-tight. Because such a wide range of materials are used in building, we need always to bear in mind the fact that relative or differential movement can occur. Relative movement between materials is principally caused by:

Table 6.4 Thermal expansion coefficients of some common building materials

Material	Approximate coeff. of linear expansion per °C ($\times 10^{-6}$)	Unrestrained movement for 50 °C change (mm/mm)
Clay bricks and tiles	5–6	0.25–0.30
Limestone	6–9	0.30–0.45
Glass	7–8	0.35–0.40
Marble	8	0.40
Slates	8	0.40
Granite	8–10	0.40–0.50
Asbestos cement	9–12	0.45–0.60
Concrete and mortars	9–13	0.45–0.65
Mild steel	11	0.55
Bricks (sand-lime)	13–15	0.65–0.75
Stainless steel (austenitic)	17	0.85
Copper	17	0.85
Glass-reinforced plastics	20	1.00
Aluminium	24	1.20
Lead	29	1.45
Zinc (pure)	31	1.55
PVC (rigid)	50	2.50
PVC (plasticised)	70	3.50
Polycarbonate	70	3.70

- temperature changes
- humidity changes
- stress.

All three of these factors will operate at the same time. For example, a load-bearing structure such as a concrete wall will be under stress because of the weight of the building that it is supporting, and at the same time it will experience changes in ambient temperature and in humidity. To understand exactly what happens, it will be useful to look at each of the above factors in turn.

6.7.1 Temperature effects

We have already seen that temperature changes bring about expansive effects. Table 6.4 gives expansion coefficient data for a range of building materials. If two different materials are in contact, and there is a temperature rise, then they will expand by differing amounts. The same is true if cooling occurs; there will be differential contraction. This means that one material will move with respect to its neighbour, and this means relative movement.

6.7.2 Humidity/moisture effects

Moisture will affect porous materials which are able to absorb moisture from the air or other surroundings. When porous materials absorb moisture, they expand; when they dry out again, they shrink or contract. Non-porous materials, e.g. steel and glass, are not so affected. Concrete, clay bricks and timber, being porous, will be affected by moisture/humidity variations. Table 6.5 gives the average values of moisture movement for a range of building materials.

It is important to remember that movements can be both reversible and non-reversible. For example, newly fired clay bricks will emerge from the brick kiln absolutely bone-dry. However, because they are porous, they will begin to absorb moisture from the air and so they will begin to expand. Once they have reached the same moisture content as their surrounding atmosphere, they will stop absorbing moisture, and they will stop expanding. This initial expansion happens once, and is irreversible. During their service lives, the bricks will then expand and contract as they follow the trends in air moisture content (humidity). This is reversible shrinkage (see Figure 6.5).

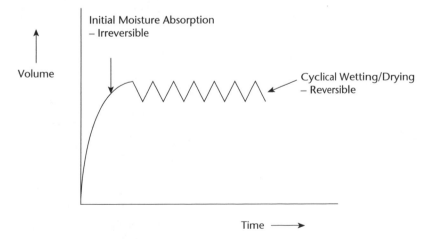

Figure 6.5 Expansive behaviour of bricks after removal from a kiln

Table 6.5 Wetting/drying movements for a range of building materials (approximately reversible)

Building material	Approximate movement (%)
Asphalt, Bitumen	Nil or negligible
Glass	Nil or negligible
Granite	Nil or negligible
Metals	Nil or negligible
Plastics	Nil or negligible
Clay bricks	0.02
Calcium silicate bricks	0.03–0.04
Concrete blocks	0.04–0.06
Dense concrete	0.03–0.04
Structural lightweight aggregate concrete	0.03–0.08
Concrete with shrinkable aggregates	0.05–0.09
Limestone	0.06–0.08
Sandstone	0.06–0.08
European spruce, Baltic whitewood	1.50* 0.7#
European larch	1.70* 0.8#
Douglas fir, Oregon pine	1.50* 1.2#
Western hemlock	1.90* 0.9#
Scots pine, Baltic redwood	2.20* 1.0#
English oak	2.50* 1.5#

* Tangential movement
# Radial movement

Freshly made concrete will start to dry out once the formwork is removed. This drying will be accompanied by shrinkage, and the initial shrinkage as the concrete comes to equilibrium with its surroundings will be irreversible. However, in service the concrete will expand and contract as it absorbs and loses moisture, tracking the local air conditions; this movement is reversible. So with concrete, the initial irreversible movement is shrinkage, whereas with bricks it is expansion. See Figure 6.6 for the behaviour of concrete.

The diagrams above illustrate the behaviour of clay bricks and concrete, but all porous materials will behave similarly, including gypsum plaster and timber. All porous materials tend to come into equilibrium

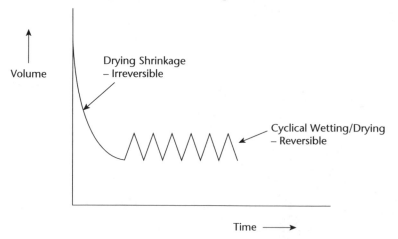

Figure 6.6   Shrinkage/expansion behaviour of concrete after initial placement

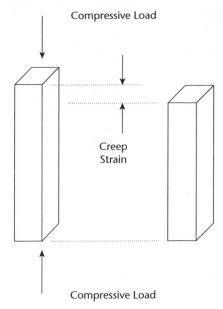

Compressive Load

Creep
Strain

Compressive Load

Figure 6.7   Creep of a column under compressive load

with their surroundings in terms of moisture content. So they absorb moisture from the air when the weather is damp and humid (and expand) and lose moisture when the weather is warm and dry (and shrink).

6.7.3 Stress

Many materials are subject to *creep* when they are put under continuous loads for protracted time periods. We have seen that creep is a permanent shape change brought about by the continuous application of load. For example, a concrete column under compressive load will suffer a small, permanent but measurable reduction in height. The column will finish up slightly shorter and slightly fatter than it was when it was first made. This is creep. Steel will also creep under load but only at elevated temperatures – *not* at room temperature. Lead, on the other hand, will creep at room temperature. To creep, metals have to be heated to a temperature of around 0.5 T_m in degrees absolute (T_m is the melting temperature K).

Figure 6.7 illustrates how a concrete column would creep under a compressive load, continuously applied over a long period of time (several years).

What is the practical significance of this? Let us consider a concrete column in a tall residential building. We have three causes of relative movement; temperature, moisture/humidity and stress. Of these three, we have seen that humidity/moisture effects and stress both give rise to permanent negative strains, i.e. they both lead to shrinkages. Any temperature effect is transient, and reversible. So we can get a permanent reduction in length of our column, and we need to evaluate this.

In concrete, the creep occurs over prolonged periods of time, although most of it occurs during the first five years of the concrete's life. Within normal design stresses for concrete for 20–35 N/mm² at 28 days, the average value of creep can be taken as 30×10^{-6} mm/mm/N/mm². The creep of concrete at 30 N/mm² may result in deformation of 0.09 per cent, which, added to irreversible drying shrinkage, gives a total shrinkage of around 0.12–1.17 per cent for a structural concrete frame.

Therefore the shrinkage movement in a three-storey height could be as much as 5 mm (or 50 mm in a ten-storey building), i.e.:

$$\text{Shrinkage Strain} = \frac{3{,}000 \text{ mm} \times 0.17}{100} = 5.1 \text{ mm}$$

In a single 3 m high storey, we can see a reduction of over 5 mm. This is significant, and it undoubtedly led to the premature failure of many of the high-rise buildings put up in the 1960s. Many of these buildings had steel-reinforced concrete frames, to which were fixed pre-fabricated wall panels, which were designed to keep the weather out, but not to be load bearing. Relative movement was not allowed for in most cases; however, if one floor deck moved 5 mm closer to the one below, then this sometimes resulted in load bearing down on the wall panel, and hence in its failure.

At least one swimming pool designed for international competition was not able to be used because the designer failed to take movement (initial permanent shrinkage) of the concrete into account. These were expensive failures, and serve to emphasise the importance of relative movement in materials.

6.8 Critical thinking and concept review

1. What is the difference between *heat* and *temperature*?
2. Name the *three* mechanisms of heat transfer that operate within and between solid bodies Explain how each mechanisms works.
3. What is the most important property needed to determine a U-value?
4. What is meant by *thermal mass*?
5. What is meant by the *relative movement* of materials in a building? List the *three* factors that determine the extent of such relative movement.
6. Define *thermal diffusivity* and explain how it is calculated.
7. In which situations is thermal diffusivity of significance?
8. Define *thermal inertia* and explain how it is calculated.
9. Define *thermal resistivity*, and give the correct units
10. How is thermal resistance calculated?
11. In which situations is thermal inertia of significance?
12. Calculate the thermal resistance of a wall 105 mm thick built with facing bricks with a thermal conductivity value of 0.50 W/m K.
13. An existing wall has a U-value of 0.6 W/m² K. By how much must the thermal resistance of the wall be increased to reduce the U-value to 0.25 W/m² K?
14. Calculate the thermal mass of a concrete wall 2.5 m high by 3 m long and 100 mm thick, if the specific heat capacity of concrete is 880 J/kg K and its density is 2,400 kg/m³.
15. What *three* factors can cause differential movement effects in building structures?

6.9 References and further reading

CARSLAW, H.S. & JAEGER, J.C. (1959), *Conduction of Heat in Solids*, Oxford University Press, Oxford.
CHARTERED INSTITUTION OF BUILDING SERVICES ENGINEERS (CIBSE) (1999), *Environmental Design, CIBSE Guide A*, chapter 3, London.
HOLMAN, J.P. (1981), *Heat Transfer*, McGraw-Hill, New York.
SHIELDS, T.J. & SILCOCK, G.W.H. (1987), *Buildings and Fire*, Longman Scientific & Technical, Harlow.
SZOKOLAY, S.V. (1980), *Environmental Science Handbook,* Construction Press, Lancaster.

7

Structures: shear force and bending moment diagrams

Contents

This chapter is concerned with structures; that is, how do buildings stand up and stay standing up? As compression and tension and their associated stresses are dealt with elsewhere in the book, in this chapter we'll be dealing with one particular type of structural member: those which span horizontally, specifically beams and slabs.

7.1 Beams or slabs?

Whereas beams are linear in nature, have a fixed width and span between two (or more) supports, slabs normally have a large width compared with their depth, cover a significant area and, if rectangular, may be supported on all four edges rather than two opposite edges. However, slabs are subjected to the same effects as beams, so whenever the word 'beam' is used in the following discussion, what is said can be assumed to apply to slabs as well.

7.2 To what effects are beams subjected?

We've all had the experience of standing on a plank of wood and feeling it deflect under the burden of our weight. The wood is *bending*. Another effect that beams experience when subjected to weights (or 'loading', as it is called) is *shear*. The term shear is harder to define and its effect more difficult to visualise, but it nonetheless exists so we have to take it seriously. The rest of this chapter is concerned with analysing the effects of bending and shear.

7.3 Bending or shear: which is more critical?

Any given beam will fail either in bending or in shear; the question is: which will occur first? What a shame there is no easy answer to this question: it would make the lives of structural engineers – and those of students doing coursework or exams in structures – much easier if there was. It depends on the magnitude and nature of the loading, and the material(s) from which the beam is made. There is no shortcut: analysis on the lines discussed below has to be carried out.

7.4 The basics

At this stage we'll have to discuss a few basic structural concepts, namely force, moment and equilibrium. These underpin all structural work, including the body of knowledge addressed in this chapter. To carry on without first reviewing these things would be like trying to build a building with no foundations (highly inadvisable). Nevertheless, if you feel confidently familiar with forces, moments and equilibrium, feel free to skip the next few sections as far as Section 7.9.

7.5 Forces

A 16-stone man standing on a floor in a building is exerting a downwards *force* (due to the effects of gravity) of 1 kN (1 kilonewton) on the supporting floor. Clearly any building is going to be subjected to forces from the people, furniture and other equipment it supports, and these forces need to be calculated.

Similarly, a retaining wall might experience a horizontal force of 50 kN from the earth that it is holding back. The wall must be designed to sustain this force, otherwise it will collapse.

7.6 Moments

A moment is a turning effect; it is concerned with rotation about a given fixed point. The use of a spanner to tighten (or loosen) a nut, or the use of a screwdriver blade to lever the lid off a can of paint, are both

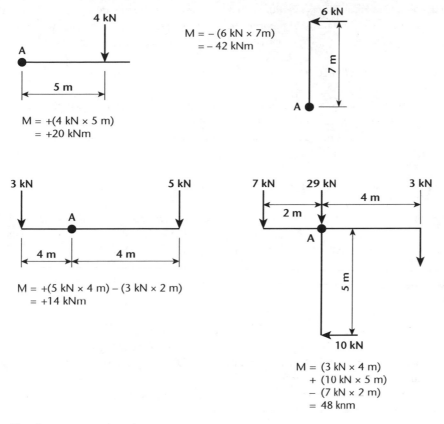

Figure 7.1 Simple moment examples

everyday examples of the application of a moment. It is significant whether the moment is clockwise or anticlockwise, as these are very different (in fact, opposite) effects. Normally we represent clockwise moments as positive (+) and anticlockwise moments as negative (−).

Numerically, a moment is the magnitude of the force (in newtons or kilonewtons) multiplied by the perpendicular distance (in millimetres or metres) from the force's line of action to the point about which rotation is taking place. The units of force are kNm or Nmm.

Some simple examples of moment are given in Figure 7.1.

7.7 Equilibrium

A 16-stone man standing on a floor of a building is, with the assistance of gravity, imposing a downward force of 1 kN on the floor. Assuming the man is stationary, the floor will react to this downward force by exerting an upward force of 1 kN. So the downward force creates an equal and opposite upward force, and the vertical forces are in *equilibrium*. This is always true of any stationary object:

Total force up = total force down. (↑ = ↓) (vertical equilibrium)

The same principle applies to horizontal forces. In a tug-of-war competition, two teams pull in opposing directions on a long piece of rope. It is easy to see if one of the two teams is winning, as any marker placed on the rope will move in the direction of the stronger team. However, if both teams are pulling hard and the marker is not moving at all, an observer would conclude that the two teams are evenly matched and neither is winning. So if the marker is stationary, the force to the left (exerted by the left-hand team) is equal to the force to the right (produced by the right-hand team). Again, this is true of any stationary object.

Total force right = Total force left (→ = ←) (horizontal equilibrium)

The same is also true of moments. If someone is using a spanner to attempt to tighten a nut by turning it clockwise and, at the same time, a second person uses another spanner on the same nut but is turning it anticlockwise, the nut will not move at all if both people are applying the same moment (turning effort). So if any object is stationary:

Total clockwise moment = Total anticlockwise moment (moment equilibrium)

These three forms of equilibrium are fundamental in solving many basic structural problems. The rules of equilibrium apply to any stationary object. That object could be a whole building, any individual part of that building (e.g. a beam or column), or any point within that beam or column.

7.8 Types of loading

There are two basic types of loading: *point loads* and *uniformly distributed loads*, the latter generally known by the acronym *udl*. As the name suggests, a point load acts at a point, or, in practice, over a very small area. Examples might include the 16-stone man above standing on a floor slab, or a column directly supported by a beam. Much more common in practice are udls. A typical steel beam in a building has its own weight evenly spread along its length. It might be directly supporting a concrete floor slab, whose load would be evenly spread along the steel beam. In turn, the floor is supporting the loads of people and furniture which, again, would be evenly spaced along the beam. So the steel beam is supporting a udl over its entire length. Point loads and udls are represented diagrammatically as shown in Figure 7.2.

7.9 Calculation of reactions

The rules of equilibrium, discussed above, tell us that the total load down on a beam (or any other part of a structure) must be opposed by a balancing upward force. This balancing upward force can only be provided by the supports of the beam (which may be columns or walls). These upward forces are called *reactions*, because they are reacting to the presence and the intensity of the downward loads.

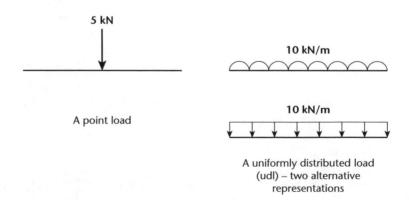

Figure 7.2   Point loads and uniformly distributed loads (udls)

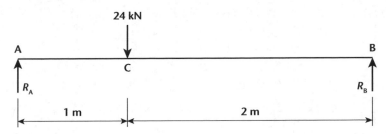

Figure 7.3 Beam for Example 1

Before we can proceed further we need to learn how to calculate reactions, and we'll see that we do that by applying the rules of equilibrium (or two of them, at least) outlined above.

7.9.1 Example 1

Consider Figure 7.3. The solid horizontal black line represents a beam, which spans between A and B. This beam is subjected to only one load, a downward point load of 24 kN located at point C, as shown. The upward arrows at A and B, where the beam is supported, represent the reactions at these points. (It is not strictly correct to represent supports as upward arrows, but it will do for the purposes of this chapter. For further information about the different types of support see my book *Basic Structures for Engineers and Architects* (Blackwell 2005).) We'll use the notation R_A to represent the reaction at A and R_B to represent the reaction at B. We wish to calculate the values of these reactions.

From vertical equilibrium (total force up = total force down) we know that:

$$R_A + R_B = 24 \text{ kN}$$

Unfortunately, this doesn't tell us what R_A is or what R_B is, so to calculate these reactions we have to do something further. Specifically, we'll use our newfound knowledge of moment equilibrium.

We found out above that, at any point in a stationary object, moment equilibrium occurs. So, at any point in this beam (e.g. point A), the total clockwise moment about the point is equal to the total anticlockwise moment about the point.

Total clockwise moment = total anticlockwise moment
24 kN × 1 m = R_B × 3 m
So R_B = 8 kN

Applying the same principle to point B:

Total clockwise moment = total anticlockwise moment
R_A × 3 m = 24 kN × 2 m
So R_A = 16 kN

Now $R_A + R_B = 16 + 8 = 24$ kN, as we expected.

7.9.2 Example 2

Consider the beam shown in Figure 7.4. There are more loads than in Example 1, and, just to spice things up, there is an overhang – in other words, one of the supports is not at the end of the beam. But the same principles apply.

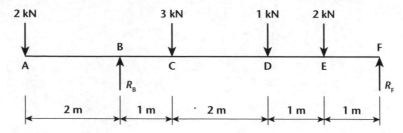

Figure 7.4   Beam for Example 2

For vertical equilibrium:

$$R_B + R_F = 2 \text{ kN} + 3 \text{ kN} + 1 \text{ kN} + 2 \text{ kN} = 8 \text{ kN}$$

Taking moments about B:

Total clockwise moment = total anticlockwise moment
$(3 \text{ kN} \times 1 \text{ m}) + (1 \text{ kN} \times 3 \text{ m}) + (2 \text{ kN} \times 4 \text{ m}) = (R_F \times 5 \text{ m}) + (2 \text{ kN} \times 2 \text{ m})$
$3 + 3 + 8 = 5R_F \times 4$
So $R_F = 2 \text{ kN}$

Taking moments about *F*:

Total clockwise moment = total anticlockwise moment
$(R_B \times 5 \text{ m}) = (2 \text{ kN} \times 7 \text{ m}) + (3 \text{ kN} \times 4 \text{ m}) + (1 \text{ kN} \times 2 \text{ m}) + (2 \text{ kN} \times 1 \text{ m})$
$5R_B = 14 + 12 + 2 + 2 = 30 \text{ kN}$
So $R_B = 6 \text{ kN}$

Check: $R_B + R_F = 6 + 2 = 8 \text{ kN}$, as expected

In the above two examples, only point loads have been considered. However, udls (uniformly distributed loads) are far more common in practice, so it's important to be able to calculate reactions for cases including udls as well. The general principle for calculating a moment due to a udl is shown in Figure 7.5.

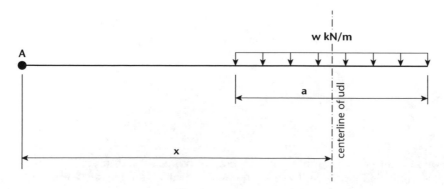

Figure 7.5   General rule for calculating moments due to uniformly distributed loads

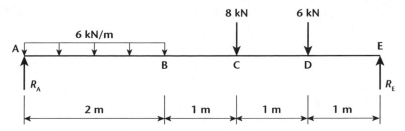

Figure 7.6 Beam for Example 3

7.9.3 Example 3

We are now going to calculate the end reactions for the beam shown in Figure 7.6, which contains both point loads and a udl.

Of course, the same general principles apply as in the earlier examples. Vertical equilibrium:

Total force up = total force down
$R_A + R_E = (6 \text{ kN m}^{-1} \times 2 \text{ m}) + 8 \text{ kN} + 6 \text{ kN} = 26 \text{ kN}$

Taking moments about A:

Total clockwise moment = total anticlockwise moment
$(6 \text{ kN m}^{-1} \times 2 \text{ m}) \times 1 \text{ m} + (8 \text{ kN} \times 3 \text{ m}) + (6 \text{ kN} \times 4 \text{ m}) = (R_E \times 5 \text{ m})$
$12 + 24 + 24 = 5R_E$
So $R_E = 12 \text{ kN}$

Taking moments about E:

Total clockwise moment = total anticlockwise moment
$(R_A \times 5 \text{ m}) = (6 \text{ kN m}^{-1} \times 2 \text{ m}) \times 4 \text{ m} + (8 \text{ kN} \times 2 \text{ m}) + (6 \text{ kN} \times 1 \text{ m})$
$5R_A = 48 + 16 + 6 = 70 \text{ kN}$
So $R_A = 14 \text{ kN}$

Check: $R_A + R_E = 12 + 14 = 26 \text{ kN}$, as expected.

7.9.4 Example 4

Have a go at calculating the end reactions for the beam shown in Figure 7.7. The solution, in brief, is given in italics below.

Vertical equilibrium: $R_A + R_E = (8 \text{ kN m}^{-1} \times 2 \text{ m}) + 14 \text{ kN} + (2 \text{ kN m}^{-1} \times 2 \text{ m}) = 34 \text{ kN}$

Taking moments about A:

$(8 \text{ kN m}^{-1} \times 2 \text{ m}) \times 1 \text{ m} + (14 \text{ kN} \times 3 \text{ m}) + (2 \text{ kN m}^{-1} \times 2 \text{ m}) \times 5 \text{ m} = 6R_E$
So $R_E = 13 \text{ kN}$

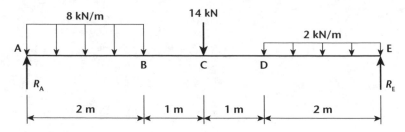

Figure 7.7   Beam for Example 4

Taking moments about E:

$$6R_A = (8 \text{ kN m}^{-1} \times 2 \text{ m}) \times 5 \text{ m} + (14 \text{ kN} \times 3 \text{ m}) + (2 \text{ kN m}^{-1} \times 2 \text{ m}) \times 1 \text{ m}$$
$$\text{So } R_A = 21 \text{ kN}$$

7.10 So what exactly is bending?

Consider a plank of wood supported at its two ends. If you were to stand on the plank, midway between the two supports, the plank would bend under your weight. In fact, it would *sag*, as shown in Figure 7.8. As the plank sags, the fibres of wood at the bottom of the plank would stretch, as they are in tension. Meanwhile, the fibres in the top of the plank would be compressed, or squashed together – they are in compression.

So any structural member that is *sagging* will experience *compression* in the top and *tension* in the bottom.

Now imagine you are sitting on the branch of a tree. Again, the branch will tend to bend under your weight. In fact, the heavier you are, the lighter the branch, and the further you are from the tree trunk, the more the branch will bend. But the branch is only supported at one end (the end where it joins the trunk of the tree) so the nature of the bending in this case would be *hogging*, as shown in Figure 7.9. In terms of forces within the member, hogging is the opposite to sagging. In the case of hogging, it is the upper part of the member which is being stretched – that is, in tension – and the lower part of the member is being squashed, that is, in compression.

So any structural member that is *hogging* will experience *tension* in the top and *compression* in the bottom. It is important to appreciate the implications of tension and compression in sagging and hogging, as some materials are poor in tension or compression – or both.

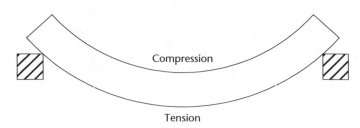

Figure 7.8   A sagging plank

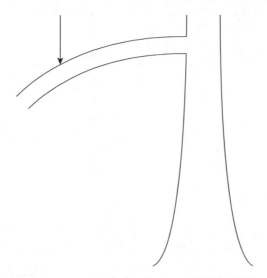

Figure 7.9   A bending tree branch

7.11 And what is shear?

No doubt you've heard others talking about a bolt 'shearing off'; maybe you've even used the expression yourself without fully understanding what it means. The term *shear* is not an easy one to explain, so I'm going to use an analogy related to those sheared-off bolts.

Figure 7.10 shows an edge-on view of two overlapping steel plates connected together by a steel bolt that passes through both plates at the overlap. Imagine applying a force of 1 kN to the left on the top plate. As we know from the discussion of equilibrium above, this system will only stay stationary if we have an equal and opposite balancing force, so we'll apply a rightwards force, also of 1 kN, to the lower plate.

Now imagine that each of these forces is increased to 2 kN, then 3 kN, then 5 kN, then 10 kN, then 20 kN. Eventually something has got to give, and, if the bolt is less strong than the plate, that something will be the bolt. It's likely that the bolt will fail (i.e. break) along a line at the level of the interface between the two plates. After failure the top part of the bolt will move off to the left with the top plate, and the bottom part of the bolt will be dragged to the right with the bottom plate.

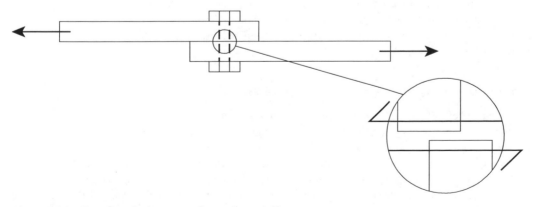

Figure 7.10   Shear in a bolt connecting two steel plates

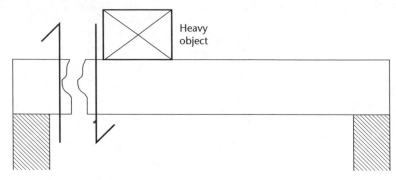

Figure 7.11   Shear in a timber joist

It can be envisaged that, in the process of failing and subsequently moving as described in the last paragraph, the new surfaces of the bolt created by the failure will slide past each other. This is characteristic of a shear failure. A shear failure involves two failed surfaces sliding past each other as failure takes place.

Let's look at another example, more closely related to building. Figure 7.11 shows a typical timber floor joist supporting the upper floor of a conventional two-storey building. The joist spans between supporting masonry walls as indicated.

Imagine the occupant of the building is foolish enough to install a very heavy object on the upper floor – a large piece of weighty machinery, for instance – relatively close to the support as shown. If the floor joists are structurally inadequate – for example, they may not be deep enough and/or wide enough to sustain heavy loads – they may simply break. If this happens the heavy object and the joist supporting it will, of course, fall. At the same time the short piece of joist remaining on the left-hand side of the break (as shown in Figure 7.11) may stay in place, which is the same as moving upwards relative to the downward-moving right-hand part of the now-broken joist. Once again we have an example of two failure surfaces sliding past each other – this time in a vertical plane.

(Incidentally, the one-sided arrowhead symbols in Figures 7.10 and 7.11 are generally used to specifically represent shear forces: watch out for them in lecture notes and other textbooks.)

7.12 Now we know what bending and shear are, how do we apply this knowledge?

We now know that a beam might fail as a result of either (a) excessive bending or (b) shear. To proceed, we need to develop some method of quantifying these bending or shear effects. These quantifications are called *bending moment* and *shear force*, respectively. The bending moment value (in kNm) tells us how much bending is taking place at a particular point in a beam, while the shear force (in kN) is the value of the shear effect at the point concerned.

7.13 How do we calculate the shear force at any given point in a beam?

The *shear force* is defined as the force tending to produce a shear failure at a point. The *value* of the shear force at any point in a beam is defined as the algebraic sum (that is, the sum taking account of plus and minus signs) of all upward and downward forces to the left of the point.

The technique is simple. Suppose we want to calculate the shear force, V, at point B in the beam shown in Figure 7.12, for which the reactions have already been calculated (check that these are correct!).

Ignore everything to the right of point B. For the part of the beam to the left of point B, the shear force at B is the total force up minus the total force down.

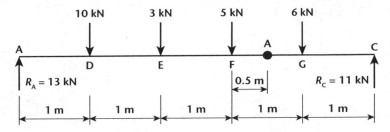

Figure 7.12 Shear force calculation example

The standard symbol for shear force is V, so we'll express the shear force at B as V_B.

$$V_B = 13 - 10 - 3 - 5 = -5 \text{ kN}$$

7.14 How do we calculate the bending moment at any given point in a beam?

The *bending moment* indicates the magnitude of the bending effect at any point in a beam. The *value* of the bending moment at any point on a beam is the sum of all moments (taken about the point) to the left of the point. Regard clockwise moments as positive and anticlockwise moments as negative.

Using the same example as above, let's now calculate the bending moment at point B.

As before, ignore everything to the right of point B. For the part of the beam to the left of point B, the bending moment at B is the total clockwise moment about B minus the total anticlockwise moment about B. We'll express the moment at B as M_B.

$$M_B = (13 \text{ kN} \times 3.5 \text{ m}) - (10 \text{ kN} \times 2.5 \text{ m}) - (3 \text{ kN} \times 1.5 \text{ m}) - (5 \text{ kN} \times 0.5 \text{ m})$$
$$= 13.5 \text{ kNm}$$

7.15 Introduction to shear force and bending moment diagrams

We have just learned how to calculate the bending moment and shear force at any given point in a beam. However, we are normally more interested in the *variation* of bending moment and shear force along the length of a beam. After all, as engineers we are most interested in the values of maximum bending moments and shear forces in a beam, the positions at which these occur and whether they are associated with hogging or sagging in the beam at these positions. This information would enable us to design a beam in a given material (e.g. timber, steel, reinforced concrete) that is capable of resisting these maximum bending or shear effects.

A shear force or bending moment diagram is a graphical representation of how shear force (or bending moment) varies in value along a given beam.

7.16 How to draw shear force diagrams

Take a piece of graph paper. Roughly half way down the sheet, draw a horizontal line the length of the beam (to some suitable scale) and label this with a number 0 (zero) at each end. This is the zero line of the shear force diagram.

Now consider the example shown in Figure 7.12 that we looked at earlier. We'll start from the left-hand end and move to the right. At point A, there is an upward force (reaction) of 13 kN. The shear force diagram reflects this upward force by moving vertically upwards from the zero line by 13 kN, taking us to a value of +13 kN. Between A and D nothing happens in terms of forces, so the shear force diagram is a

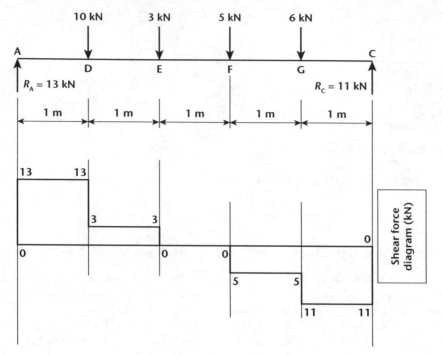

Figure 7.13   Shear force diagram for Figure 7.12

horizontal straight line. At D there is a 10 kN downward force, so the shear force diagram drops sharply downwards by 10 kN. As we're already at +13 kN, the shear force diagram drops to a value of +3 kN. There are no forces between D and E, so the shear force diagram is a horizontal straight line. At E there is a 3 kN downward force, so the shear force diagram drops by 3 kN, taking it to zero. There are no forces between E and F, so the shear force diagram stays at zero. At F there is a downward force of 5 kN, so the shear force diagram drops by 5 kN, to −5 kN. As there are no forces anywhere between F and G, the value of shear force stays at −5 kN. At G there is a downward force of 6 kN, which takes the shear force diagram down to −11 kN (i.e. −5 − 6 = −11). As there are no forces between G and C, the value of shear force remains constant at −11 kN. At C, the upward reaction of 11 kN takes the shear force diagram back to zero. The resultant shear force diagram is illustrated in Figure 7.13.

There are a few points to note from this exercise. First, the shear force diagram ends up at zero exactly. This will always be the case (unless a mistake has been made). Second, look at the shape of the shear force diagram. You will see that it is 'stepped', i.e. it comprises a series of 'steps' in the form of vertical and horizontal lines. If a beam is loaded with point loads only, its shear force diagram will always be stepped in this way.

7.17 Shear force diagrams involving udls

Now we are going to sketch the shear force diagram for an example that includes udls. Look again at Figure 7.7 from Example 4.

Again, we'll draw a horizontal line representing zero shear force, and we'll start from the left-hand end – point A.

Earlier we calculated that the reaction at A (R_A) in Example 4 was 21 kN. So the shear force diagram will jump upwards at A from zero to a value of +21 kN.

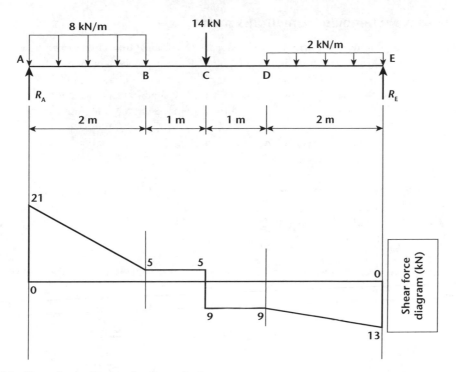

Figure 7.14 Shear force diagram for Example 4

Between A and B, the beam experiences a downward load of 8 kN m⁻¹ over a length of 2 m, so the shear force reduces by a total of $(8 \times 2) = 16$ kN over this distance. The value at A is +21 kN, so by the time B is reached, the shear force has reduced to $(21 - 16) = 5$ kN. The reduction in shear force over this length is uniform, i.e. linear, as indicated by the sloping straight line on the shear force diagram shown in Figure 7.14.

Between B and C there are no forces, so the shear force diagram remains constant at a value of + 5 kN. At C the beam experiences a downward force of 14 kN, so the shear force diagram jumps downwards by 14 kN, taking it from +5 kN to −9 kN. Between C and D there are no forces, so the shear force diagram remains constant at a value of −9 kN.

Between D and E, the beam experiences a downward load of 2 kN m⁻¹ over a length of 2 m, so the shear force reduces by a total of $(2 \times 2) = 4$ kN over this distance. The value at D is −9 kN, so by the time E is reached, the shear force has reduced to $(−9 − 4) = −13$ kN. Again, the reduction in shear force over this length is uniform, i.e. linear, as indicated by the sloping straight line on the shear force diagram shown in Figure 7.14.

At E there is an upward force (reaction) of 13 kN, which brings the shear force diagram up from −13 kN to zero.

Again, there are a few points to note from inspection of Figure 7.14. As before, the shear force diagram ends up at zero exactly. Once again, this will always be the case unless a mistake has been made. Second, look at the shape of the shear force diagram. In the region where there are only point loads (between B and D) the shear force diagram is 'stepped', i.e. it comprises a series of 'steps' in the form of vertical and horizontal lines, as we found in the above example. But between points A and B, and again between points D and E, the beam is subjected to udls and these are expressed as sloping straight lines on the shear force diagram.

7.18 How to draw bending moment diagrams

Before drawing a bending moment diagram it is necessary to calculate the values of bending moments at significant points along the beam. By 'significant points' I mean points at which a point load occurs (or a udl starts or ends) or where there is a support.

Again, we'll refer to the example shown in Figure 7.12. The significant points are A, D, E, F, G and C. We'll now calculate the bending moment at each of those points using the rule outlined above.

Moment at A $= 0$ kNm (there is nothing to the left of A)
Moment at D $= (13 \text{ kN} \times 1 \text{ m}) = 13$ kNm
Moment at E $= (13 \text{ kN} \times 2 \text{ m}) - (10 \text{ kN} \times 1 \text{ m}) = 16$ kNm
Moment at F $= (13 \text{ kN} \times 3 \text{ m}) - (10 \text{ kN} \times 2 \text{ m}) - (3 \text{ kN} \times 1 \text{ m}) = 16$ kNm
Moment at G $= (13 \text{ kN} \times 4 \text{ m}) - (10 \text{ kN} \times 3 \text{ m}) - (3 \text{ kN} \times 2 \text{ m}) - (5 \text{ kN} \times 1 \text{ m}) = 11$ kNm
Moment at C $= (13 \text{ kN} \times 5 \text{ m}) - (10 \text{ kN} \times 4 \text{ m}) - (3 \text{ kN} \times 3 \text{ m}) - (5 \text{ kN} \times 2 \text{ m}) - (6 \text{ kN} \times 1 \text{ m})$
$\qquad = 0$ kNm

As a general rule, the bending moments at the ends of the beams (A and C in this case) will always be zero, unless there are moments directly applied at these points.

Once we have calculated these values we can plot them on a sheet of graph paper. The general rule for doing this is that we draw the bending moment diagram on the side of the beam that experiences *tension*. A glance at Figure 7.12 suggests that the beam will *sag* along its entire length due to the nature of its loading and the position of its supports, so tension will occur on the *underside* of the beam. We therefore plot the bending moment diagram *below* its zero line, as indicated in Figure 7.15. Note that we simply join the dots with straight lines to obtain the bending moment diagram.

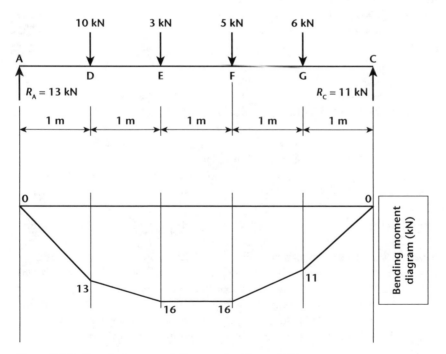

Figure 7.15   Bending moment diagram for Figure 7.12

There are a few points to note from this exercise. First, the bending moment is zero at the two ends of the beam. This is usually the case, as discussed above. Second, the bending moment diagram comprises sloping straight lines, and this is always the case for bending moment diagrams for beams which contain point loads only.

7.19 Bending moment diagrams involving udls

Now we are going to draw the bending moment diagram for an example that includes udls. Look again at Figure 7.7 from Example 4.

The significant points are A, B, C, D and E. We'll now calculate the bending moment at each of those points using the rule outlined above. (At this stage you might find it useful to look again at Figure 7.5, which shows the general principle for calculating a moment due to a udl.)

Moment at A = 0 kNm (there is nothing to the left of A).
Moment at B = (21 kN × 2 m) − (8 kNm × 2 m × 1 m) = 26 kNm
Moment at C = (21 kN × 3 m) − (8 kNm × 2 m × 2 m) = 31 kNm
Moment at D = (21 kN × 4 m) − (8 kNm × 2 m × 4 m) − (14 kN × 1 m) = 22 kNm
Moment at E = (21 kN × 6 m) − (8 kNm × 2 m × 5 m) − (14 kN × 3 m) − (2 kNm × 2 m × 1 m)
= 0 kNm

As in the previous example, the beam will be sagging throughout its length, so the underside of the beam is in tension and therefore the bending moment diagram is plotted below the zero line. These points can now be plotted on a sheet of graph paper.

This time we must be cautious about plotting the diagram itself as it is not the simple 'join the dots' exercise of the above example. Where there is no udl (i.e. between points B and D) we can indeed connect the points we've plotted with straight lines. However, where there is a udl (that is, between A and B, and again between D and E) the lines connecting the points will be parabolic (curved) in shape. If you wish to check this you could calculate the bending moments at 0.5 m intervals between A and B, and you will find that the values are 10.5, 17 and 22.5 kNm, respectively: a line drawn through these points will be a smooth curve. The full bending moment diagram is shown in Figure 7.16.

7.20 Shear force and bending moment diagrams together

It is conventional to draw a loaded beam along with its shear force and bending moment diagrams lining up with each other on the same diagram. We'll do this for the two examples calculated so far. Figure 7.17 combines Figures 7.12, 7.13 and 7.15, while Figure 7.18 – from Example 4 above – combines Figures 7.7, 7.14 and 7.16. There is a good reason for this convention, as we shall discuss.

7.21 Where does the maximum bending moment occur?

As discussed earlier in this chapter, engineers are normally most interested in the *maximum* shear forces and bending moments in a beam, as we have to design beams (of steel, concrete or timber) that are capable of sustaining these maximum effects. The value and position of maximum shear force are readily determined from inspection of a shear force diagram, but determination of maximum bending moments is not always so simple. We plot bending moment values at specific points, but how can we be sure that a maximum (or peak) value of bending moment doesn't occur somewhere *between* the points we've selected?

It can be demonstrated mathematically that where shear force is zero, the bending moment is always either (a) zero or (b) a local maximum or minimum (i.e. a 'peak' or 'trough' point).

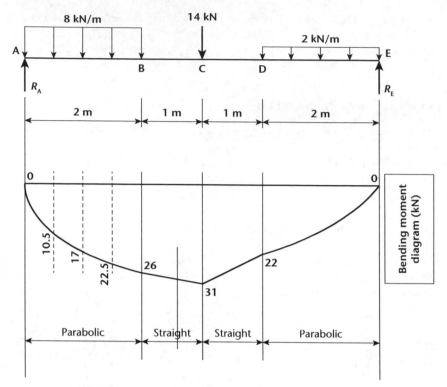

Figure 7.16　Bending moment diagram for Example 4

Looking at Figure 7.17, we see that the shear force is zero (i.e. the graph touches or cuts through the zero line) at points A, E–F and C. Looking at the bending moment diagram, we see that the bending moment is zero at points A and C and a maximum (16 kNm) all the way from E to F. If we look at Figure 7.18 we see that the shear force is zero at A, C and E. The bending moment is also zero at A and C and a maximum (31 kNm) at point C.

7.21.1 Example 5

Find the position and value of the maximum bending moment for the beam shown in Figure 7.19. To do this, we first need to draw the shear force and bending moment diagrams. The calculations are given below.

Calculation of reactions:

Vertical equilibrium:
Total force up = total force down
$R_A + R_B = (6 \text{ kN m}^{-1} \times 4 \text{ m}) = 24 \text{ kN}$

Taking moments about A:

Total clockwise moment = total anticlockwise moment
$(6 \text{ kN m}^{-1} \times 4 \text{ m}) \times 2 \text{ m} = (R_B \times 6 \text{ m})$
So $R_B = 8 \text{ kN}$

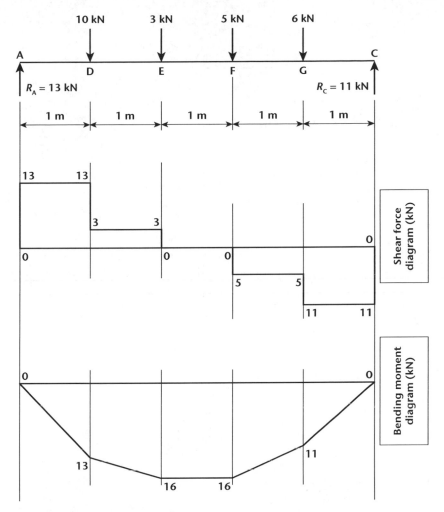

Figure 7.17 Shear force and bending moment diagrams for Figure 7.12

Taking moments about B:

> Total clockwise moment = total anticlockwise moment
> $(R_A \times 6\text{ m}) = (6\text{ kN m}^{-1} \times 4\text{ m}) \times 4\text{ m}$
> $R_A = 16\text{ kN}$

> Moment at A = 0 kNm (there is nothing to the left of A)
> Moment at B = $(16\text{ kN} \times 6\text{ m}) - (6\text{ kN m}^{-1} \times 4\text{ m} \times 4\text{ m}) = 0$ kNm
> Moment at C = $(16\text{ kN} \times 4\text{ m}) - (6\text{ kN m}^{-1} \times 4\text{ m} \times 2\text{ m}) = 16$ kNm

Now the problem is we don't know the value of the maximum bending moment. It would be dangerous (and, as we shall see, incorrect) to assume that it is 16 kNm simply because that is the maximum value we obtained from the points we selected. There is a good chance that the maximum bending moment is greater than this, but how do we determine if it is, and, if so, what value it has? The key lies in the fact, outlined above, that the maximum bending moment will always occur at a point where the shear force is zero.

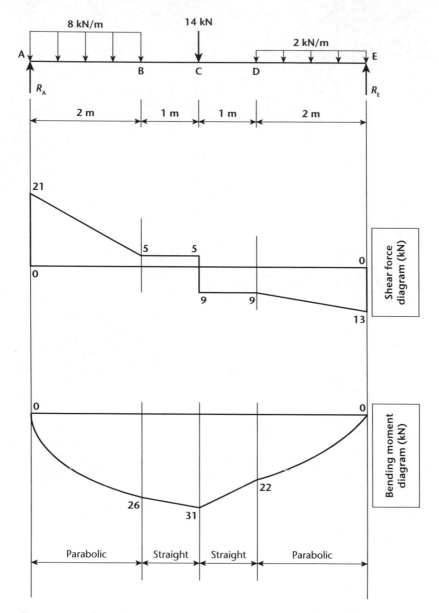

Figure 7.18   Shear force and bending moment diagrams for Example 4

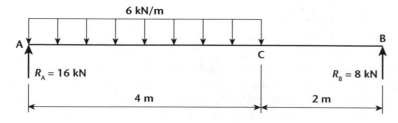

Figure 7.19   Beam for Example 5

The shear force and bending moment diagrams are shown in Figure 7.20. From this we see that the shear force is zero at a position some distance to the left of point C. Using similar triangles (Figure 7.21) or the simple technique shown in italics below, we can show that the position of zero shear force is located 2.67 m to the right of point A.

The udl between A and C is 6 kN m^{-1}. In other words, the shear force reduces by 6 kN for every metre we move rightwards along the shear force diagram. We wish to locate the point where the shear force is zero. As the shear force is 16 kN at A, it has to reduce by 16 kN to attain a value of zero. If the shear force reduces by 6 kN for every metre travelled, the distance required to 'lose' 16 kN is 16/6 = 2.67 m. We can therefore conclude that the shear force is zero (and therefore the bending moment is a maximum) at a point 2.67 m to the right of point A.

This must also be the position of maximum bending moment. Now we know this, we can work out the value of bending moment at this point in the usual way:

$$\text{Max moment} = (16 \text{ kN} \times 2.67 \text{ m}) - (6 \text{ kN m}^{-1} \times 2.67 \text{ m} \times 2.67/2) = 21.33 \text{ kNm}$$

For completeness, we'll also work out the bending moments at points D, E and F, which are 1, 2 and 3 metres, respectively, to the right of point A:

$$\text{Moment at D} = (16 \text{ kN} \times 1 \text{ m}) - (6 \text{ kN m}^{-1} \times 1 \text{ m} \times 0.5 \text{ m}) = 16 - 3 = 13 \text{ kNm}$$
$$\text{Moment at E} = (16 \text{ kN} \times 2 \text{ m}) - (6 \text{ kN m}^{-1} \times 2 \text{ m} \times 1.0 \text{ m}) = 32 - 12 = 20 \text{ kNm}$$
$$\text{Moment at F} = (16 \text{ kN} \times 3 \text{ m}) - (6 \text{ kN m}^{-1} \times 3 \text{ m} \times 1.5 \text{ m}) = 48 - 27 = 21 \text{ kNm}$$

7.22 Exercises

Draw the shear force and bending moment diagrams for the beams given in Examples 1, 2 and 3 earlier in this chapter. The answers are shown in Figures 7.22, 7.23 and 7.24, respectively.

7.23 Critical thinking and concept review

1. Explain the difference between a beam and a slab.
2. What is meant by the term deflection?
3. What is a moment? Give everyday examples of situations where a moment is applied.
4. Which is more common, a point load or a UDL? Give examples of each.
5. Explain what is meant by the term reaction, and explain how the principle of equilibrium can be used to determine reactions.
6. Draw diagrams illustrating sagging and hogging in beams, and explain in each case in which parts of the beam tension and compression will occur.
7. Explain the mechanism of shear failure.
8. How can the position of maximum bending moment be determined from a shear force diagram?
9. What is the purpose of shear force and bending moment diagrams?

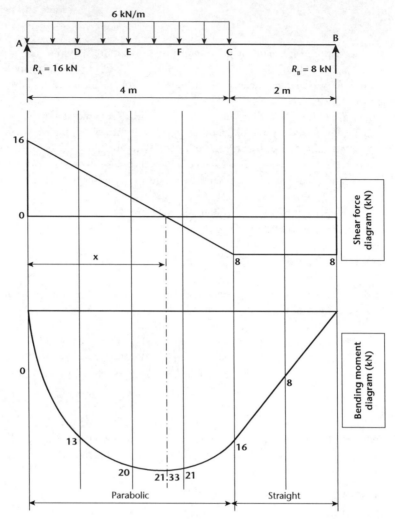

Figure 7.20 Shear force and bending moment diagrams for Example 5

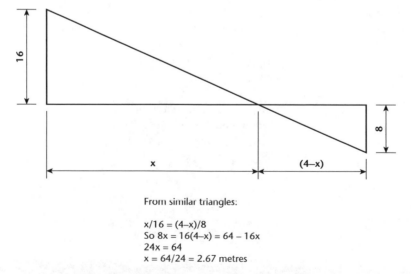

From similar triangles:

x/16 = (4–x)/8
So 8x = 16(4–x) = 64 – 16x
24x = 64
x = 64/24 = 2.67 metres

Figure 7.21 Determination of position of zero shear force in Example 5

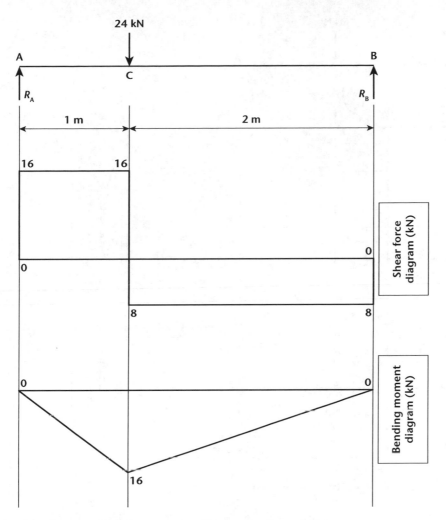

Figure 7.22 Shear force and bending moment diagrams for Example 1

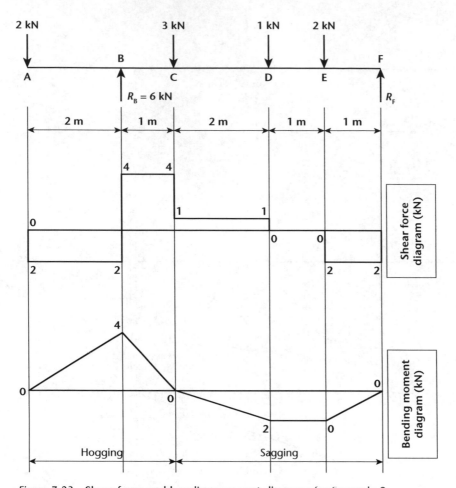

Figure 7.23   Shear force and bending moment diagrams for Example 2

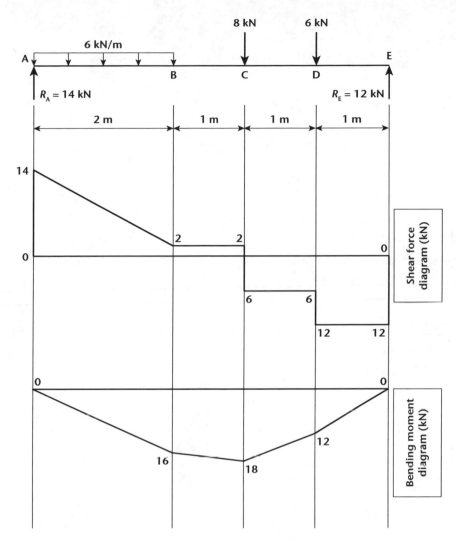

Figure 7.24 Shear force and bending moment diagrams for Example 3

Part II

Individual types and classes of materials

8

Ferrous metals

Contents

8.1 Introduction

There are about 75 metallic elements known to science, and they are divided into two groups:

1. ferrous metals – iron and steel, and all metals and alloys *based* on iron
2. non-ferrous metals – all the metals and alloys *not* based on iron.

Total world production of metals is around 1,000,000,000 tonnes per year. Of this total, 90 per cent is iron and steel, the other 10 per cent is all the other metals combined. The most widely used ferrous metal is steel so it is easily the most important metal in use in the world today for all major industries, especially construction. Steel is the major metal used as illustrated below.

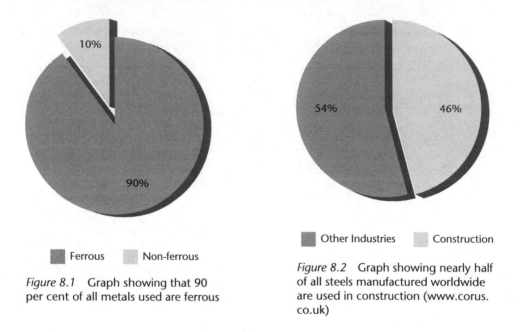

Figure 8.1 Graph showing that 90 per cent of all metals used are ferrous

Figure 8.2 Graph showing nearly half of all steels manufactured worldwide are used in construction (www.corus. co.uk)

This chapter will look into the basic chemistry, different types, properties and usage of ferrous metals. Non-ferrous metals are dealt with in Chapter 9.

8.2 What is a ferrous metal?

We need to clarify the difference between ferrous and non-ferrous metals. As mentioned above, ferrous metals are *based* on iron or have iron as the main constituent; however, non-ferrous alloys are *not* based on iron or do not have iron as the main constituent. However, most ferrous metals do contain many non-ferrous alloying elements. There are three primary reasons why ferrous metals are so widely used. First, iron-containing compounds exist in abundant quantities in the earth's crust, thus iron is cheap in comparison to other metals. Second, ferrous alloys are very versatile in that they can be tailored to have a range of physical and mechanical properties. Third, metallic iron and steel alloys can be extracted, produced or fabricated using economical techniques. Furthermore, another major advantage of steels is that they are comparatively easy to recycle and hence confer desirable sustainability properties, which is discussed in more detail in Chapter 22.

Of all materials used in construction, steel (along with concrete) is the most important and widely used. Indeed 50 per cent of all steel manufactured is utilised in construction, as illustrated in Figure 8.2.

Figures 8.3 and 8.4 are typical examples of famous landmarks in the UK, both constructed using steel structures; in the case of the football stadium, the cantilever roof is also made from steel. These examples typify the importance of steel as construction material.

8.3 Classification of ferrous alloys

There are several types of ferrous metals. The best known are steel, wrought iron, cast iron and stainless steel, of which ordinary steel is the most important in terms of usage and application. Essentially steels are iron–carbon alloys that may contain other alloying elements; in fact, it is possible to have thousands of different types of steels with different chemical compositions. As we shall see later in Section 8.5, the mechanical properties of steel are sensitive to the addition of minute amounts of carbon, which is usually

Figure 8.3 Hilton Hotel, Manchester, constructed with a core steel structure, by far the tallest building in the city (2010)

less than 1 wt%; such steels are often referred to as *plain carbon steels*. In general, many common steels are categorised in accordance with their carbon content, namely low, medium and high carbon types, as discussed later in this chapter. As there are so many different types of ferrous metals, for simplification these have been classified in Figure 8.5. It is also vitally important to grasp the chemical differences and phases that exist in ferrous alloys, namely steels and cast irons; the iron–iron carbide diagram, commonly known as the Fe–Fe$_3$C phase diagram, provides the basis for understanding the chemical composition (what distinguishes them) and mechanical properties of steels and cast irons. This is illustrated in Figure 8.6. As can be seen, there are many different possible combinations of steels and cast irons.

The iron–iron carbide diagram is undoubtedly the most important phase diagram in metallurgy. This is so as steels are the primary structural metals in practically every technological and engineering application. The possible information to be deduced from this diagram is vast – primarily for metallurgists/materials scientists. Although civil engineers and architects need not delve deeply into the metallurgical theories, it is still important to grasp the fundamental science. From the Fe–Fe$_3$C phase diagram it is important to note that steels contain up to 2.1 wt% carbon (C), although it is usually less than 1 wt%. Cast irons contain 2.1–6.67 wt% C, though most cast irons contain 3.0–4.5 wt% C. Therefore, the chemical compositions of both steel and cast iron are almost identical (iron + carbon (up to a few per cent)); thus it is imperative that we are able to differentiate and distinguish between the two.

The phase diagram above is only a portion of the iron–iron carbide phase diagram which extends to 100 wt% C; however, the portion up to 6.7 per cent is of importance as it covers all steels (up to 2.1 wt%) and cast irons (2.1–6.7 wt%). Given the majority of steels are either low or medium carbon (as discussed in Section 8.3.1) at least 70 per cent of all metals (steels) used worldwide contains less than 1.0 wt% carbon – in fact in most cases less than 0.7 wt%.

Figure 8.4 Old Trafford Stadium, home to Manchester United Football Club, and the cantilever roof structure, which is made from steel (www.manutd.com)

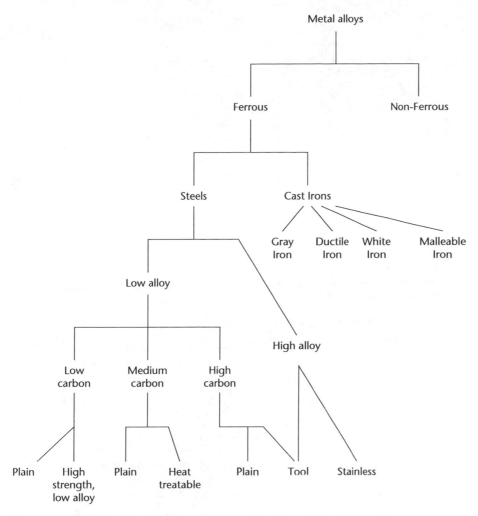

Figure 8.5 Classification of ferrous alloys

At room temperature most steels have a chemical structure/phase present called α-ferrite or α-iron, which has a BCC crystal structure. This phase exists as part of α-*ferrite* + *Fe*$_3$*C (cementite)* – this dual phase is known as *pearlite*, which is present in all steels and cast irons at room temperature. At 727–912 °C the α-*ferrite* transforms to γ-*iron* or γ-*austenite*, which has an FCC structure. However, γ-austenite starting to form alloys containing 0–0.76 wt% C as a phase of $\alpha + \gamma$ is present on the left-hand side of the phase diagram. With an increase in temperature, eventually there is a complete transformation to γ-*austenite*. The transformation temperature range is 727–912 °C as this is dependent on the carbon content, as shown on the bottom left-hand side of Figure 8.6 – we can see that as we increase the carbon from 0 to 0.76 wt% the temperature at which α-ferrite transforms to γ-iron or austenite decreases from 912 to 727 °C as it follows the solid tie line. As the temperature is increased further, the γ-austenite transforms to δ-*ferrite* (at about 1,400 °C) which has a BCC crystal structure. As the temperature reaches about 1,500 °C for steels containing up to 1 wt% carbon, we enter the liquid phase, thus the alloy melts at about 1,500 °C.

125

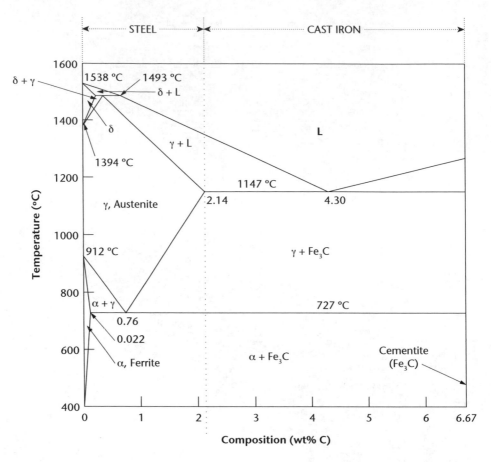

Figure 8.6 The iron–iron carbide phase diagram showing the chemical phases that exist in steels and cast irons

If we now increase the carbon content to 1.0–2.14 wt%, above 727 °C the metals transforms from pearlite to $\gamma + Fe_3C$, then as the temperature is raised there is a transformation to γ, then to *liquid* + γ (at this stage the metal starts to melt), then finally to complete liquid.

Cast irons (2.14–6.67 wt% C) have similar reactions/transformations, with the difference being that there is a larger amount of cementite (Fe_3C) present. Another important point to note from Figure 8.6 is that as the carbon content increases from 2.14 to 4.30 wt%, the melting point falls sharply from 1,300 to 1,147 °C. Thus, in general, cast irons have melting points which are considerably lower than steels (about 1,500 °C), therefore they are easily melted and amenable to casting, hence, the name.

There is a myriad of information which can be deduced from the iron–iron carbide diagram. However, it is not expected that students or professionals from the civil engineering or architectural disciplines will be required to acquaint themselves with much more than the fundamentals covered in this chapter; nevertheless, a list of indicative sources for further reading is included at the end of this chapter.

Having looked at the chemical properties of ferrous alloys and, more importantly, what distinguishes between steels and cast irons, the remainder of this section further elaborates on the different types of ferrous alloys. The following are the most common types.

8.3.1 Steel

An alloy of iron (Fe) and carbon (C). It consists of up to 1.0 per cent carbon in iron. Such steels are often referred to as plain carbon steels, which can be subdivided into three categories.

Plain carbon steels

This group of steel can be sub-categorised into low-, medium- and high-carbon steels, dependent on their carbon content.

LOW-CARBON STEELS

Of all the different types of steels those produced in the highest quantity are low-carbon steels. Typically, these contain anything up to 0.25 wt% C. Chemically, they are composed of ferrite and pearlite. Low-carbon steels are relatively soft and weak in comparison to other types of steel but have excellent ductility. In construction they are used as structural I-beams, as sheetform in buildings, bridges and pipelines. They have a yield strength of about 280 MPa, tensile strengths of around 400–550 MPa and a ductility of 25 per cent elongation (at failure they stretch up to a quarter of their original length, i.e. a 100 mm sample will stretch to 125 mm at failure). Another group of low-carbon steels are the high-strength low-alloy steels (HSLAs); in addition to carbon, other alloying elements include vanadium, nickel, copper and molybdenum – the combined alloyed content is up to 10 wt%. As a result HSLAs have higher strengths than plain low-carbon steels (of around 500 MPa) without compromising on ductility and machinability. These steels have applications in construction/civil engineering where structural strength is of great importance in bridges, support columns in multi-storey buildings and towers.

MEDIUM-CARBON STEELS

Medium-carbon steels have carbon of 0.25–0.60 wt%. These alloys have higher strength than low-carbon steels but lower ductility and toughness. However, they have a good compromise and balance between strength and ductility. The main application in civil engineering is in high-strength structural components where a combination of high strength, ductility and toughness are critical. Medium-carbon steels are also extensively used in the railway industry. They have a yield strength of about 450 MPa, tensile strengths of around 600–800 MPa and a ductility of about 15–20 per cent elongation (at failure they stretch up to one-fifth of their original length, i.e. a 100 mm sample will stretch to 115–120 mm at failure).

Most constructional steels contain between 0.15 and 0.4 wt% of carbon. Thus depending on application and strength requirements mainly medium- or low-carbon steels are used in construction (structural steel).

HIGH-CARBON STEELS

High-carbon steels usually have carbon content of 0.60–1.4 wt%. These alloys are the strongest and hardest of the plain carbon steels. However, the disadvantage is that they have the lowest ductility. As ductility and toughness are such important properties required for structural steels, high-carbon steels have limited application or usage in construction. They are mainly used as cutting tools, as well as high-strength wire. High-carbon steels can have yield strengths of up to 1,700 MPa, tensile strengths of up to 1,800 MPa and ductilities as low as 5 per cent.

8.3.2 Cast iron

Cast iron is also an alloy of iron and carbon, consisting of between 2.1–4.0 per cent carbon in iron. Therefore cast iron contains up to ten times more carbon than steel. This carbon gives it excellent casting properties, making it possible to produce many complex items and shapes by casting. This is explained by referring to the phase diagram in Figure 8.6. We can see that as the carbon content increases, especially from 2.0 to 4.2 wt%, the melting temperature reduces (refer to the liquid phase at the top of the figure). Thus, they are easily melted and more amenable to casting. In general, cast irons tend to be brittle with the exception of malleable cast iron. They have a tensile strength of 150–400 MPa. Cast iron has limited application in construction.

8.3.3 Wrought iron

Essentially pure iron, wrought implies the alloy is relatively ductile. It used to be made and used in very large quantities for construction and general engineering purposes in Victorian times (1800s and earlier), e.g. bridges, metal fencing, etc., but today it is not produced any more, except for tiny amounts made for demonstration purposes in industrial museums.

8.3.4 Galvanised steel

As we shall see in Section 8.4, iron, hence steel, is highly prone to both wet and atmospheric (attack by oxygen) corrosion; therefore, it is vital to protect steel in any application. One method is by coating the material with a protective or corrosion resistant layer. As zinc has excellent resistance to corrosion, is chemically compatible with steel and very cheap, it is the ideal material. The zinc is applied to the steel metal by a process known as galvanising. Galvanised steel has gone through a chemical process to keep it from corroding. The steel gets coated in layers of zinc because rust will not attack this protective metal. During the galvanising process the steel is submerged in melted zinc, the chemical reaction permanently bonds the zinc to the steel. Therefore, the zinc isn't exactly a sealer, like paint, because it doesn't just coat the steel; it actually permanently becomes a part of it. A chemical reaction takes place between the zinc and iron molecules, thus, forming a homogenised material. The most external layer is all zinc, but successive layers are a mixture of zinc and iron, with an interior of pure steel. These multiple layers are responsible for the amazing property of the metal to withstand corrosion-inducing circumstances, such as moisture, air or saltwater. The zinc also protects the steel by acting as a sacrificial layer. If for some reason rust does take hold on the surface of galvanised steel, the zinc will be corroded first. This allows the zinc to prevent rust from reaching the steel. The major advantage of this process is cost; thus, given the large scale of steel structures required in construction, the vast majority of steel used in construction is galvanised.

8.3.5 Stainless steel

Stainless steel is an alloy of iron, carbon, chromium (Cr) and nickel (Ni). Stainless steel was discovered by accident in 1913 in Sheffield, and the most common form contains 18 per cent Cr and 8 per cent Ni (and is usually referred to as 18/8 stainless steel). Chromium is very expensive and so stainless steel is also expensive (around ten times the price of ordinary plain carbon steel). The material is highly resistant to corrosion (rusting) in a variety of environments, especially the ambient atmosphere. Chromium is the predominant alloying element and a concentration of at least 11 wt% is required. The addition of nickel and molybdenum also enhances the corrosion resistance. As chromium has such a strong affinity for oxygen, it rapidly reacts with oxygen to form a protective passive layer (chromium oxide). As a result the stainless steel has excellent resistance to corrosion; indeed, it provides superior protection to galvanised steel, but due to cost it would be too expensive to consider using stainless steel for structural applications in civil engineering. As a result stainless steels are utilised as cutlery, gas turbines, boilers, nuclear power stations, etc.

8.4 Properties of steel

Why is steel such a popular material? The reason is that it has a remarkable combination of properties, and it also has the merit of being inexpensive. It is cheaper than any other metal except cast iron, with steel prices beginning at about £700 per tonne (2011).

In terms of its mechanical properties, it has:

1. High stiffness (high Young's modulus). This means that a steel structure placed under load will not distort or deflect elastically very much.
2. High strength (high tensile strength). This means that the steel will not break under high loads. Metals possess tensile strength, unlike stone, concrete, brick, etc. A large suspension bridge such as the Humber Bridge, or a cable stay bridge like the Dartford Bridge over the Thames would be impossible to build without steel.
3. Good ductility (capable of undergoing considerable plastic deformation before it breaks). Its ductility means that it is capable of being shaped by rolling, forging, drawing, etc. into any desired shapes (angles, channels, I-beams, etc.). Steel also has quite a high fracture toughness.

Table 8.1 summarises the main mechanical properties of the common plain carbon steels. Although, steels have in general highly desirable mechanical properties, they also have some properties that are not so good, including the following.

8.4.1 Corrosion

Steel is prone to wet corrosion. The reason is that it is a two-phase material, i.e. its microstructure consists of two phases – ferrite (virtually pure iron) and pearlite (a mixture of pure iron and iron carbide, see Figure 8.6). Wrought iron, being virtually pure iron, is single phase, and so is much more corrosion resistant than steel.

8.4.2 High density

Steel has quite a high density (7,800 kg m^{-3}), and so structures and components made from steel can be rather heavy. This is not usually a problem with buildings, but occasionally weight is important, and this can rule out the use of steel.

8.4.3 Poor fire performance

As steel is so dense and is such a good conductor of heat, it does not perform well in fire unless it is protected. Steel loses much of its strength if heated to 1,000°C. Most building fires do not reach temperatures higher than 700–800°C, but if part of a steel structure is heated to this temperature the steel will buckle or collapse under the load it is carrying. Steel also expands when heated, and expansion stresses are also responsible for some of the problems that occur with unprotected steelwork in fires.

Table 8.1 Mechanical properties of the common plain carbon steels

Type of steel	Carbon content (wt%)	Young's modulus (GPa)	Yield strength (MPa)	Tensile strength (MPa)	Ductility (%)
Low carbon	Up to 0.25	~210	~250	400–550	20–25
Medium carbon	0.25–0.60	~210	~450	600–800	15–20
High carbon	0.60–1.40	~210	Up to 1,700	Up to 1,800	5–10

8.5 Effect of carbon on steel

Carbon has a very marked effect on the properties of steel. Although we usually only add a 1.0 per cent fraction, its effect is very great. This ought not to surprise us too much, if we remember that there are more than 10×10^{22} atoms per cubic centimetre of metal; 0.1 per cent represents 1 part in 1,000, so 0.1 per cent carbon steel contains 10×10^{20} carbon atoms. Expressed like this, we can see that 0.1 per cent carbon is really quite a lot of carbon. As discussed in Section 8.3 and as shown in Table 8.1 and Figure 8.7, increasing the carbon content increases:

- the strength of the steel
- the yield strength of the steel.

Furthermore, the hardness of the steel increases with an increase in carbon. Figure 8.7 shows the tensile test plots for two low–carbon steels (0.1 and 0.2 wt% C), a medium–carbon steel (0.4 wt% C) and a high–carbon steel (0.8 wt% C). It can be seen that the higher the carbon content, the higher the load required to fail the material – indicating a rise in tensile strength with carbon content. However, we can also see that the extension at failure for the samples decreases with increasing carbon content. Thus, increasing the carbon content decreases the ductility of the steel. It affects the fracture behaviour of the steel – higher–carbon steel necks less than lower carbon steel when it fails in tension. Finally, the carbon content has no real effect on the stiffness of the steel (no effect on Young's modulus E), as shown in Table 8.1 and Figure 8.7.

Thus we can see why steel is such an attractive and the most important material for all engineering applications. It is one of the few metals which has a good compromise between strength and ductility. It is also very flexible to modify, i.e. it is possible to greatly alter its strength and/or ductility, hardness, etc. Furthermore, it is available in abundance, relatively easy to fabricate and, most importantly, very economical. As a result it is the exclusive choice for structural applications in civil engineering.

8.6 Environmental implications of steel

A major benefit of steel is its recyclability. It is a relatively easy metal to recycle in a way which requires little energy. The sustainability of steel and other materials are dealt with in greater detail in Chapter 22.

In general, the average life cycle of a building is around 60 years; Figure 8.8 shows the typical life cycle of an office building over the 60-year cycle.

When it comes to environmental considerations, the most important parts are the 'embodied energy' at the beginning and the 'end of life' at the end.

In modern-day construction (of buildings with steel) certain steps are being taken to reduce the environmental impact of steel; these include the following.

8.6.1 Demountable buildings

This is where the structures are erected in such a manner that it is possible to take apart the building structure (demount) with minimum of fuss. The main advantage is that, provided the steel structure has not

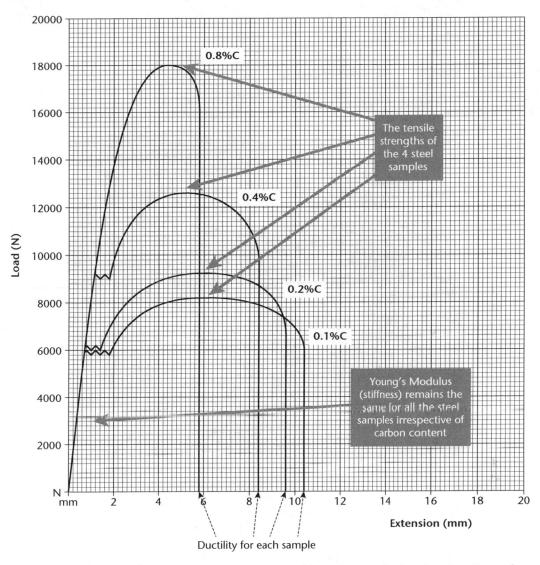

Figure 8.7 Load vs extension graphs for tension tests on plain carbon steels, showing the effects of increasing carbon content

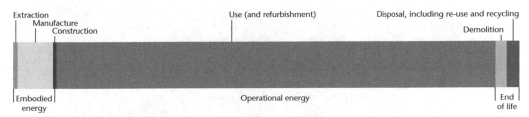

Figure 8.8 Typical life cycle of steel in construction applications

(After www.corus.co.uk)

suffered any damage, it can be re-erected and hence re-used. For this kind of technology the steel beams are erected using mechanical fixings such as bolting, as illustrated in Figure 8.9.

8.6.2 Manufacturing off-site and site erection

Another advantage of steel is that it can be manufactured or fabricated off-site and then erected on-site. The advantages of such a process are:

Figure 8.9 Prefabricated steel structures at the Emirates Old Trafford Cricket Ground, Manchester

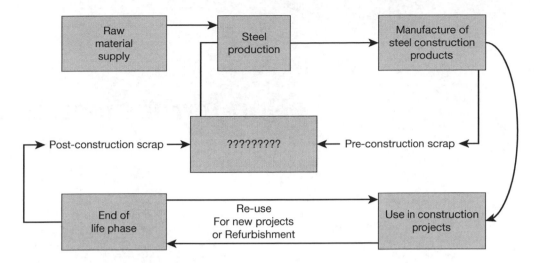

Figure 8.10 The typical life cycle for steel used in civil engineering applications, showing how sustainable the material is

- large-scale volume production;
- pre-engineered to required dimensions;
- precision made to manufacturing tolerances;
- choice of shape and size, ductile material;
- zero waste;
- quality assured, low–defect products;
- fast erection time;
- reduced snagging on site.

Figure 8.10 shows the life cycle of steel in civil engineering applications. Thus, in theory all steel used should be either recycled or re-used, which is a huge environmental and economical advantage (see Chapter 22). Currently in the UK at least 85 per cent of steel used in construction is re-used or recycled; furthermore, 99 per cent of structural applications are recycled or re-used as shown in Table 8.2. Thus, only *1 per cent* of structural steel is lost to landfill or rust.

References and further reading

ASKELAND, D.R. (2013), *The Science and Engineering of Materials*, fourth edition, Wiley.
CALISTER, W.D. (2007), *Materials Science & Engineering*, seventh edition, Wiley.

Table 8.2 Recycling and re-use of construction steel in the UK

	Structural sections	Purlins and rails	Cladding	Decking	Rebar	Internal non-structural steel
Recycling %	86	89	79	79	91	85
Reuse %	13	10	15	6	1	2
Total %	99	99	94	85	92	87

Source: www.corus.co.uk.

9

Non-ferrous metals

Contents

9.1 Introduction

In the previous chapter we looked at ferrous metals. Although non-ferrous metals account for only 10 per cent of all metals utilised in construction, nevertheless they are a very important group of materials in construction and other industries, as we shall see in this chapter.

9.2 What is a non-ferrous metal?

First, we need to clarify the difference between ferrous and non-ferrous metals. Ferrous metals are based on iron or have iron as the main constituent; however, non-ferrous alloys are *not* based on iron or do *not* have iron as the main constituent. This does not mean that non-ferrous metals do not contain iron – several non-ferrous metals contain iron as an alloying element (less than 10 per cent content), thus it is vital to distinguish between difference between these two types of metal.

The main advantage with non-ferrous metals is that generally they are relatively light in comparison to ferrous alloys and in general they have better corrosion properties. Table 9.1 gives a list of various non-ferrous alloys (mainly those used in construction) with the densities for comparison with steel.

As can be seen, both aluminium and titanium are substantially lighter than iron (and therefore also steel), and thus are extremely useful for low-density, high-strength applications. Also, the tensile strengths (and

Table 9.1 Non-ferrous alloys with chemical formula and density and comparison with iron

Metal	Chemical formula	Density (kg m⁻³)	Ultimate tensile strength (MPa)
Aluminium (alloys)	Al	2,700	Up to 600
Copper (alloys)	Cu	8,900	Up to 600
Lead	Pb	11,300	15
Nickel	Ni	8,900	500
Titanium	Ti	4,500	Up to 1200
Zinc	Zn	7,100	Up to 200
Iron (steel)	Fe	7,900	Up to 1000 (steels)

other properties) can vary greatly for any metal as this can be modified by treating the metal, i.e. heating and subsequent cooling, modifying the microstructure, etc. Therefore, in Table 9.1 there is a wide range of ultimate tensile strength values for some metals.

9.3 Aluminium alloys

Aluminium is one of the most abundantly available metals on earth. Given its low density, impressive strength and, more importantly, excellent resistance to wet corrosion, aluminium is one of the most widely used and inexpensive engineering materials.

In construction, aluminium alloys are used externally for window frames and structural glazing systems, roofing and claddings, flashings, rain-water goods, etc. It is used internally for ceilings, panelling, ducting, light fittings, vapour barriers (as aluminium foil), architectural hardware, walkways, handrails, etc. Aluminium has the merit of being light (density 2,700 kg m⁻³ as compared to 7,900 kg m⁻³ for steel) and ductile, and easily rolled into sheet and thin strip, and extruded into the complex sections required for window frame manufacture. It resists corrosion, especially if it is anodised. However, it has only one-third of the stiffness of steel, and it melts at the relatively low temperature of 550–600 °C (in comparison to approximately 1,500 °C for steels). Its performance in fire is therefore not good, and so it cannot be used structurally in buildings.

Aluminium has high electrical and thermal conductivities; as cables it can be utilised for the transmission and distribution of electrical power. Since the late 1990s aluminium has replaced copper conductors for these applications, and is the standard material for electrical conductors. These conductor designs have consistently provided a superior combination of strength and conductivity for distribution and transmission applications. Aluminium has a distinct advantage over copper as it is nearly 70 per cent less dense.

Aluminium has an FCC crystal structure, thus its ductility is retained even at very low temperatures. Aluminium can be alloyed with lithium to further enhance its properties. These materials have lower densities (2,500–2,600 kg m⁻³) in comparison to other metals, especially ferrous alloys, high-specific moduli (elastic modulus–specific gravity ratios), and excellent fatigue and low temperature toughness properties. Aluminium–lithium alloys are mainly utilised in the aerospace industry.

Due to its immense versatility aluminium is also used in paints as a primer for use on all types of wood and aged creosoted or bitumen-coated surfaces. It is ideal for use on fire damaged wood after appropriate treatment. An aluminium–zinc coating is also used as an anti-corrosion treatment of metal-coated steel. The coating applied to steel provides enhanced corrosion protection and can improve steel lifetime by up to four times that of galvanised steel under the same conditions. Furthermore, aluminium paint is commonly used for interior and exterior use where a durable and bright reflective finish is required, ideal for use on heated pipework. There are many different types of aluminium alloys – Table 9.2 shows a selection.

Table 9.2 A selection of various aluminium alloys and their general application in construction and other industries

Composition of aluminium alloy (wt%)	Ultimate tensile strength (MPa)	Application
99.9 Al/0.1 Cu	90	Heat exchangers, light reflectors
98.58 Al/0.12 Cu/1.2 Mn/0.1 Zn	110	Pressure vessels and piping
97.25 Al/2.5 Mg/0.25 Cr	230	Wires (electrical) and rivets
97.9 Al/1.0 Mg/0.6 Si/0.3 Cu/0.2 Cr	240	Pipelines and furniture
95.65 Al/1.3 Cu/0.95 Mg/2.0 Li/0.1 Zr	465	Aeroplane structures

9.3.1 Environmental implications of aluminium

A major benefit of aluminium is its recyclability. It is a relatively easy metal to recycle and requires little energy. The sustainability of aluminium and other materials are dealt with in greater detail in Chapter 22.

9.4 Copper and copper alloys

Copper is used extensively for a variety of applications, especially in the construction industry; copper is used in building for plumbing pipes and fittings, and sometimes in sheet form for roofing. Pure copper melts at around 1,080 °C, and is used in pure sheet form to cover roofs. It has a density of 8,900 kg m^{-3}. The weight of copper is rarely a problem. Copper is used primarily because of its excellent electrical conductivity and its high thermal conductivity, as well as its good corrosion resistance.

Copper is a reddish-coloured metal; it has its characteristic colour because of its band structure (chemistry). In its liquefied state, a pure copper surface without ambient light appears somewhat greenish, a characteristic shared with gold. When liquid copper is in bright ambient light it retains some of its pinkish lustre. Due to its chemical structure, copper has very high thermal and electrical conductivity, and is malleable. It has resistance to wet (water) corrosion. Copper is a metal that does not react with water, but the oxygen of the air will react slowly at room temperature to form a layer of brown–black copper oxide on copper metal. It is important to note that in contrast to the oxidation of iron by wet air that the layer formed by the reaction of air with copper has a protective effect against further corrosion.

Copper can be alloyed with zinc to form the alloy brass or with tin to form the alloy bronze. These alloys, especially brass, find many applications in plumbing and electrical fittings, e.g. bathroom taps. There are many different types of copper alloys – Table 9.3 shows a selection.

9.5 Lead

Lead is a soft, silvery-white or greyish metal. It is very malleable and ductile; lead melts at a very low temperature of 327 °C, and is a very dense, heavy metal (density 11,300 kg m^{-3}). Chemically it is not very

Table 9.3 A selection of copper alloys and their general application

Composition of copper alloy (wt%)	Ultimate tensile strength (MPa)	Application
99.9 Cu	~230	Plumbing pipes and electrical wiring
70 Cu/30 Ni	~400	Salt-water piping, heat exchangers
~70 Cu/~ 30 Zn (brass)	230	Radiator fittings, bathroom fittings, plumbing applications
~90 Cu/10 Sn/2 Zn (bronze)	~300	Steam fittings, piston rings, bearings

reactive. Lead is commonly used in construction in sheet form for roofing, flashings and lead sheet cladding. Lead is dense and heavy, but it is quite corrosion resistant. It can be repeatedly bent and folded without failing, a property which is useful when forming flashings for roofing purposes. It is also alloyed with tin to form solder; this is used mainly for making electrical and plumbing joints. The microstructure of solder (lead–tin) is discussed in detail in Chapter 5.

9.6 Zinc

Zinc is a relatively soft metal with a low melting temperature and a density of $7,100 \text{ kg m}^{-3}$. In construction zinc is used to a small extent for roofing and cladding, but its main importance in construction is as corrosion protection for steel as galvanising. In galvanised steels, a thin coating of zinc is applied to steel items to protect them against wet corrosion. This process is cheap, and it significantly extends the life of steel items in service, i.e. it improves the durability of steel. This is especially economical in comparison to stainless steel, which is expensive due to the high price of chromium (more than 10 per cent of this metal is required for stainless steel). Thus, for large-scale production galvanising is always preferred; hence, nearly all structural steel beams used in construction are galvanised. Other non-construction applications of zinc include automotive parts (door handles and grilles), padlocks and office equipment.

Zinc is also a very important constituent in paints. Zinc chromate and zinc silicate primers are used as a protective coating in metal components, providing superior corrosion resistance and additionally protecting the surface from proliferation of organic matter.

Zinc oxide powder, ZnO, is used as a pigment, in compounding rubber and in the manufacture of plastics. Other primers include zinc phosphate primer, which is a versatile, fast-drying, high-build primer with excellent rust-inhibiting performance.

9.7 Titanium

Titanium is a light, strong, lustrous, corrosion-resistant (including to sea water and chlorine) transition metal with a greyish colour. Titanium can be alloyed with iron, aluminium, vanadium, molybdenum, among other elements, to produce strong, lightweight alloys for a variety of applications, especially aerospace. Titanium is recognised for its high strength-to-weight ratio. In addition to being light and strong with low density, it has good ductility. The relatively high melting point (about 1,700 °C) makes it useful as a refractory metal. Commercial (99.2 per cent pure) grades of titanium have ultimate tensile strength of about 435 MPa, equal to that of some steel alloys, but are 45 per cent lighter (density is $4,400 \text{ kg m}^{-3}$, whereas steel is nearly 8,000). Titanium is 60 per cent heavier than aluminium, but more than twice as strong as the most commonly used aluminium alloy. Certain titanium alloys achieve tensile strengths of over 1,380 MPa. It is a fairly hard material (although not as hard as some grades of heat-treated steel), non-magnetic and a poor conductor of heat. Like those made from steel, titanium structures have a fatigue limit which guarantees longevity in some applications. Although titanium is about half as light as steel, with similar strength and highly superior resistance to corrosion, it is very expensive. Therefore, its application is restricted to high-risk areas, e.g. aerospace industry. Otherwise it would replace steel in most construction and automobile applications.

9.8 Critical thinking and concept review

1. What is the difference between ferrous and non-ferrous metals?
2. How do the properties of non-ferrous metals compare with ferritic metals?
3. What are the advantages of non-ferrous metals, especially aluminium and titanium?
4. Why are copper and copper alloys so useful?
5. Which non-ferrous metal is used in galvanised steel?

This chapter deals with glass materials and glazing, which is widely used in building construction. Glass has been used for glazing for many centuries, as well as for bottles and containers, drinking vessels and decorative objects. In the twentieth century, the use of glass widened to include fibres for reinforcing polymers, glass wool for thermal insulation and, following advances in electronics, the use of long fibres for fibre-optic devices. The chemistry and structure of glass is dealt with here, as well as the processes of glassmaking and the shaping and forming of glass into the various products used in buildings. The durability of glass is discussed, as well as the various problems that arise with the use of glass. While glass is transparent, it also has low fracture toughness, and this brings problems in its wake. These problems and the ways in which they may be mitigated are discussed here. The mechanical properties of glass which make it useful as a reinforcement material for composites, and the production of fibres for this purpose, are also discussed.

Contents

10.1 Introduction

Glass artefacts were first made in ancient Egypt, and in Europe since the Iron Age. All these glasses were made from naturally occurring raw materials and contained a major percentage of silica (SiO_2), a glass-forming oxide. These early artefacts were vessels and containers of various types, with some being made for utilitarian purposes and others for artistic/decorative reasons.

Today, glass is made for many purposes: glazing for buildings, containers, ovenware, for optical purposes and for the production of glass fibres – both for optical reasons and for the reinforcement of composites. Decorative and artistic objects are still made in large numbers. Glass is a hard, transparent solid at room temperature, although it is often described as a super-cooled liquid, and sometimes as an inorganic polymer. The important property of glass is its transparency, and it finds its use in building construction primarily for this reason. Glass is used in windows to admit daylight while keeping out the weather, and also retaining the internal heat of the building.

Glass can also be given a lustrous appearance by the addition of lead to the mix, and so can be used to make attractive drinking vessels and containers. It can be coloured by the addition of other materials, and there is virtually no limit to the range of aesthetically pleasing objects and works of art that can be made from glass. A visit to any Venetian gift shop will quickly confirm this assertion.

However, it is not only the transparency of glass that dictates its usage; it has useful mechanical properties too. The physical properties of glass are given later, in Table 10.2. It can be seen that glass possesses a combination of high stiffness and strength with a modest density value, and this makes it attractive to use as reinforcement in the production of light but stiff articles.

10.2 Glassmaking

The raw materials for making flat glass would be as follows:

> sand, soda ash, limestone ($CaCO_3$), dolomite (mixture of $CaCO_3$ and $MgCO_3$), feldspar (contains potassium, sodium, aluminium, silicon and oxygen), sodium sulphate (Na_2SO_4) and cullet (broken/scrap glass).

The raw materials are put into a furnace and heated until they have fused to form a glassy mass of material. This material can then be fed into the appropriate process for the manufacture of flat glass, glass fibres (both short and long), glass wool or matting for insulation, light bulbs, containers such as bottles and jars, etc. Technological advances have changed the market for glass products quite dramatically over recent years – for example, much glass was used in the manufacture of cathode ray tube envelopes for use in television sets, computer monitors, oscilloscopes, etc. Flat-screen technology and high-speed analogue-to-digital conversion devices have rendered the cathode ray tube obsolete. Polymeric materials have taken other markets for transparent materials. Aircraft windows are made from polymethylmethacrylate (Perspex), because it is significantly lighter in weight than glass. In situations requiring high security, such as for glazing in banks, transparent forms of polycarbonates are now used because they have higher fracture toughness than glass.

On the other hand, certain developments have created new markets for glass products. One good example is the requirement for long, continuous fibres for fibre-optic applications. Copper is a metal of strategic importance, and it is also expensive and in short supply. Many years ago, electrically controlled servo-motors replaced metal cables, levers and bell-cranks for the control of aircraft in flight. Opto-electronic devices have now replaced many electrical systems, thus obviating the need for copper wiring to connect the servo-motors to the flight controls. Glass fibres are lighter and cheaper than copper wires, and the signals transmitted consist of light pulses instead of pulses of electricity. Optical control systems are

Table 10.1 Composition and characteristics of the main commercial glass types

Glass type	Composition (% by weight)						Characteristics & Applications
	SiO_2	Na_2O	CaO	Al_2O_3	B_2O_3	Other	
Fused silica	99						Very low thermal expansion, very high viscosity
96% silica (Vycor)	96				4		Very low thermal expansion, very high viscosity
Borosilicate (Pyrex)	81			2	12		Low thermal expansion
Container glass	74	15	5	1		MgO 4	Easy workability, high durability
Float glass	73	14	9			MgO 4	High durability
Glass fibres	54		16	14	10	MgO 4	Low alkalinity
Lead glass tableware	67	6				PbO 17	High refractive index
Optical flint	50	1				PbO 19	Specific index and dispersion values
						BaO 13	
						K_2O 8	
						ZnO 8	

currently finding applications in many different areas, including buildings, and so the demand for glass fibres is likely to increase.

Since we are considering materials for construction, we shall consider flat glass, glass fibres and glass matting for insulation here. Flat glass is required for glazing, and glass wool for insulation, whereas short glass fibres are needed for glass reinforced plastic (GRP) materials and long fibres for control systems. Glass is made for a wide range of applications, and the composition is varied depending on the properties required and the intended application. The compositions of the various commercial glass types are given in Table 10.1.

10.3 Structure of glass

Glass for building is made from two of the most abundant elements in the Earth's crust, namely *silicon (Si)* and *oxygen (O)*, in the form of silica (SiO_2). Glass was invented a very long time ago, but it is also a component of some of our most recently developed technology, e.g. lasers and fibre-optic systems. Although it is strictly a ceramic material, it is also referred to as an *inorganic polymer*, and it has both ceramic-like and polymer-like properties. It is sometimes referred to as a 'super-cooled liquid', and sometimes as 'an inorganic product of fusion which has cooled to a rigid condition without crystallisation'. These names tell us that it is not a true solid with a fixed melting point, but rather a material which gradually softens as it is heated, and eventually becomes liquid at high temperature.

In fact, silica is only one of a group of so-called 'glass formers', refractory oxides which will form glasses when heated to elevated temperature. Other glass formers include germanium dioxide (GeO_2), boron oxide (B_2O_3) and phosphorus oxide (P_2O_5). Germanium dioxide is rare and therefore is not often used, but boron oxide is added to silica in the production of Pyrex ovenware. This addition confers a low coefficient of thermal expansion on the glass, thereby reducing its liability to thermal shock. Pyrex is eminently suitable for making ovenware because of this. Pyrex was also used to make the 200-inch mirror fitted to the Mount Palomar reflecting telescope, because of its low coefficient of thermal expansion. The highest possible dimensional stability was required for this telescope so that it can form accurate, distortion-free optical images of the night sky and of very distant stellar objects

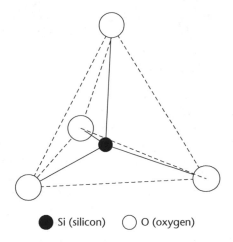

Figure 10.1 The SiO$_4$ tetrahedron – the building block of the glass structure

Silicon is tetravalent (it forms four bonds), and when heated in the furnace to make glass it forms a tetrahedral structure with oxygen, which is the basic building block (or monomer) of the glass structure (Figure 10.1).

Silica can form a truly crystalline structure – e.g. quartz, which is the structure shown in Figure 10.2a. *Extremely slow* cooling is required to form quartz crystals. Such slow cooling rates occur under the Earth's crust. The times involved are measured in millions of years in some cases.

If we cool silica in a furnace to make glass (where the cooling rate is much faster than that required to produce quartz), we obtain an amorphous structure as shown in Figure 10.2b. The 'melting' temperatures of both crystalline and amorphous quartz are so high that they would not be of much practical use, as they would need heating to very high temperatures (well in excess of 1,000 °C) to be worked or shaped. This would make glass too expensive for many purposes. Therefore, so-called 'network modifiers' are added to the mix to make it possible to work and shape glass at much lower temperatures. Sodium and calcium atoms can act in this way, as illustrated in Figure 10.2c. The modifiers break up the silica structure network,

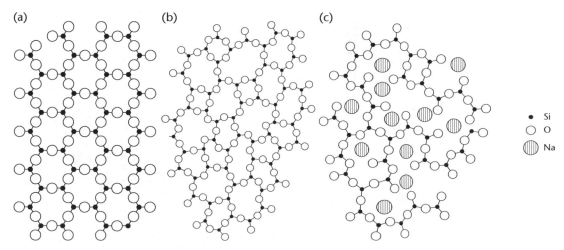

Figure 10.2 (a) Quartz (pure silica); (b) quartz glass (amorphous); (c) Soda glass structure, with network modifiers (Na)

reducing the number of bonds per unit volume, and so making it easier to soften the material. We shall see the practical effects of this in Section 10.4, which deals with the rheology of glass.

10.4 Effect of temperature on the rheology of glass

As has been mentioned already, glass is really a super-cooled liquid, and as such it does not have a defined melting temperature. Indeed, it does not melt in the conventional sense of the term. Rather, it softens as it is heated; its viscosity decreasing with increasing temperature until it becomes truly liquid. Figure 10.3 illustrates how viscosity reduces as glass is heated.

The four curves in Figure 10.3 are for pure silica, Pyrex (borosilicate glass), container glass and lead glass. The trend shown is quite clear – as the temperature is increased the viscosity of the four glass types reduces. In everyday parlance – the glasses become more 'runny' or more liquid. The *strain point* is the temperature at which any internal stresses are reduced significantly in a matter of hours. The *annealing point* is the temperature at which any internal stresses are reduced to a commercially acceptable level in a matter of minutes.

Below the strain and annealing point lines lies the working range, bounded by two horizontal dashed lines. In some texts the upper and lower lines of this range are designated the softening point and the working point. However, from our point of view, the working range is the important temperature range for shaping glass products.

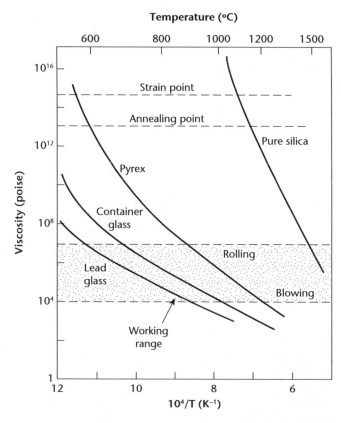

Figure 10.3 How viscosity (η) of glass reduces as it is heated. It obeys an Arrhenius law ($\eta \propto \exp(Q/RT)$) at high temperature

(After Ashby & Jones, 1980)

10.5 Shaping and forming of glass

In order to shape glass into the various products with which we are familiar, the glass must be heated to an elevated temperature. The temperature chosen depends upon the composition of the glass. It is important to recognise that, when heated, glass does *not* melt in the conventional sense. As has already been mentioned, there is no fixed temperature at which glass passes from solid to liquid – rather the glass progressively softens as the temperature is raised. This is why glass is sometimes called a 'super-cooled liquid'.

We can distinguish a number of glass-shaping processes:

- production of flat glass for glazing;
- production of glass fibres, short and long;
- production of glass wool;
- production of bottles and containers.

Of these, the first three are of interest to the construction industry.

10.5.1 Production of flat glass

The transparency of glass has made its use desirable in the construction of buildings for centuries. However, the difficulties in producing flat glass for glazing purposes have led to many different production methods over the years.

The earliest methods of making flat glass involved blowing the glass into cylinders or muffs. These were then split along their length, and while still hot they were gradually opened up and flattened out. The glass so produced would not be exactly flat or of uniform thickness, but this did not matter as the sheets were cut up into small diamond shapes, squares and rectangles. The pane of glass to fit the window opening was then assembled like a jig-saw from the shaped pieces, which were held together by strips of lead with grooves on each side (these were called 'cames'). The cames were cut to length and could be soldered together to produce the window pane. The traditional lattice window panes were made his way. At this time (fifteenth and early sixteenth centuries), glass was very expensive indeed, and only very rich people could afford to build houses with large windows.

Most of the glazing in Georgian windows was not produced from blown cylinders but by spinning. The glass-worker would first get a lump of molten glass on the end of his long iron blowing tube. He would blow an initial bubble and then rest the tube on a special forked rest and spin it round as fast as he could while flattening out the bubble using a wooden bat held in his other hand. By this means he could produce a disc of anything between 1.0 and 1.35 metres in diameter, and occasionally of 1.5 metres. The glass disc would have a thick boss at its centre where the iron tube was attached; the tube would be broken off the boss, and the disc could then be cut into square shapes, diamond shapes, etc., for making up the window pane with cames. The boss was usually sold cheaply to someone who could not afford to have windows of any size made for them.

The next method used for making flat glass enabled wider plates to be produced. In this process a steel plate was lowered into a furnace full of molten glass, until the glass had wetted the lower edge of the steel. The steel plate was slowly raised from the furnace, drawing a film of hot glass out of the furnace. The rate of withdrawal had to be carefully adjusted to allow the glass time to cool and stiffen enough to carry the weight of the film.

This process allowed larger glass plates to be made, but the glass so produced was of uneven thickness, and therefore its optical properties were less than satisfactory. As retail activity developed during the latter part of the nineteenth century, demand grew for superior glass of uniform thickness so that goods could be clearly displayed in shop windows to potential purchasers.

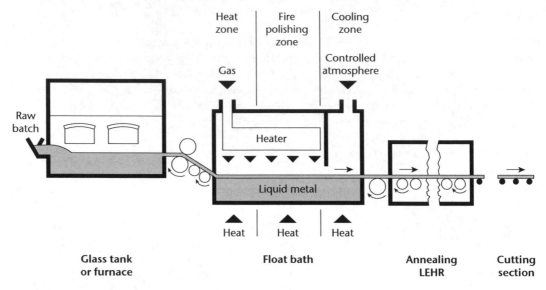

Figure 10.4 Diagram of the Pilkington float glass process

Plate glass used to be made by softening and rolling the glass to a rough cast ribbon. The ribbon was then annealed to remove all residual strains and then the two surfaces were simultaneously ground and polished using sand and emery suspensions of progressively finer grain sizes. The twin grinders obviously had to extend over the full width of the ribbon, and this process was expensive.

However, in 1959 the Pilkington Glass Company in the UK came up with a seminal invention; this was the *float glass* process. This makes it possible to produce perfectly flat, uniform thickness plates of glass in large widths without the need for the expensive grinding and polishing that was always necessary previously. The float glass process just uses heat and gravity. Figure 10.4 shows a schematic diagram of the process.

The first part of the process is the same as for plate glass. The ribbon of glass is then drawn across the surface of a bath of molten tin. The hot glass 'floats' on the tin and therefore becomes perfectly flat, smooth and of uniform thickness. As the heat is applied to both faces of the glass – this is 'fire polishing' – it gives perfectly flat and parallel faces to the glass sheet. An inert atmosphere protects the tin from oxidation.

The selection of tin as the liquid metal on which to 'float' the glass was crucial. It had to be a metal with a sufficiently low melting temperature for it to stay liquid as the glass cooled to a point where it began to stiffen. The material had to have a significantly higher density than glass, and it was also essential that the glass did not react chemically with the liquid metal. Tin melts at 505 K (232 °C) and has a density of 7,300 kg/m^3 (nearly three times that of glass). It meets all the technical requirements for a substrate material and has proved to be excellent for the purpose.

The Pilkington Company protected the process with patents, but an American company quickly developed another version, and the float glass process is now the worldwide standard method for the production of flat glass.

10.5.2 Production of short glass fibres

This is an important field of use for glass as glass fibres are used for the reinforcement of plastics to make light but stiff composite materials. The combination of physical and mechanical properties of glass that make it suitable for use in composites is set out in Table 10.2.

Table 10.2 Representative physical properties of glass

Property	Representative value and units
Density (ρ)	2,600 kg/m^3
Thermal conductivity (k)	1 W/m K
Specific heat capacity (C_p)	670 J/kg K
Coefficient of thermal expansion (α)	9×10^{-6}
Tensile strength (σ_{TS})	100 MN/m^2
Young's modulus of elasticity (E)	71 GN/m^2
Melting temperature (T_m)	1,400 K

To make fibres it is necessary to thermally soften the glass and then to stretch it into filaments which are then rapidly cooled to give the fibre material. Short fibres up to 5–10 mm in length for reinforcing plastics are made in this way. Softened glass is poured into a centrifugal spinner. The glass emerges in filaments from the holes in the spinner, and jets of cold air play on these filaments, quickly cooling them. They fall onto a conveyor and are collected; this is shown in Figure 10.5.

To be really satisfactory for use as reinforcement material, the fibre surfaces must be pristine and free from even the tiniest scratches and defects. Although a scratch may be invisible to the naked eye, such scratches represent stress raisers, and are therefore potential sites for failure. Therefore, short fibres are best used when freshly made, and before there is time for any surface degradation to occur.

10.5.3 Production of long, continuous fibres

Optical fibres can be made from almost any continuous glass fibre if it is not too long and the light losses are tolerable. Modern fibre-optic systems are multi-channel, using modulated-light, and these systems need

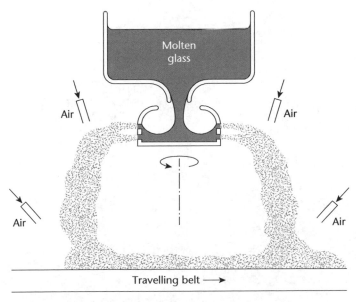

Figure 10.5 Discontinuous glass fibre production by centrifugal spraying

(After Van Vlack, 1982)

to be capable of transmitting light over great distances. This means that the composition, structure and processing of these fibres must be closely controlled. The glass used must be free from all oxides that will absorb light. The transition metal oxides such as FeO tend to absorb light to an unacceptable degree. Absorption can also be caused by chemically combined water, present as −OH groups in the glass. This problem can be reduced by making sure that the gas streams are as dry as possible, and by excluding the atmosphere during the pre-form and fibre-drawing processes.

The comments made above about the need for the fibre surfaces to be free from scratches and other defects applies to optical fibres too. Such scratches will cause dispersion effects, resulting in transmission losses.

Continuous fibres are made by drawing molten glass through a spinnerette (a steel plate with an array of fine holes through it). The glass filaments emerge and are wound onto a take-up drum or winder at high speed (300–500 m/sec); as shown in Figure 10.6, several hundred filaments can be drawn together.

10.5.4 Production of glass wool insulation

Glass wool can be made using a centrifugal spinner. Jets of air can be directed downwards onto the emerging filaments so that they remain continuous, and then the fibrous material can be sprayed with an organic binder so that the familiar glass wool blanket material is produced. A schematic of the process is given in Figure 10.7. Glass wool works as insulation because of the pockets of air caught in the mass of fibres; air is a poor conductor of heat, and so the wool acts as a glass–air composite and provides good thermal insulation. It is easy and convenient to apply inside buildings, roof structures, etc.

The manufacture of bottles and containers from glass is a large industrial sector, but it is not relevant to construction, and so it will not be considered here.

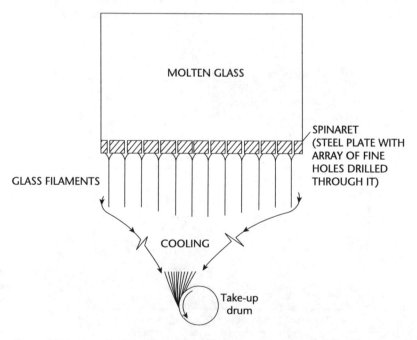

Figure 10.6 Continuous fibre production by viscous drawing through a spinnerette at high velocities (After Van Vlack, 1982)

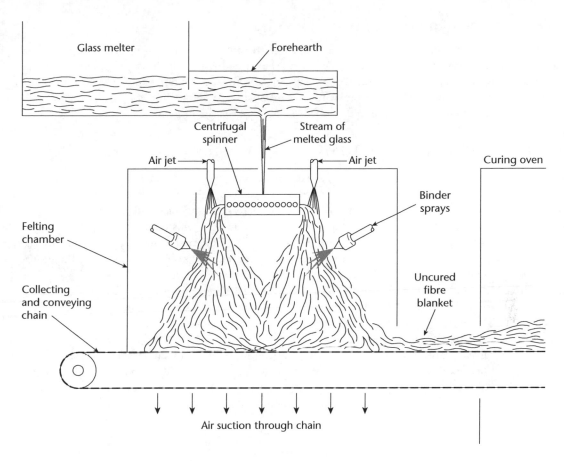

Figure 10.7 Rotary process for glass wool production

(After Mohr & Rowe, 1978)

10.6 Durability of glass

While glass is strong and stiff, it has a low fracture toughness, which means it is brittle and reacts poorly to impact situations. However, it resists weathering very well, as can be seen by the survival of so much medieval stained glass in old churches in the UK and Europe. Glass in general has a very high resistance to the corroding action of water and atmospheric agencies. Old glass may not be so resistant. Glass windows will resist the action of the atmosphere and its contents for many years. Most glazing is fixed in a vertical position, and this helps because rainwater (and its contents) runs off immediately. However, under certain conditions, glass can be corroded, so we must not think that glass is an insoluble material.

Glass is readily dissolved by hydrofluoric acid (HF), which is commonly used in the etching of glass. Glass has good resistance to attack by most other acids. However, it can be attacked by alkaline solutions, quite severely in some cases. Glass is attacked by lime and by cementitious water. This means that water that has been in contact with fresh cement, mortar or concrete can be quite corrosive to glass. It is important to prevent such water from flowing over glass windows and glazing during construction.

10.7 Tempering of glass

One of the major problems with glass is its inherently low fracture toughness, or put another way, it is a naturally brittle material. Glass breaks with a chonchoidal fracture surface, which is another indication that it is in fact a super-cooled liquid. If a sheet of glass is subject to flexure (bending), the highest stress will be at the surface or extreme fibre. Therefore, it follows that we can improve things by making glass with residual stress of a compressive nature in the surface layers of our sheets of glass.

10.8 Problems with glass and glazing

Glass is an essential material in building due to its combination of *transparency* and *durability*. Most buildings are designed and constructed with windows, and beside admitting daylight, they can enhance the appearance of the building from both inside and out. Glass has been increasingly popular with architects, and it has been used extensively in many urban buildings, including some very tall structures. Many modern and striking buildings now incorporate large areas of fenestration. While offering many attractive and beneficial aesthetic and technical advantages, the use of glass does bring certain problems in its wake. Traditional single-glazed windows can be regarded as areas of weakness in many respects.

We have looked at the physical properties of glass in Table 10.2. From this we can see that while glass is not a good conductor of heat, at least compared with metals, it does have a modest value of thermal conductivity. Therefore, heat losses through large areas of single glazing will be unacceptably large, as we now live in an energy-conscious world.

Another thermal problem is that of solar gain, or the greenhouse effect. Incoming sunlight is short-wave radiation, and heat radiated back to the windows from warm objects inside the building will be long-

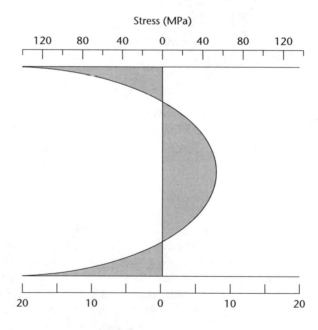

Figure 10.8   Room temperature residual stress distribution over the cross-section of a tempered glass plate

(After Kingery et al., 1976).

wave radiation. Ordinary glass is transparent to incoming sunlight, but relatively opaque to the heat energy trying to exit the building interior. If we have large areas of windows, then solar gain can make the building interior uncomfortable for the occupants. Energy-consuming air-conditioning is not the desired solution.

Large panes of glass will have quite high natural frequency values, and sound transmission will potentially pose problems, especially in urban areas. The glass is able to vibrate in synchrony with incident external sound waves and so transmit the sound into the interior of the building. These thermal and acoustic problems are the continuous day-to-day issues that the building designers have to contend with. An exceptional hazard is that of fire, and there are potential difficulties with single-glazed windows in fire situations.

Glass has a low value of fracture toughness (not shown in Table 10.2), which makes it susceptible to impact by flying objects, and to the blast waves generated by nearby explosions, should one occur. Therefore, the principal problems with glass windows come under the following headings:

- thermal problems;
- sound or acoustic problems;
- fire and impact problems.

10.8.1 Thermal problems

Single glazing has a very high U-value at around 5.0 W/m² °C, and therefore a large amount of heat can be lost through single-glazed windows. However, new buildings are required to be fitted with double-glazing, and this reduces the heat losses by around 50 per cent. The current need for energy conservation means that even U-values of 2.5 W/m² °C are too high (Dept for Communities and Local Govt, 2008). To achieve the very low glazing U-values required by the *Code for Sustainable Homes* (Dept for Communities and Local Govt, 2008), we must have recourse to triple-glazing, with Krypton gas infill and low-emissivity glass. By these means we can reduce the U-values to 1.0–1.2 W/m² °C, or perhaps a little lower.

In summer, the interior of buildings can become uncomfortably warm because of solar gain (a form of the well-known greenhouse effect). This leads building designers to install air-conditioning systems, which increase energy consumption, which is not a good thing. This problem can be alleviated by the appropriate use of low-emissivity glass.

Visible light is only a proportion of the total energy received from the sun; part of the insolation is heat energy. Low emissivity glass ensures that more of the thermal component is kept out. The diagram in Figure 10.9 illustrates this. It plots light transmission percentage against total radiation transmission percentage. The performance envelopes for body-tinted glass, reflective glass and low-emissivity glass are shown.

10.8.2 Sound and acoustic problems

In urban areas traffic noise can be very intrusive through single-glazed windows, particularly if the window frame is ill-fitting. Sound insulation can be markedly improved by double-glazing. Strictly, the spacing of the two panes of glass for sound insulation is wider than that required for thermal insulation, but nevertheless, double-glazing for the purposes of thermal insulation will deliver an impressive improvement in sound insulation too.

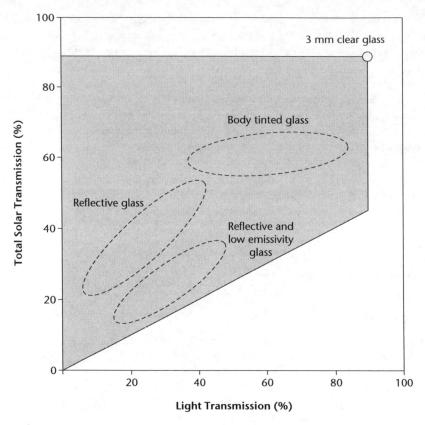

Figure 10.9 Light transmission percentage vs total solar radiation transmission percentage

10.8.3 Fire and impact problems

Windows are vulnerable in fire situations because of thermal shock, which causes cracking and shattering of the glass. If a window is lost, this will enable two things to happen:

1 Flames can escape from the room or enclosure, and thus spread the fire.
2 Air can enter the room or enclosure, and thereby ventilate the fire (i.e. allow oxygen to reach and therefore sustain the fire).

If the pane of glass does not shatter, heating by the fire may cause the glass to thermally soften, and the pane of glass can sag out of the frame, with similar effects to those mentioned above, i.e. ventilation and escape of the fire.

To prevent a cracked glass from falling out of a window pane, wired glass is commonly used. This ensures that a cracked window pane remains in place, preventing both escape and ventilation of the fire. To prevent softening glass from falling out of the window frame, strips of intumescent material can be located along the edges of the glass inside the window frame rebate. As the fire heats the intumescent strips, they will swell and therefore exert a positive grip on the edges of the pane, thus preventing it from falling out (Figure 10.10).

Windows are particularly vulnerable in the event of an explosion. Explosions produce blast waves, which move outwards from the centre of the explosion at the speed of sound (330 m/sec). The shock front

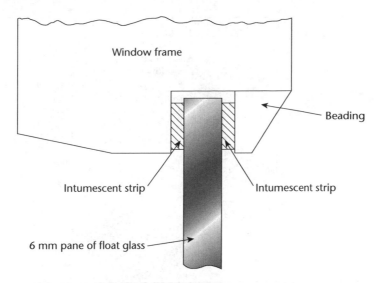

Figure 10.10 Use of intumescent material to hold glass in the event of fire

of a blast wave is a very thin zone of raised pressure (twice atmospheric or more), followed by a rarefaction wave, which is a region of low pressure (below atmospheric). When a blast wave strikes a window, it may crack the glass, and in a few cases blow the broken glass into the building. In many cases, however, the pane absorbs the pressure and bulges inwards like a tennis racket. As this inward flexing occurs, the width of the pane may be effectively reduced by a few millimetres, and when the pane flexes outwards, the rarefaction wave arrives just in time to 'suck' the pane of glass out of the frame, so that the glass ends up in the street outside the building. In the Baltic Exchange and Bishopgate bombing attacks in the early 1990s much of the glass from these buildings ended up in the streets outside.

Windows are also susceptible to impact by flying debris. Glass is very stiff and strong at room temperature, but it has no ductility which, coupled with its very low fracture toughness, means that it tends to shatter when struck by flying objects and other missiles. Flying glass represents a very serious safety hazard, and so steps must obviously be taken to minimise the problem.

To prevent glass from being sucked out of the window frame, improved glazing systems have been designed that hold the glass to a greater depth at its edges. Typically, a 30 mm holding depth would be used rather than the 4–5 mm common to normal glazing systems using putty as sealant (Figure 10.11).

To improve performance against blast waves, laminated glass is used as part of a double-glazed unit. The outer pane will be of toughened glass. The inner pane will be laminated glass containing a sheet of clear transparent polymeric material sandwiched between two sheets of glass, at least one of which will also be toughened glass (Figure 10.11). The importance of the laminated glass is the way it prevents the outer envelope of the building from being penetrated. In the London bombings of the early 1990s referred to above, the single-glazed windows were shattered and either sucked out or blown in, and the blast waves then created havoc within the offices inside. Papers were blown from peoples' desks, so it took weeks to restore normal service, with the consequent economic costs to the businesses concerned. Modern windows are designed to prevent this scale of disruption.

10.9 Critical thinking and concept review

1.   Explain why glass is called a super-cooled liquid.
2.   What are network modifiers, and why are they added to glass?

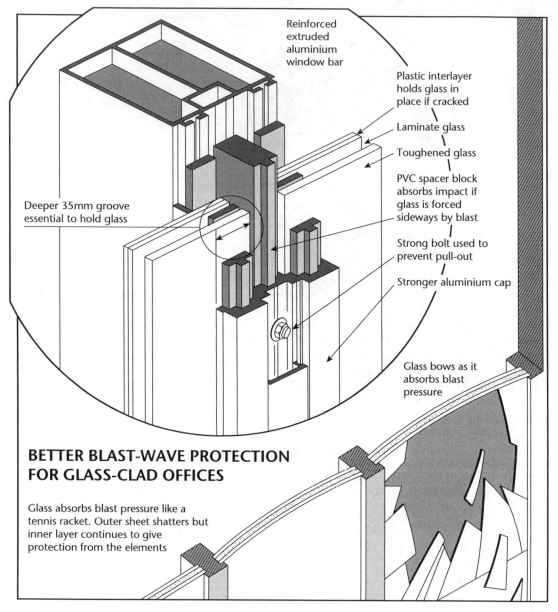

Reinforced extruded aluminium window bar

Plastic interlayer holds glass in place if cracked

Laminate glass

Toughened glass

PVC spacer block absorbs impact if glass is forced sideways by blast

Deeper 35mm groove essential to hold glass

Strong bolt used to prevent pull-out

Stronger aluminium cap

Glass bows as it absorbs blast pressure

BETTER BLAST-WAVE PROTECTION FOR GLASS-CLAD OFFICES

Glass absorbs blast pressure like a tennis racket. Outer sheet shatters but inner layer continues to give protection from the elements

Figure 10.11 Design of window 'hardened' against damage by blast waves

3. Why is borosilicate glass resistant to thermal shock?
4. What advantages did the float glass process bring to the mass production of flat glass for glazing?
5. Name two liquids that can corrode window glass.
6. How is plate glass tempered?
7. Why does tempering have the effect of toughening the glass?
8. Explain how glazing can be made to counter the problem of solar gain.
9. Briefly explain how low-emissivity glass works.

10. Why are glass fibres such an attractive material for the reinforcement of plastics in the manufacture of composites?
11. Explain how glass wool works as an insulating material.
12. Briefly explain why long glass fibres have become materials of such technological importance.
13. Single-glazing gives very high I-values of around 5.0 W/m^2 °C. Briefly discuss the methods used to reduce the U-value of windows for low-energy buildings.
14. Explain how panes of glass can be sucked out of a window frame when impacted by blast waves.
15. Briefly explain the measures that can be taken to 'harden' windows against the effects of blast waves caused by explosions outside a building.

10.10 References and further reading

ASHBY, M.F. & JONES, D.R.H. (1980), *Engineering Materials 1*, Pergamon Press, Oxford.
ASHBY, M.F. & JONES, D.R.H. (1986), *Engineering Materials 2*, Pergamon Press, Oxford.
BUTTON, D. & PYE, B. (eds.) (1993) *Glass in Building*, Butterworth Architecture, Oxford.
CALLISTER, W.D. (1994), *Materials Science and Engineering*, John Wiley, New York.
CLIFTON-TAYLOR, A. (1987), *The Pattern of English Building*, 4th edition, Faber & Faber, London.
DEPT FOR COMMUNITIES AND LOCAL GOVERNMENT (2008), *Code for Sustainable Homes*, Drafted by the BREEAM Centre, BRE, Watford.
KINGERY ET AL. (1976) [TO BE COMPLETED AT PROOF STAGE]
LYONS, A.R. (1997), *Materials for Architects and Builders*, Arnold, London.
MOHR, J.G. & ROWE, W.P. (1978), *Fibre Glass*, Van Nostrand-Reinhold, New York and London.
RAWSON, H. (1991), *Glasses and their Applications*, The Institute of Metals, London.
VAN VLACK, L.H. (1982), *Materials for Engineering*, Addison-Wesley, Reading.

11

Clay brickwork

Contents

11.1 Introduction

This chapter introduces clay brickwork as a structural material. A clay brick, clay unit or masonry clay unit can be defined as a rectangular, cuboid-shaped object made of clay which could be solid or contain from three to over 30 holes through the unit. This composite is one of the oldest construction materials, having been used for up to 10,000 years. In most forms of construction in the UK, clay brickwork is formed of units usually 103 × 65 × 215 mm in size which are bonded together using various forms of mortar. The precise dimensions of clay bricks can vary and in some parts of the world much larger clay units are manufactured. To ensure clay brickwork is fit for purpose, the material needs to be designed to ensure it is strong enough to carry the loads imposed on it, durable enough to survive for the design period of the building and it should be relatively cheap and easy to maintain.

11.2 History of clay brickwork

Clay bricks have been used to construct numerous famous and long lasting structures (Campbell & Price, 2003), including:

- the Pyramids of Giza
- the Great Wall of China
- the Coliseum
- medieval castles
- the 2,000 temples in Pagan in Burma
- Brunelleschi's Dome in Florence
- the structure of the Taj Mahal
- the 1,200 miles of Victorian sewers under London
- the Chrysler Building in New York City

In the Near East and India, clay bricks have been in use for more than 5,000 years. The Tigris–Euphrates plain lacks rocks and trees. Sumerian structures were thus built of plano-convex mudbricks. These bricks are somewhat rounded and unstable in behaviour so Sumerian bricklayers would lay a row of bricks perpendicular to the rest every few rows. The gaps were filled with bitumen, straw, marsh reeds and weeds.

The Ancient Egyptians and the Indus Valley civilisation also used mudbrick extensively, as can be seen in the ruins of Buhen, Mohenjo-Daro and Harappa, for example. In the Indus Valley civilisation particularly, all bricks corresponded to sizes in a perfect ratio of 4:2:1. Interestingly, today this ratio is still considered optimal for effective bonding.

The Romans made use of fired bricks, and the Roman legions, which operated mobile kilns, introduced bricks to many parts of the empire. Roman bricks are often stamped with the mark of the legion that supervised its production so the bricks used in Southern and Western Germany, for example, can be traced back to traditions already described by the Roman architect Vitruvius. The skill to fire bricks was lost as the Roman Empire declined, but revived in the thirteenth century

During the Renaissance and Baroque periods, visible brick walls were unpopular and brickwork was often covered with plaster. It was only during the mid eighteenth century that visible brick walls regained some degree of popularity, as illustrated by the Dutch Quarter of Potsdam, for example. Today, one of the reasons brickwork is still utilised is its aesthetic appeal.

The transport in bulk of building materials such as bricks over long distances was rare before the age of canals, railways, good roads and large goods vehicles. Before this time bricks were generally made as close as possible to their point of intended use (it has been estimated that in England in the eighteenth century carrying bricks by horse and cart for ten miles over the poor roads then existing could more than double their price). Bricks were made using locally available materials in regions that lacked stone and other materials suitable for building close at hand, including, for example, much of south-eastern England, large parts of the American south-west and the Netherlands – all places lacking stone but possessing the essential requisites for brick-making; suitable clays and fuel for firing.

The use of clay brickwork has declined in recent decades, reaching its peak during the Victorian building booms and between the wars when factories, houses, sewers, bridges and retaining walls were all usually constructed using clay brickwork. During those times up to ten billion bricks were produced annually.

11.3 The codification of masonry

For thousands of years (Figure 11.1) construction has been controlled using a series of codes and standards which are developed for particular materials in order to ensure consistency of materials and construction. In the UK British Codes and Standards governed construction up until March 2010, when they were superseded by a unified set of European codes. This section briefly outlines the current codes of practice and regulations which govern the construction of clay brickwork.

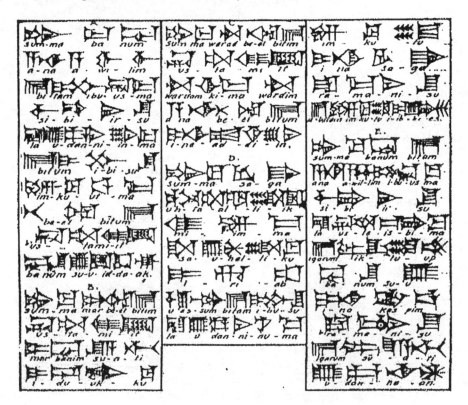

Figure 11.1 The code of Hammurabi

The masonry code, Eurocode 6, is a series of guides which control the construction of all types of masonry, including clay masonry. The code is divided into four parts:

1   BS EN 1996-1-1:2005: General rules for reinforced and unreinforced masonry structures
2   BS EN 1996-1-2:2005: Structural fire design
3   BS EN 1996-2:2006: Design considerations, selection of materials and execution of masonry
4   BS EN 1996-3:2006: Simplified calculation methods for unreinforced masonry structures

Eurocode 6 enables engineers to design all types of masonry buildings and structures, and in addition gives guidance on materials and the execution or construction of buildings. Many other European standards support Eurocode 6. For example, there are standards which describe how to test brickwork for strength, frost resistance or resistance to aggressive chemicals. Additional standards include test procedures for the individual clay units or mortar. Then there are standards which give advice and test procedures for items regularly associated with clay masonry, such as wall ties required when building cavity walls or lintels used to span over windows.

The Building Regulations are a legal document as required in the *Building Act of 1984*, and apply in England and Wales. The current edition of the regulations is *The Building Regulations 2000* and the majority of building projects are required to comply with them. They exist to ensure the health and safety of people in and around all types of buildings (i.e. domestic, commercial and industrial). They also provide for energy conservation, and access to and use of buildings.

11.4 The manufacture of clay bricks

The earliest clay bricks were formed by hand and dried in the sun to produce a meaningful building material. Often the wet clay had straw added, which reinforced and improved a brick's durability.

11.4.1 Clay for brickwork

Clay used to make bricks can vary in physical properties, colour and mineralogical content, but all clay used to make bricks must be crushable, and be able to be mixed with water to form a plastic material which can be moulded into various shapes. The precise mineral content of the raw material is also important and will affect a brick's properties. The source of clay is primary igneous rocks which are weathered and transported by water, wind or ice and re-deposited elsewhere. In the process the eroded igneous rock picks up impurities, such as quartz, mica, calcium carbonate (lime) and iron oxide. The final material is deposited in layers and becomes a sedimentary rock. Variations in age, the method of deposition, the weight of rock above and the type of impurities included will occur and may affect the brick–making process and the properties of the finished product.

11.4.2 Winning clay for brickwork

Gathering the raw material for brick manufacture is termed clay winning. How this is achieved will depend on the depth, thickness, hardness and physical geology of the clay beds. Good clay for brick manufacture is not always adjacent to the factory and in this instance suitable quarries are mined once or twice a year using heavy excavators, back actors, and a large volume of clay is then stockpiled nearer the factory. The advantages of bulk winning are that it can take place during good weather, and a large reserve of clay close to the factory means that breakdowns at the quarry are not critical. The process of layering the stockpile means the stockpiled clay is mixed prior to brick manufacture, a process which will eliminate localised variations which may be evident in the quarry. Laboratory testing of clay from different parts of a quarry determines the characteristics of the layers and clay is mixed according to the required properties of the finished item. Preventing environmental pollution at the quarry and restoring quarries to acceptable environmental standards after closure form part of the process and should be included in the cost.

11.4.3 Clay preparation

Clay preparation consists of transforming the clay raw materials into a plastic mouldable product by a process of grinding and mixing with water. A typical factory would include a primary crusher to break down large lumps of clay to manageable size, and a secondary crusher where the clay is reduced in size further. At this stage water can be added, but in other instances the clay is reduced to dust and water added later. Further crushing takes place through conveyor rollers which reduce the clay particles to about 1–2 mm. At this stage the clay is a plastic material.

11.4.4 Shaping clay bricks

Forming regular units enables walls to be built quicker and with pleasing regular patterns. Modern brick manufacturing utilises two main procedures to achieve this.

Clay extrusion. With extrusion the clay is mixed to a stiff texture and is then loaded into an extruder, where a worm screw pushes it along a barrel into a vacuum chamber which compresses it through a taper and out through a die. The die is machined to a precise size and shape slightly larger than the finished size of the brick, such that with shrinkage during the firing process the brick ends up the correct size. The clay emerges as a continuous brick-shaped column. Initially this is smooth but it can be modified by removing a thin sliver from the top and sides using a taught wire to produce a 'wiredrag' effect or by placing textured rollers over the column to create a rusticated effect, or even by blasting the column with sand. The clay column is then cut into single bricks and palletised ready either for a dryer prior to entering the kiln or, in some instances, for direct entry into the kiln. Extruded bricks are generally perforated, can be solid but cannot be frogged.

Soft mud moulding. This covers a number of techniques which may be mechanical or simple hand processes. Soft clay is thrown into a mould; a mould-release medium, which is usually oil, sand or water, prevents the clay from sticking to the mould box. Excess clay is stuck off from the top of the mould and the bricks are turned out. In its most simple form this is done by hand when one brick is produced at a time, but this is obviously labour intensive, slow and expensive and is usually only used for making special shapes or decorative bricks. For standard bricks large automated machines can replicate the hand-making process much quicker by using banks of mould boxes on a circuit. The washing, sanding, throwing, compressing, striking off and turning out of the wet units is undertaken automatically. Moulded bricks, unlike extruded bricks, can have frogs included.

11.4.5 Drying

Before bricks can be fired, as much moisture as possible must be removed or they will explode in the kilns. Drying involves the removal of water from the wet brick in such a way as to dry them out evenly from inside out. If the outer skin of the brick dries first it becomes impossible for the moisture inside to escape. In the kiln the extreme temperatures force out any remaining moisture and some cracking may occur. To minimise this the dryers are kept at temperatures of about 80–120 °C and the atmosphere is very humid, keeping the exterior of the brick as moist as possible. Bricks shrink in the dryers as moisture evaporates and the clay particles pack closer together. Drying schedules vary but 18–40 hours is typical for an automated plant. Special shapes and large units can take longer, sometimes up to a week or more. The dry bricks are then set onto kiln cars ready to be fired.

11.4.6 Firing

About 4,500 years ago, in the Tigris–Euphrates Valley, kiln-fired pottery developed and was also applied to bricks. The technique enabled bricks formed in moulds to be subsequently fired. Obviously bricks cannot suddenly be subjected to excessive temperatures so firing occurs in three stages. First, there is pre-heating which ensures total dryness of the brick and utilises combustion gases in the kiln to raise the brick temperature. This is followed by firing proper in which fuel, usually natural gas, liquefied petroleum gas, oil or coal is used to raise and maintain the temperature to the required level over a few hours. Finally, cooling occurs when cold air is carefully drawn into the kiln to cool the bricks slowly, ready for sorting and packing. This air becomes hot and can be drawn off and recycled for re-use in the drying process

The important difference between fired and sun-dried clay units is that the firing process alters the chemistry of the clay, forming a material with the particles fused together, termed a ceramic, which has properties more akin to glass than to a dried mud. This new vitrified material is much stronger and more durable than the original unfired clay would have been. Colour changes also result during the vitrification process. During the firing process temperatures vary considerably depending on clay type, but are generally in the range of 900–1200 °C.

The vitrification process occurs in a kiln, of which there are several types. Intermittent kilns are usually small, static kilns and are used for firing low-volume batches such as when relatively few bricks of an unusual shape are required. The kiln is loaded with bricks, taken through the firing process then unloaded. Continuous kilns are used for large-scale production of standard-shape bricks. They are more economical and produce large quantities of bricks at a constant rate. The continuous chamber kiln is an annular tunnel divided off into chambers. At any one time one section representing about 20–25 per cent of the kiln is being fired. The firing rotates round the kiln with chambers being lit and quenched in a specific sequence. Bricks are loaded into the kiln in front of the fire and pre-heated for 1–2 days, then fired for 2–3 days, after which they are cooled before being removed from the kiln and replaced with fresh dry bricks awaiting the fire's next circuit. In the tunnel kiln-dry bricks are loaded onto a fireproof trolley or kiln car. This then travels very slowly through the kiln, typically taking 3–4 days from end to end. Tunnel kilns are more expensive to build than chamber kilns but more economical to run and lend themselves to high degrees of automatic control. This is important because the performance of a brick depends on the firing procedure it is subjected to. It is usual for these kilns to run for 3–4 years between shutdowns. Production is maximised with long runs and few changes.

11.4.7 Packaging

Following firing, bricks are selected and packaged. This may be carried out manually or by machine. Mechanised packing is limited to regular production types where all bricks are of the same size and shape. Special shapes are packed manually. Figure 11.2 illustrates the manufacture of clay bricks using the extrusion process.

11.5 Construction formats using clay bricks

11.5.1 Bonding patterns for clay brickwork

Clay brickwork comes in many different colours and textures, but the way it is assembled on site determines the final appearance. Figure 11.3 indicates five common bonding patterns, only one of which can be used for a wall one brick wide.

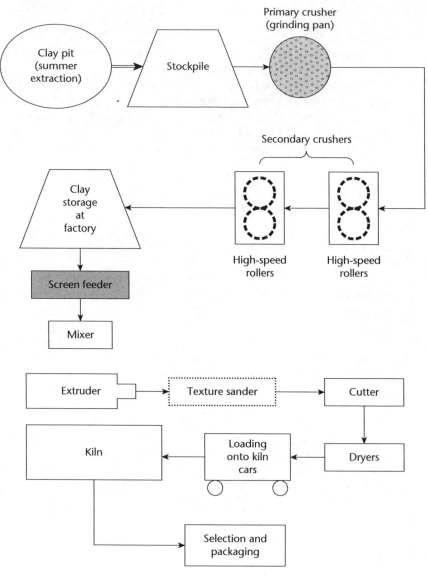

Figure 11.2 Manufacturing clay bricks using the extrusion process

11.5.2 Solid and cavity wall construction

In Victorian times solid walls were the usual way to construct masonry. Most walls were 225 mm thick. One problem with solid walls is rain penetration, so from the middle of the last century cavity wall construction became established. This utilises two leaves of masonry with a gap between. Since the 1950s the inner skin has been formed using concrete blocks and only the outer leaf uses clay bricks. The two leaves of masonry are joined using wire wall ties which keep the walls apart and improve the stability and stiffness of the construction. Wire ties are designed with a drip so any water penetrating from the outside will be shed down the centre of the cavity and drained out of the building

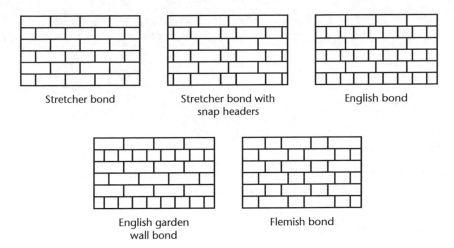

Figure 11.3   Common bonding patterns

when there are storms. A modern trend is to increase thermal performance of cavity walls by including insulation in the cavity, which means the width of cavities has increased from the original 50 mm to 100 mm or more.

11.5.3 Masonry diaphragm walls

Masonry diaphragm walls are an efficient structural solution which combine the attractive appearance of masonry with the need for efficient on-site construction. Structural efficiency is achieved because the diaphragm wall comprises two leaves linked with solid masonry diaphragms, resulting in a very stiff wall (Figure 11.4). Their use enables elegant and imaginative structures to be constructed using simple foundations and the continuity and speed of a single trade operation. They are ideally suited to applications which require an attractive high-quality finish allied to robustness, low maintenance, good sound insulation and inherently good fire resistance. Very high standards of thermal performance can be achieved by incorporating insulation within the cavities of the wall.

These walls use conventional brick or block masonry units and do not require new skills in their construction. Commercial centres, sports halls, auditoria, churches, assembly halls, etc. are all particularly appropriate applications.

The structural cross-section provided by diaphragm walls performs efficiently under both vertical loads and lateral wind loading, hence using diaphragm walls can often eliminate the need for structural steel or reinforced concrete columns, necessary to support the roof structure when using conventional solid or cavity walls. Figure 11.4 is a section through a diaphragm wall.

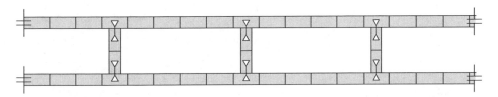

Figure 11.4   Brickwork diaphragm

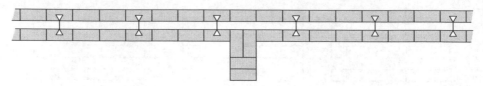

Figure 11.5 Brickwork fin cavity

11.5.4 Masonry fin walls

Masonry fin walls are usually a traditional cavity wall but with the outer skin stiffened by piers placed at regular intervals along the wall. Figure 11.5 illustrates a typical fin wall.

11.6 Clay units for masonry

Clay masonry units are specified according to BS EN 771-1: 2003. Most clay units used in the UK are termed HD units according to this code, but the code does, however, include what are known as LD units. LD or low-density clay units should have a density below 1,000 kg m^{-3}, whereas HD ones should exceed this value. Figures 11.6 and 11.7 illustrate typical LD and HD units.

BS EN 771-1:2003 specifies a range of requirements for clay bricks for specific uses, and it is up to the engineer or architect to ensure the units comply. Dimensional tolerances are specified and it is usual for the manufacturer to ensure his units comply. The initial rate of water absorption affects how the mortar and unit bond. This is reflected in the values of flexural strength of masonry as shown in Table NA.6 of the UK National Annex to BS EN 1996-1-1:2005. Masonry density affects both the acoustic and thermal behaviour of walls and rigorous techniques for determining the dry density of masonry are included in BS EN 772-13:1999. Thermal effectiveness can then be evaluated using techniques based on unit dry density as indicated in BS EN 1745:2002.

Of greater importance is the freeze–thaw resistance of clay units. Many clay bricks come out of the kiln with an open texture and as such absorb water into the brick when wetted. This can be a problem if the water then freezes and expands. The surface of the unit can be damaged and crumble away. To overcome this, units need to be impermeable or very strong to resist these bursting forces. The European Standard BS EN771-1:2003 notes that the manufacturer should test and grade his units to show their resistance to frost. Units may be:

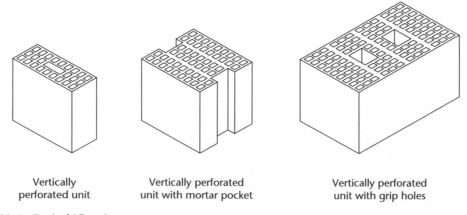

Vertically
perforated unit

Vertically perforated
unit with mortar pocket

Vertically perforated
unit with grip holes

Figure 11.6 Typical LD units

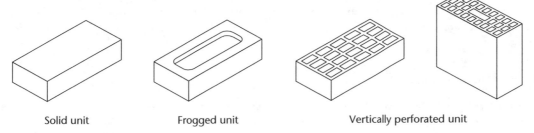

| Solid unit | Frogged unit | Vertically perforated unit |

Figure 11.7 Typical HD units

F0   suitable for passive exposure
F1   suitable for moderate exposure
F2   suitable for severe exposure.

Of equal importance to frost resistance is the active soluble salt content of clay units. Salts within the clay can be leached out and when they come into contact with cement mortars can degrade the mortar. Again, it is the manufacturer who is required to grade his units in accordance with Table 11.1

Probably the most important property of clay bricks is their compressive strength, which BS EN 771-1:2003 requires manufacturers to declare. The test to determine this property requires the faces of the unit either be ground smooth or bedded into mortar, but in both instances a flat plane through which the compressive force can be applied should be achieved. The orientation of the unit should be as it would be when built into a wall and the location of frogs should be positively noted. i.e. frog-up or frog-down. Class A and B engineering bricks (traditional UK brick types) are no longer specifically specified but obviously units with corresponding strength and water absorption could be specified as equivalent.

11.7 Movement in masonry

11.7.1 Technical background

All materials move during their lives and clay masonry is no exception. An indication of the various factors which affect movement in masonry, together with typical values for unrestrained free thermal and moisture movement, are presented in this section using information provided in BS EN 1996-1-1:2005. Nevertheless, it is almost impossible to calculate with any degree of certainty the extent of movement that will actually occur. The determination of movement is a complex problem and not merely a summation or subtraction of extremes of individual values of thermal and moisture movement, creep, deflection, and so on. For example, as a material expands due to increase in temperature, it will also shrink as moisture is lost. In

Table 11.1 Active soluble salts content categories (BS EN 771-1:2003)

Category	Total % by mass not greater than	
	Na^+ and K^+	Mg^+
S0	No requirement	No requirement
S1	0.17	0.08
S2	0.06	0.03

addition each movement will be controlled to some extent by the degree of restraint to which the masonry is subjected and its orientation and location within the structure. The code indicates movements due to different causes separately. It is the task of the engineer to evaluate how these interact and affect a structure.

11.7.2 Thermal movement

The total range of free movement due to thermal effect, which is generally reversible, is equal to the temperature range multiplied by the appropriate coefficient of thermal expansion. The movement, however, that actually occurs within a wall after construction depends not only on the range of temperature but also on the initial temperature of the units as laid and on their moisture content. This will vary with the time of year and the weather conditions during the construction period and may, with some materials, be influenced by the age of units. For example, certain steam or similarly cured units have sufficient strength to be delivered relatively fresh from the curing chamber, and may have a higher initial temperature than air-cured units. To determine the effective free movement that could occur, therefore, some estimation of the initial temperature and temperature range must be made. The effective free movement, so derived, must still be modified to allow for the effects of restraints.

Table 11.2 indicates the recommended code (BS EN 1996-1-1:2005) values and typical ranges of coefficients of thermal movement. Based on these some estimation must be made of the actual value for the material being used, although most manufacturers should be able to supply more precise values for their own materials.

The differences between mortars and units can largely be neglected when considering movement along the wall, since the effect is controlled by the adhesion of the mortar to the units, although some slight adjustment may be necessary for brick walls due to the greater quantity of mortar present. The units do not restrain the mortar in the vertical direction and therefore the movement up the height of the wall may be determined by multiplying the dimensions of the units and the mortar by the respective coefficients.

11.7.3 Moisture movement

The movement occurring along and up a wall as a result of changes in moisture content is controlled in the same way as thermal movement, except that the minimum, initial and maximum moisture content rather than temperatures are considered. The effective movement within a wall is related to the moisture content of the units at the time of laying, and it is clear that keeping the units as dry as possible before and during construction will reduce subsequent movement. The effective free movement will need to be modified to take restraints into account, which when at the end of a wall, will increase the tensile stresses in the wall.

Table 11.2 Coefficient of linear thermal movement of units and mortar

Materials	Coeff. of linear thermal movement $\times 10^{-6}$ per °C	
	Code value	Range
Fired clay bricks and blocks	6	4–8 (depending on type of clay)
Concrete bricks and blocks	10	7–14 (depending on aggregate and mix)
Calcium silicate bricks	10	10–15
Mortars		11–30

The free moisture movement of fired clay units is generally less than 0.02 per cent and is usually neglected. Attention should, however, be given to the long-term expansive movement of fired clay units due to adsorption of moisture, especially when concrete and clay units are used together as, for example, in a cavity wall. As clay bricks expand and concrete blocks contract, differential movement will result. The total long-term unrestrained expansion is typically 0.10 per cent, but a lower value may be appropriate depending upon the type of clay. With this in mind it should also be noted that wire ties have greater flexibility than flat, twisted ties.

The free moisture of mortar is similar to that of concrete units, although the effective free movement is likely to be greater, since initial moisture loss will not take place. Reversible movement of internal walls may generally be neglected since they are unlikely to become wet after initial drying out.

11.7.4 Determination of total movement within a wall

To determine the movement likely to take place in a wall it is necessary to combine the individual effective movements due to thermal, moisture and other effects. The effective thermal and moisture movements are not directly additive since a wall is unlikely to be at both its maximum temperature and its saturated condition at the same time, so to estimate maximum movement it is necessary to consider carefully the temperature range over which the moisture movement occurs and make some attempt to combine the thermal and moisture movements on a rational basis rather than just considering the extremes. With many variables it is difficult to determine with any degree of certainty the actual movement that will occur. Practical rules of thumb are usually adopted.

With fired clay walls long-term expansion usually predominates and a simple solution would be to adopt an effective global movement. A good general rule is to assume that unrestrained or lightly restrained unreinforced walls, built off membrane-type DPCs, can expand up to 1 mm per metre during the life of a building. The expansion may be less in normal-storey height walls.

To cater for this movement in clay brickwork, expansion joints are usually recommended, typically at about 12 m but not exceeding 15 m intervals along clay walls. Where bed joint reinforcement is included expert advice should be sought.

11.7.5 Accommodation of movement

Defects such as cracking are undesirable and difficult to deal with after the event, and it is most important to consider provision for movement at the design stage.

Discrete panels. One way of ensuring that the masonry is able to accommodate small seasonal movements is to design the building so that the masonry is separated into discrete panels, for example by use of feature panels at window openings and storey-height door openings.

Movement joints. The provision of movement joints (control joints) has the same effect as the discrete panel method in that joints divide the wall into defined lengths which are able to accommodate the strains arising from temperature and moisture variations. This method is suitable in situations where long masonry walls occur.

Specification of mortar. It is important to note that mortar influences the way in which a wall accommodates movements. A masonry wall built with a relatively low-strength mortar is better able to accommodate the stresses developing in a wall as a result of movement than a wall with a very strong mortar, as the weaker mortar relieves the stresses that may otherwise cause problems. As a result it may be taken that the general recommendations on the accommodation of movement, and joint spacings, apply to situations where the appropriate mortar is employed. In most situations an M4 (designation (iii)) mortar will suffice. Where stronger mortars are required for structural or durability reasons, then some modification to the recommendations may be necessary. Higher-strength mortars can normally be tolerated in vertically reinforced walls due to the presence of the reinforcement.

Storage and protection of masonry. For cracking to occur within a masonry wall tensile stresses must be present. These stresses chiefly occur as a result of moisture loss, differential movement and thermal contraction. The higher the moisture content at the time of laying, the greater the shrinkage will be, and to reduce subsequent shrinkage it is desirable to protect units from rain while they are stored and during construction.

Reinforcement. Bed joint reinforcement can be used to modify joint spacings and to assist in situations where high stress concentrations occur as, for example, at window openings. The reinforcement should be of the tram-line type, preferably with an effective diameter of 3–5 mm, with a cover of at least 20 mm from the face of the mortar, and should be of austenitic stainless steel where it is to be used in a wall which is exposed to the weather. Galvanised mild steel may be appropriate for use on internal walls. Such reinforcement should extend at least 600 mm past openings. Reinforcement may also be used to increase the distance between movement joints. The provision of reinforcement will not ensure that cracking does not occur but it will considerably lessen the risk.

Bonding pattern. Although maintaining the bond pattern is an automatic criterion for facing work, it is often found that little attention is paid to this aspect in work that is to be rendered or plastered. This may seem to be of little consequence but in practice the lack of supervision, together with poor workmanship, can result in a series of virtually aligning perpendicular joints with the result that the wall is considerably weakened and cracking is more likely to occur. Similarly, it is of benefit to design the building so that wherever possible lintels can be supported by a full-length block.

11.7.6 Differential movement

Substantial differential movement may occur when one leaf is of fired clay masonry and the other is of concrete masonry. It is essential that the effects of differential movement are considered since serious problems may otherwise arise. In principle the wall details should be checked to ensure that any differential movement is free to take place. When designing with both fired clay and concrete masonry the essential point to remember is that fired clay walling has a general tendency to an expansive movement while concrete masonry has a tendency to shrink.

11.8 Mortar for clay brickwork

Mortar for clay masonry can be defined as a cementitious material used to bond individual clay units together. In its basic form this comprises cement, sand and water which, when mixed together, forms a plastic material which bonds units together and then hardens over several hours. To assist in the building process, the mortar may include lime or chemical admixtures which improve workability on site.

The effect of mortar joints on the appearance of brickwork has always been neglected or overlooked. Although mortar joints appear on brickwork as narrow lines, they actually take up a significant proportion (about 15–25 per cent) of the total wall surface, depending on the bonding pattern. This explains the surprising effect of mortar on the look of finished brickwork. A dark mortar tends to make the bricks look darker and richer in colour, whereas a lighter mortar tends to make bricks appear a lighter tone.

Tooling of the joints tends to assist in bringing the mortar and units into close contact. Mortar joints should be finished with a consistently shaped profile to enhance the characteristic appearance of brickwork. Four common types of joint are illustrated in Figure 11.8.

Flush. This profile is suitable for moderate and sheltered exposure. No attempt is made to iron the joint surface, resulting in an open texture. The bond between the mortar and units is dependent upon the skill

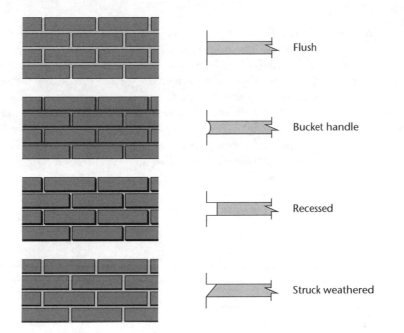

Figure 11.8 Mortar joints in clay brickwork

used during the unit laying. This type of joint has the potential to produce a uniformly coloured and textured appearance.

Struck or weathered. When used with the appropriate mortar this profile is suitable for every degree of exposure. The surface of the mortar is closed by ironing and inclined to the lower edge to shed off water. The ironing produces a good bond between the mortar and units. Masonry constructed with this joint type requires good workmanship to obtain a uniform colour and texture.

Bucket handle. This profile is also potentially suitable for every degree of exposure. The surface of the mortar is closed by ironing and the curved profile tends to allow water to run from the face. The ironing produces a good bond between the mortar and units. The use of this joint type requires good workmanship to obtain a uniform colour and texture.

Recessed. This profile is only suitable for sheltered and moderate exposure. A tool should be used to rake the joint to a consistent depth. Deep raking should be avoided and should never be more than half way between the face of a unit and its perforations. Consideration should be given prior to using recessed joints in applications where durability is critical, such as exposed and coastal environments. This type of joint has the potential to produce excellent uniformity of colour and shade.

Modern masonry utilises Portland cement as the mortar binder, although mortars using a blend of cement and lime are also important. In this latter instance the lime is primarily included as a workability aid. Many other cements and binders can be used to produce mortar, but these are generally for specialist applications. For example, phosphate cements, magnesium oxychloride cements, calcium aluminates and a variety of other materials are used for very special applications. Gypsum–based mortars, although only used for restricted applications, will be briefly discussed as they have a long history of use.

The most ancient mortar discovered to date is from Galilee, Israel near Yiftah'el and is reputed to be 10,000 years old. Surviving examples of the built environment dating back about 5,000 years to the time of Mesopotamia still survive today in the form of masonry ruins. Much of the mortar in those ruins remains durable today, a testament to the long-lasting nature of masonry.

11.8.1 Gypsum mortar

The Egyptians used gypsum mortars in the construction of the pyramids. More recently, this type of mortar was used to rebuild much of the city of Paris, which gave its name to gypsum, sometimes called plaster of Paris, as a result of the natural gypsum deposits found under the district of Montmartre.

Some of this gypsum-bonded masonry still exists in the older parts of Paris. More recently, an external render, based on gypsum and originally developed in France, was imported into the UK. However, after a short period of time and a bad history of failures it was withdrawn from the market. In addition, many of the existing historic dwellings in Paris, which were originally built with gypsum mortar and have not been restored, suffer badly from damp and deterioration. As a result of these adverse experiences, and because gypsum does not form hydrates that are stable in the presence of water, these mortars are not used today and are certainly not recommended for external work.

In very dry climates, or where there is protection from the elements, these mortars could be used without problems, such as when used to bond internal blocks in continental Europe.

11.8.2 Lime mortars

The Egyptians also used lime mortars, with literature on this subject dating back over 2,000 years. Lime mortars harden and gain strength by the evaporation of water and the absorption of carbon dioxide from the atmosphere, a process known as carbonation. With pure lime this results in the gradual conversion of lime into calcium carbonate, but strength gain in the mortar is too slow for practical building. To obtain lime mortar which can be practically useful it is helpful to understand how quicklime, hydrated lime and hydraulic lime are produced, and their properties.

Quicklime. Pure lime is produced by burning mineral raw materials, usually those based on calcium although magnesium-based limes also exist. In the UK the raw materials burned in the kiln to produce pure quicklime are chalk or limestone (usually calcium carbonate), but in theory any calcareous feedstock could be used. In some countries, shells, corals and other sources of calcium are used satisfactorily. When the raw material is heated to about 950 °C, the combined carbon dioxide is driven off in the form of gas, and calcium oxide or quicklime remains.

Hydrated lime. Quicklimes are not used directly as mortar, but first reacted with water to produce calcium hydroxide, known as 'slaked' or 'hydrated' lime, in which form the mixture is ready to be incorporated in a mortar mix. However, pure calcium hydroxide mortars, although capable of hardening and strength development, in reality react extremely slowly. Typically, it would take about 100 years for a mortar joint to carbonate to a depth of 6–10 mm. Ancient craftsmen understood this, knowing that for building pure lime was inferior to one that had some impurities.

Hydraulic lime. When cement is manufactured, lime is burned in a kiln with clay to produce a cementing compound that reacts with water to produce a hardened hydrate. This has similarities with lime, when the presence of clay impurities produces weak cement. So it is necessary to include some clay during the burning of lime to produce lime mortars which harden at a meaningful rate, although this rate is still much slower than would be achieved using Portland cement mortars. Nevertheless, lime mortars produce acceptable working properties for masons, but even a high-quality lime, with a good ultimate strength, could prove very problematic for winter usage. Indeed, it is probable that the majority of masonry construction continued little, if at all, during the winter months. Most successful historic structures were built with hydraulic lime.

Many examples of Roman mortar survive, some within the UK. For example, Hadrian's Wall still contains unrestored areas of original materials. A more recent example, the Tower of London, some 900 years old, provides further evidence of the durability of masonry materials and historic lime mortar.

Ash lime mortar. Ash lime mortar utilises the pozzolanic properties of ash or other material containing reactive silica mixed with lime. Using ground brick dust as a source of silica and mixing with lime produces a binder which can be used in masonry construction.

Early bricks were fired at low temperatures and the resultant product was highly reactive in the presence of free lime. Modern bricks, however, are fired at a much higher temperature and are not nearly as reactive. Therefore the specification of ground brick dust in conjunction with lime to form mortar should be undertaken cautiously. Using furnace ash to make ash lime mortar was successfully undertaken from the Industrial Revolution, but there are reservations over using these mortars when steel is bonded into the material as corrosion may occur. Nevertheless many examples of buildings constructed in Victorian times using ash lime mortar exist.

11.8.3 Modern cement mortars

Most modern mortar for masonry includes Portland cement, sand and some lime, the latter material usually added to improve workability.

Cement–lime–sand mortars. As a result of the development of Portland cements, the potential for masonry construction greatly increased. This was attributed to the reliable strength development and much increased rate of strength gain when compared to lime mortars, which enabled construction to be planned and executed far more rapidly. Further, construction could continue in most weathers with the exception of freezing and wet conditions.

With lime mortars masons used mix proportions of between one part binder to two or two and a half parts sand (1:2 and 1:2.5). With cement mortars the mix was altered and a general cement mix contains only one part cement to six parts sand or, if lime is included, one part cement, one part lime and six parts sand. Including higher proportions of cement produced mortar which was too strong for the clay and was brittle and prone to cracking. Interestingly, when both lime and cement are included in the mortar, the proportion of binder (cement and lime) to sand is similar to the traditional lime mortar.

Masonry cement mortars. Masonry cement is produced in a cement works and is an alternative concept to that of mixing cement and lime on site to obtain a blend of the properties of each. The cement includes an inert material instead of lime, either ground limestone or fine silica, and Portland cement. In practice, ground limestone is used as this is often available at a cement works, which means that haulage is not a factor.

Recent research work has shown that limestone may also provide an enhanced strength development in the medium and long term as it contributes to a slow continuation of the cement hydration. In addition to the mixture of cement and other material, masonry cement mortars also include an air-entraining admixture which improves frost resistance and workability.

Current practice. Site-made mortars are less common today, partly due to an increased emphasis on accurate gauging and mixing. Factory-produced mortars manufactured under controlled conditions have advantages in both quality and consistency when compared to site-mixed materials. Hence, factory-produced mortars now control a larger share of the market to the extent that on large sites today it is rare to find site mixing taking place. Factory-made mortar may be one of the following types:

- wet ready to use
- dry ready to use (delivered in silos or bags)
- lime sand for mortar.

Wet ready to use. This is cement and sand, sometimes with the addition of lime, and contains a cement set retarder. It is gauged with water ready to use at the time of mixing in the factory, and remains workable for a specified period of time, generally two working days. It is delivered in bulk using specialist vehicles and discharged into designated containers on site ready for use.

Dry ready to use. This is cement and dried sand, sometimes with the addition of lime and/or admixtures. It is stored on site in bulk silos or bags. The silos incorporate their own integral mixer, with provision for the connection of water and power. This means that only the desired amount may be mixed at any one time.

Lime sand for mortar. This is sand and lime, sometimes containing admixtures which require the addition of cement and water on site. The material is usually delivered to site in tippers or skips.

Construction using mortar. Mortar is one of the most basic elements of masonry construction. However, in order to achieve excellent brick and blockwork, a thorough knowledge of good site practices and a skilled workforce is required. The influence of site practice and procedures on the final properties of mortar is profound. The appearance, mechanical properties and durability are all affected by site operations. Correct practice is needed through all construction stages.

11.8.4 Good practice

Cold weather. Mortar stored in containers should be adequately covered to provide protection against rain, frost and snow. During prolonged periods of very cold weather, it is best practice to also protect containers storing mortar.

If mortar freezes during storage, remove and discard the frozen material prior to using the remainder. Do not use mortar that includes any ice particles or lay mortar on frozen surfaces. The inclusion of anti-freeze agents for mortars is not recommended.

Hot weather. In hot weather mortar will tend to lose its plasticity more rapidly due to evaporation of the water from the mix and the increased rate of hydration of the cement. Mortars mixed at high temperatures may have a higher water content, a lower air content and a shorter board life than those mixed at normal temperatures, unless compensatory measures are taken. Therefore, mortar with a high lime content and high water retention characteristics is sometimes considered for use in these conditions.

In hot summer conditions materials and mixing equipment should be shaded from direct sunlight prior to use. Mortar tubs and mortar boards should be rinsed with cool water before they come into contact with mortar.

11.9 Durability of clay brickwork masonry

Clay brickwork is generally very durable. In the UK many examples of centuries-old brickwork can easily be seen. Bricks are porous, and it is this porosity that leads to problems, as water can enter the bricks by capillarity. Problems are best avoided by careful selection of the bricks for any given situation. The effects of the environment on brickwork, such as moisture and frost, are dealt with in Chapter 20.

The other factor that can affect brickwork is the incidence of fire. Here again, bricks generally survive in fire better than most other construction materials, and the effects of fire are dealt with in Chapter 21.

11.10 Critical thinking and concept review

1. What is brick made from?
2. What are the different types of bricks?
3. How are bricks manufactured?
4. What are the mechanical properties of bricks?
5. What makes bricks so durable?
6. Why is it disadvantageous to use very dry bricks with mortar?

11.11 References and further reading

BRITISH STANDARDS INSTITUTION, *BS EN 1996-1-1:2005. Incorporating Corrigenda February 2006 and July 2009. Eurocode 6 − Design of Masonry Structures − Part 1-1: General Rules for Reinforced and Unreinforced Masonry Structures*, British Standards Institution.

BRITISH STANDARDS INSTITUTION, *BS EN 1996-1-2:2005: Incorporating Corrigendum October 2010. Eurocode 6 − Design of Masonry Structures − Part 1-2: General Rules − Structural Fire Design*, British Standards Institution.

BRITISH STANDARDS INSTITUTION, *BS EN 1996-2:2006: Incorporating Corrigendum September 2009. Eurocode 6 − Design of Masonry Structures − Part 2: Design Considerations, Selection of Materials and Execution of Masonry*, British Standards Institution

BRITISH STANDARDS INSTITUTION, *BS EN 1996-3:2006: Incorporating Corrigendum October 2009. Eurocode 6 − Design of Masonry Structures − Part 3: Simplified Calculation Methods for Unreinforced Masonry Structures*, British Standards Institution.

BRITISH STANDARDS INSTITUTION, *BS EN 771-1:2003. Incorporating Amendment No 1: Specification for Masonry Units − Part 1: Clay Masonry Units*, British Standards Institution.

BRITISH STANDARDS INSTITUTION, *NA to BS EN 1996-1-1:2005. UK National Annex to Eurocode 6 − Design of Masonry Structures − Part 1-1: General Rules for Reinforced and Unreinforced Masonry Structures*, British Standards Institution.

BRITISH STANDARDS INSTITUTION, *BS EN 772-13:2000. Methods of Test for Masonry Units. Part 13: Determination of Net and Gross Dry Density of Masonry Units (Except for Natural Stone)*, British Standards Institution.

BRITISH STANDARDS INSTITUTION, *BS EN 1745:2002. Masonry and Masonry Products − Methods for Determining Design Thermal Values*, British Standards Institution.

CAMPBELL, J.W.P. and PRYCE, W. (2003), *Brick: A World History*, Thames and Hudson, London.

The Building Regulations 2000: Including Approved documents Part A−H and J−P, Stationery Office Books.

Contents

12.1 Introduction

Concrete in its simplest form is usually a mixture of cement powder, water, sand and gravel. A range of cement powders exist on the market which impart different characteristics to the material. The water is necessary as when cement comes into contact with water a chemical reaction commences which ultimately results in the paste hardening. The sand and gravel, sometimes termed fine and coarse aggregate, form a matrix of strong particles of differing sizes which when 'glued' together by the cement paste form an extremely strong and durable engineering material. Sand and gravel can also vary in strength, shape and texture. In the production of concrete it is usual for all the materials to be added to a mixer concurrently rather than one after the other, although in certain instances the materials are added in a specific order. Modern concretes often include admixtures. These are small quantities of materials usually added into the mixing water which impart very specific properties to concrete. One example of this is the use of air entraining agents which produce millions of very tiny bubbles equally dispersed through the concrete, which prevent the concrete from being damaged if it freezes. Figure 12.1 is a schematic illustration of concrete and the various constituent parts.

While concrete in one form or another has existed for several thousand years, modern concrete can be dated from 1824 when the patent for Portland cement was first taken out. Since that time the type and nature of cements have varied both as a result of new materials being employed and as the technology used in its manufacture has altered. Since the introduction of cement, concrete as a material has been used in many different ways. Typical examples include the frames of high-rise buildings, as part of road bases, in tunnelling, bridges, dams and ports. Indeed, most of the built environment contains some fraction formed using concrete.

Concrete as a building material is very widely used and unlikely to be replaced in the future. It does, however, require furnaces which use fossil fuels in its manufacture, and these dispel CO_2 into the atmosphere. In addition, the manufacturing process includes grinding the material to a powder, a process which also uses energy. Consequently, at present there is much effort from the concrete industry to reduce the impacts it has on the environment. In addition to this, challenges to construct iconic and durable structures require concrete researchers to be continuously developing the material to meet these challenges.

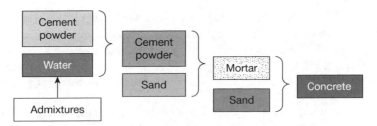

Figure 12.1   Schematic illustration of concrete ingredients and how they combine

12.2 The constituents of concrete

Basic concrete comprises a mixture of cement powder, water, sand and gravel. Modern concretes often have an admixture added.

12.2.1 Cement powder

Most modern cements, whether they are used in combination with other materials or not, contain some Portland cement. Portland cement is made by burning limestone and clay or similar materials in a kiln at temperatures of up to about 1,450 °C. During the process, the material sinters and produces balls of clinker, which are ground to form cement powder. Both the raw materials and the fineness of the material influences the properties of the cement. Portland cement is now known as Portland cement CEM I, manufactured to conform to BS EN 197-1. This cement comes in two different strength classes and can be either normal or rapid hardening. For example, CEM I 42,5N represents the normal less-strong type of this cement, whereas CEM I 52.5R represents the stronger, rapid-hardening version. To understand why a cement is rapid hardening or not, it is necessary first to understand the basic chemistry when water is added to cement powder.

A reaction termed hydration occurs when water and cement powder are combined. The reaction is exothermic and so gives off heat. When cement particles are very finely ground, they have a large surface area compared to more coarse cement, so when water is added, more of the cement is available to react with the water. Rapid-hardening cement is thus generally finer, but in addition contains slightly different proportions of some ingredients to ordinary Portland cement. So when cement powder and water are combined the hydration reaction commences and after several hours the paste hardens. From then on it continues to gain strength as a result of the hydration reaction. The strength gain is more rapid in the first few days, after which the rate of strength gain reduces. Nevertheless, strength gain continues throughout the life of the cement provided the hydration reaction continues. Hydration only ceases when there is no water available, so when concrete is placed, hydration is often promoted by keeping the young concrete wet, a process known as curing.

Two other types of Portland cement are used. These are white cement, which is combined with washed clean aggregate and produces concrete which is much whiter and used for architectural purposes. Another problem with normal or rapid-hardening Portland cements is that certain sulphates found in groundwater or some pollutants react with the cement to degrade it. Sulphate-resisting cement is a Portland cement with the component which reacts with sulphates removed, so eliminating the problem.

The manufacture of Portland cement uses considerable quantities of energy and one way of mitigating this is to combine waste materials which can be encouraged to contribute to the hydration reaction with cement. The waste materials do not require fossil fuel to burn them, this having been achieved in a previous process. They may, though, still need to be ground to an acceptable fineness. The two most common waste materials to be combined with Portland cement are ground granulated blast furnace slag (GGBS), which is a by-product of iron smelting and pulverised fuel ash (PFA), derived when pulverised coal is used in coal-powered power stations.

Ground granulated blast furnace cement has slow hydraulic activity on its own and requires being combined with Portland cement to be activated. Activation results from the lime and other alkalines which are produced from the hydration of the Portland cement. The GGBS may be ground into the Portland cement or blended with it at the time of manufacture. These blast furnace cements should conform to BS EN 197-1. If GGBS is added to a concrete which contains Portland cement at the time of mixing then the material should conform to BS EN 15167-1.

In most practical uses 36–66 per cent of a 'cement' would be formed using GGBS. In these proportions the blast furnace cement can be used for similar purposes to CEM I, but it gains strength more slowly. This means concrete formed using a blast furnace cement will generate heat from the hydration reaction more slowly and would therefore be suitable for very large concrete pours where the temperature differences from the centre of the pour to the edge can be significant. Because concrete formed using blast furnace cement gains strength more slowly, it is not suitable where formwork needs to be struck at early ages. In addition, blast furnace cements are more resistant to sulphates than ordinary Portland cement. In less common cases, the proportion of GGBS can be increased up to 80 per cent of the overall cement mass, which produces even slower heat generation during curing and better resistance to sulphate attack.

Pulverised fuel ash is gathered using electrostatic precipitation, which captures the very fine ash produced at coal-fired power stations when pulverised coal is burnt. The material, often termed fly ash, is a fine powder of glassy spheres that when mixed with cement and water combines chemically with the lime produced during the hydration of cement. As such fly ash can be used to replace part of the cement. Fly ash is less dense than ordinary cement, contributes less to the early age strength of concrete and has a lower water demand than ordinary cement and hence cements with part of the Portland cement replaced by fly ash need to be carefully designed. With fly ash replacing 25–30 per cent of the ordinary cement, the blend can be used for most purposes. When the proportion of fly ash reaches 40 per cent, the cement has sulphate-resisting properties and in addition can reduce alkali–silica reaction. Alkali–silica reaction occurs when the hydrating cement reacts with certain silicates which can be found in some aggregates. Including up to 55 per cent fly ash in cements has the additional effect of generating heat more slowly. With fly ash, the hydration reaction is more sensitive to water so the curing of these concretes is important.

Cement may be delivered to site either in bags or as a bulk material. It is important to keep cement dry during storage to prevent 'air setting' from occurring. This results when moist air causes the cement to hydrate and form localised lumps of set or partially set cement. If this is then used in concrete, it can cause significant reductions in strength. Bulk storage silos would usually accommodate 25 tonnes of cement is a weather-tight unit. The cement is pumped into the silo by compressed air. As the silo is water-tight, the risk of the cement air setting is low, but condensation against the wall of the silo could result in moisture mixing with the cement, so to abrogate the likelihood of this occurring, aeration is undertaken in which the cement is circulated within the silo. Cement silos often have scales and hoppers attached to enable concrete batching to occur. All this equipment needs to be kept clean, be regularly inspected and the accuracy of scales checked at regular intervals. Cement may be delivered to smaller sites in 25 kg bags. Bagged cement should be stored on a raised floor in a water-tight shed, but piles should not exceed 1.5 m in height to avoid warehouse set, which occurs when cement is over compacted in the dry state. If indoor storage is not possible, the bags should be raised off the ground and covered with waterproof tarpaulins. Cement should not be stored for any great length of time on site, whether in a silo or bag. On any site, the oldest cement should always be used first, and it is good management practice to store different cements in different places.

Cement testing is usually undertaken at the cost of the manufacturer but by an independent testing house at regular intervals. This is necessary to ensure the quality of the concrete produced is of an acceptable level.

12.2.2 Aggregates

Aggregates form the bulk of a concrete mix and comprise graded sand and gravel. Sometimes sand is termed as fine aggregate and the gravel fraction as coarse aggregate. These materials are combined with cement powder and water and form the basis of a concrete mix. Although inert, aggregates form a large fraction of a concrete mix and are important.

Aggregate may be formed from mined gravel or gravel reclaimed from the sea and subsequently washed. Both of these aggregate types need to be carefully managed to ensure the distribution of particles is correct for concrete manufacture. This is particularly so of marine aggregates, which tend to contain larger proportions of single-sized particles. Alternatively, rocks may be crushed to form aggregates. These are all natural aggregates and are specified to conform to BS EN 12620. It is possible to obtain recycled aggregate (RA) and recycled concrete aggregate (RCA) for use in concrete, which should conform to BS 8500-2. Further, lightweight aggregate is sometimes required and may be natural but sometimes it is man-made.

Aggregate for use in most structural concrete is graded in a particular way. This means it has a specific range of particle sizes which enable smaller particles to fit within the spaces made by larger ones, yet when wet the mix should be easily placed and compacted. Particle sizes are monitored using a set of graded sieves through which a sample of the aggregate is passed. Material retained on the different sieves is weighed and enables a particle size distribution for the aggregate to be determined. Sieving should be in accordance with BS EN 933-1. Sometimes a gap-graded aggregate is required, in which specific particle sizes are omitted. This aggregate can be used to provide architectural finishes if specified.

Aggregate is usually specified by the maximum size of aggregate included. In the UK most structural concrete is formed using 20 mm aggregate, which means the maximum stone size in the mix should rarely exceed 20 mm. If there are large pours of concrete, larger maximum aggregate sizes are sometimes used. Structural concrete usually includes a mixture of sand and gravel which are only combined at the mixer. Coarse aggregate sizes are not permitted to be below 4 mm and the maximum sand size is set at 4 mm. The reason for using two different aggregate types is that if the aggregates were combined at source, and then transported to site on a lorry, say, much of the sand would filter through the gravel voids to the bottom of the pile so concrete made using the top of a pile would have different aggregate than concrete made using aggregate recovered from the bottom of a pile.

Aggregates should be clean and able to bond to the cement paste. If materials such as clay are stuck to the aggregate surface cement paste will not bond to the stone and the strength of the mix will be compromised. Aggregate should not have excessive dust coatings, which can reduce the bond between cement and stones and increase water demand. Other undesirable ingredients include pyrite, coal and sulphate. If pyrite is in aggregate it can decompose and this results in staining of the concrete surface. Coal and lignite may swell during the life of a structure and they sometimes decompose; this can leave voids in the concrete. Some aggregates are susceptible to alkali aggregate reaction. These aggregates contain particular silicates which can react with the alkalis produced when cement hydrates. Fortunately, this occurrence is rare, but when it occurs, because the reaction is expansive, the concrete can spall and crack.

When delivered to site, aggregate should be inspected before discharge. With sand, it is important to ensure it is not contaminated with silt and clay. Rubbing a damp sample in the palms of the hand will indicate if impurities are in the mix. The field settling test will also reveal contamination, but both of these tests are only indications of the presence of contaminants and if suspicions exist detailed laboratory analysis should commence. With coarse aggregate, the material should be inspected visually and a knife used to scrape individual particles to ensure clay is not adhering to the aggregate. Coarse aggregate should be inspected to ensure it has not been placed in layers on the truck. All aggregate should be well mixed before loading. Layered loads should be discarded.

When delivered to site, aggregate should be tipped onto a concrete apron laid to falls. Partition walls to separate different aggregates are essential and a sensible means of discharging aggregate should be developed. If it is impossible to provide a concrete surface below the aggregate, the bottom 300 mm of aggregate should be discarded as contaminants from the ground can affect this zone. Aggregates should be stored in as large a pile as possible as this helps to ensure an even distribution of moisture content throughout the material. Leaving aggregate for 12 hours after any rain will usually result in a reasonably uniform moisture content except at the very bottom of the pile. With sand, though, the time may need to be extended somewhat. In any case, the moisture content of the aggregate should be regularly checked.

12.2.3 Water for concrete

Water in concrete has two important functions. First, it mixes with cement powder and results in the hydration reaction commencing, which ultimately results in the concrete hardening. But the concrete also needs to be placed and the term used to describe this is concrete workability. To be workable concrete needs more water in the mix than just for hydration. This water either remains trapped in the concrete matrix after the concrete has hardened or evaporates, resulting in a degree of permeability being imparted to the mix. Adding too much water increases the permeability, which reduces both concrete strength and durability.

Usually, if water is drinkable it will suffice for concrete, unless sweetened by sugar. But it is important to note that organic impurities in the water can cause retardation of the mix and water must be free from chlorides as these can cause steel in the mix to corrode. Acids and sulphates in the mix water can have detrimental influences but these only occur in the longer term. BS EN 1008 specifies the quality of water required but a practical means of checking that water is acceptable is to make trial mixes using the available water. Long-term effects cannot be highlighted but most other problems will be evident.

Recycled water is now being used routinely in many ready-mix factories, where a series of large linked settling ponds enable the water to be purified and re-used.

12.2.4 The use of admixtures in reinforced concrete

An admixture is a material, usually a liquid, which is added to a batch of concrete during mixing in order to modify the properties of the fresh or hardened concrete in some way. Nowadays admixtures are added to most mixes and in some cases they are essential. Because admixtures are added to concrete mixes in small quantities, their inclusion should be carefully managed. Incorrect dosage of an admixture – either too much or too little – may adversely affect the strength and other properties of the concrete.

BS EN 934-2 specifies the requirements for the common admixtures: accelerating water-reducing; retarding water-reducing; normal water-reducing (plasticising); high-range water-reducing (super-plasticising); water-resisting (water-proofing) and air entraining admixtures.

12.2.5 Accelerating water-reducing admixtures

These admixtures speed up the chemical reaction in concrete both while the mix is wet and subsequently. Hence concrete with this admixture both hardens and gains strength quicker than untreated mixes. The concrete remains workable when first mixed and achieves similar 28-day strengths to untreated concretes.

Until 1977 the most widely used accelerator was calcium chloride, either as a proprietary solution or in flake form. Because the presence of chlorides, even in small amounts, increases the risk of corrosion of embedded reinforcement the use of admixtures based on calcium chloride is now prohibited in all reinforced and pre-stressed concrete. But in plain concrete not containing any embedded metal, chloride-based admixtures may be used when early rapid hardening is required, in which case a dosage rate of 1.5 per cent anhydrous calcium chloride by weight of the cement will usually be sufficient. Chloride-based admixtures should not be used with sulphate-resisting Portland cement as the sulphate-resisting properties would be reduced.

Unfortunately, there is as yet no alternative material that has all the advantages but none of the disadvantages of calcium chloride. Some chloride-free accelerators are available, but they are generally not as effective.

Calcium chloride increases drying shrinkage and this has caused problems with high-strength granolithic toppings and screeds, particularly when laid separately or unbonded. For this reason calcium chloride should not be used in toppings. Accelerators are less effective in mortars and thin toppings because the thickness of mortar, either in a joint or on a rendering, is such that any heat generated by the faster reaction is quickly dissipated.

12.2.6 Retarding water-reducing admixtures.

Retarding admixtures inhibit the initial reaction between cement and water by reducing the rate of water penetration to the cement and hence slowing down the production of hydration products. The concrete therefore stays workable for longer than it would otherwise. Concrete workability and the time a concrete remains workable depends on water–cement ratio, ambient temperature and, if retarder is included, its dosage.

In the UK these admixtures can be used when the temperature exceeds 20–25 °C to prevent early setting of concrete. Normal concrete will harden quicker if the external temperature is high. Another application is with very large pours when it can be difficult to place new concrete on already placed concrete before the earlier concrete has set. This results in what are termed cold joints (wet concrete placed on hardened concrete). Retarders can help in these situations. Slip-forming concrete is another area where retarders can be used. This procedure is used to build towers, lift shafts or large, hollow concrete columns. The formwork is jacked up the sides of the wall as it is being built at a steady rate. Concrete emerges from the base of the formwork in a hardened state having been placed in the usual manner in the top. Sometimes slip-forming needs to very slow and retarders can prevent the concrete setting against the formwork. Another situation where retarders are used is where there is likely to be a long time between mixing and placing concrete, such as could occur with traffic delays.

The quantity of retarder can be varied, providing retardations of up to four to six hours, by altering the dosage, but longer delays of up to two days or so can be obtained for special purposes. While the early strength of concrete is reduced by using a retarder, which may affect formwork striking times, the 7- and 28-day strengths are not likely to be significantly affected.

Retarded concrete mixes need careful proportioning to minimise bleeding due to the longer period during which the concrete remains fresh. It is important for retarders to be added with the gauging water; retarders added towards the end of the mixing cycle can lead to a considerably greater retardation than if they were added at the beginning of the cycle.

12.2.7 Normal water-reducing (plasticising) admixtures

Water-reducing admixtures reduce the attraction between cement particles in concrete mixes. The main consequence of this is to produce a more workable mix. If the admixture is added to a normal concrete the mix becomes more workable and the slump will increase by about 60 mm. This assists in the placing of concrete, especially if there is complicated reinforcement to be surrounded by concrete. Alternatively, the admixture can be used to produce concrete with standard workability but less water. This material will be stronger than concrete with more water in it. More dense concrete will also be more durable. It is also possible to produce concrete of a given strength and workability by using this admixture and reducing the cement content, so economising in that way.

Overdosing may result in retardation and/or a degree of air entrainment, but does not necessarily increase the workability and therefore may not be noticed in fresh concrete.

12.2.8 High-range water reducing (super-plasticising) admixtures

These are chemical polymers which when included in concrete have a greater plasticising effect than normal water-reducing admixtures. They are used for one of two reasons: first, to increase greatly the workability of a mix so that 'flowing' concrete is produced; and second, to produce high-strength concrete by reducing the water content much more than can be achieved by using normal water-reducing admixtures.

Flowing concrete is usually obtained by first producing a mix with a slump of about 75 mm and then adding the super-plasticiser, which will increase the slump to over 200 mm. This high workability only lasts for about 30–60 minutes; stiffening and hardening then proceed normally. Because of this limited duration of increased workability, when ready-mixed concrete is used it is usual for the super-plasticiser to be added to the mix on site rather than at the batching or mixing plant.

The use of flowing concrete is likely to be restricted to work where the advantages in ease and speed of placing offset the increased cost of the concrete – considerably more than for other admixtures. Typical examples are where the reinforcement is particularly congested, making both placing and vibration difficult, and where large areas such as slabs would benefit from a flowing, easily placed concrete.

The fluidity of flowing concrete increases the pressures on formwork, which should be designed to resist full or near full hydrostatic pressure.

When used to produce high-strength concrete, a reduction in water content of as much as 30 per cent can be obtained using a super-plasticiser, compared with a water reduction of only about 10 per cent when using a normal plasticiser; 1- and 28-day strengths can be increased by as much as 50 per cent.

High-strength water-reduced concrete containing a super-plasticiser is mainly used for the manufacture of precast units where the increased early strength allows earlier de-moulding.

12.2.9 Water-resisting admixtures

There are two families of water-resisting admixtures. Hydrophobic water-resisting admixtures combine with the cement in such a way that external water is not drawn into the capillary pores. If, however, there is a head of water exerting pressure on the concrete, water will enter into the capillary voids. Hence, for concrete in splash zones or concrete above the water table this treatment is adequate. The second water-resisting admixture is termed a water blocker. This admixture is polymer-based and blocks capillaries, preventing any water ingress even under a significant head of water. If concrete is below water this admixture prevents water ingress.

12.2.10 Air entraining admixtures

These are usually organic surfactants which entrain a controlled amount (about 5 per cent by volume) of air in the concrete in the form of small air bubbles. These bubbles should be very small (about 0.5 mm in diameter, with most being <0.3 mm) and evenly dispersed.

The main reason for using an air entraining admixture is that the presence of the tiny air bubbles in the hardened concrete increases its resistance to frost action, especially when aggravated by the application of de-icing salts and fluids. Saturated concrete – as most external paving concrete will be – can be seriously affected by the freezing of the water in the capillary voids, which will expand and tend to burst it, causing spalling; but if the concrete is air entrained, the small air bubbles which intersect the capillaries and remain unfilled with water even when the concrete is saturated, act as pressure relief valves and cushion the expansive effect by providing voids into which the water can expand as it freezes without disrupting the concrete. When the ice melts, surface tension effects draw the water back out of the bubbles.

Air entrained concrete should be specified and used for all forms of external paving, from major roads and airfield runways down to garage drives and footpaths which are likely to be subjected to frost and to de-icing salts, either applied directly or dripping from the underside of vehicles.

The optimum air content for a mix with 20 mm maximum-sized aggregate is 5 per cent; for 40 mm and 10 mm maximum-sized aggregate the optimum amounts are 4 per cent and 7 per cent, respectively.

Air entrainment also affects the properties of the fresh concrete. The minute air bubbles act like ball-bearings and have a plasticising effect, resulting in increased workability. Mixes which are lacking in cohesion, harsh or which tend to bleed excessively are greatly improved by air entrainment. Air entrainment also reduces the risk of plastic settlement and plastic shrinkage cracks. Some evidence indicates that uniformity of colour is improved and surface blemishes reduced.

One factor which has to be taken into account when using air entrained concrete is that the strength of the concrete is reduced by about 5 per cent for every 1 per cent of air entrained. However, the plasticising effect of the admixture means that the water content of the mix can be reduced, which will offset most of the strength loss which would otherwise occur. The amount of air entrained for a given dosage depends on sand grading, variation in temperature and mixing time. Hence dosages may need to be altered as a job proceeds.

12.2.11 Storage of admixtures

Admixtures are generally stable, but may require protection against frost, which can permanently damage them, and may also require stirring; the manufacturer's instructions should be followed.

12.2.12 Dispensing admixtures

Because admixtures are usually added in small quantities, generally 30–1,000 ml per 50 kg of cement, accurate and uniform dispensing is essential. This is best done using manual or automatic dispensers so that the admixture is dissolved in the mixing water as it is added to the mix. Super-plasticisers for flowing concrete, however, are usually added just before discharge and the concrete is then mixed for a further two minutes.

12.2.13 Trial mixes for admixtures

Preliminary trials are a wise precaution to check that the required modification of the concrete property can be achieved. The use of an admixture is likely to require some adjustment of the mix proportions. For example, when using an air entrained mix, the additional lubrication produced by the small air bubbles permits a reduction in the water content, and it is usually advantageous at the same time to reduce the sand content by about 4 per cent. The correct adjustments can best be determined by trial mixes. In essence, the trial mixes should be used to confirm the recommended dosage of the manufacturer.

12.3 The properties of fresh concrete

As soon as water is added to cement, sand and stone, wet concrete is formed and needs to be carefully managed if a strong, durable final product is to be produced. The control of wet concrete is undertaken by measuring the consistence or workability of the mix.

12.3.1 The workability of concrete

The workability of concrete can be thought of as the ease with which concrete can be mixed, placed, compacted and finished. Nowadays it is often termed the consistence of concrete. It is a difficult property to measure yet it affects the compaction of concrete, how easy it is to shift concrete in formwork without segregation and whether concrete remains stable when pumped or transported. There is no single test which accurately measures consistence but a suite of tests are included in BS EN 12350 which enable the consistence of most concretes to be determined.

The slump test is described in BS EN 12350-2 and it is widely used around the world. In the slump test, the freshly made concrete is placed in three successive layers in a slump cone, which is an open-ended sheet-metal mould shaped like a truncated cone. Each layer is tamped 25 times with a standard rod. The mould is then removed carefully by lifting it vertically, and the reduction in height of the concrete is called the slump. The slump test is simple to use and very useful for detecting variations in the batch-to-batch uniformity (particularly in the water content) of a concrete during production. It is particularly suitable for the richer and more workable concretes. It is not suitable for dry mixes which have almost no slump; nor is it reliable for lean mixes. It is also not appropriate for very wet mixes.

For stiffer mixes, the VB consistometer test (BS EN 12350-3) developed by V. Bahrner of Sweden is more satisfactory than the slump test. The VB apparatus consists essentially of a standard slump cone placed concentrically inside a cylinder of 240 mm internal diameter and 200 mm high, the cylinder being mounted on to a vibrating table. To carry out a VB test, the slump cone is filled with concrete as in a standard slump test. The slump cone is then removed, which allows the cone of concrete to subside. A swivel arm enables a sliding transparent disc (230 mm diameter) to be swung into position to rest on the subsided cone of concrete. This is then fixed so the disc can only move vertically and vibration is started. The remoulding of the concrete in the cylinder is observed through the transparent disc. The VB time is the time in seconds between the start of vibration and the instant when the whole surface of the transparent disc is covered with cement paste, i.e. it is the time of vibration required to change the shape of the cone of concrete, left standing after lifting off the slump cone, into that of a cylinder with a level top surface. Thus, the stiffer and the less workable the concrete mix, the higher will be the VB time.

The degree of compactability test (BS EN 12350-4) is also used to measure the workability of stiff mixes, but is not common. A mould is loosely filled with concrete and then vibrated to achieve full compaction. The extent to which the concrete consolidates is recorded and gives a measure of workability. This measurement also gives an indication of the response of a concrete to vibration on site.

The flow table test (BS EN 12350-4) is used to evaluate the workability of very wet mixes which would collapse if tested in a slump cone. A truncated cone is used to position concrete at the centre of a large table which is hinged along one side but can be raised to 40 mm on the other edge. The table is lifted and allowed to fall 15 times, after which the spread of the concrete is recorded, which gives an indication of the workability.

More refined tests have been developed to measure the workability of self-compacting or self-levelling concretes. The tests are designed to measure flowability, passing ability and resistance to segregation.

12.3.2 Hardening of concrete

Concrete is used structurally to carry loads when hardened. But the material needs to be managed from the workable state to when it is sufficiently hard to carry loads. The process during which the material changes from being workable to when it is actually a solid material is termed concrete hardening. Because the chemical reaction which causes concrete to harden is exothermic, heat is given out and this obviously causes expansion. So if concrete is unrestrained while hardening then it will not crack, but if there is restraint, cracking can occur. Cracking can occur if concrete is externally restrained – for example, concrete built on an already hardened base or tied between fixed columns will expand while hardening, then set and finally cool. Hence these thermal effects can cause the weak but still solid concrete to crack when it contracts. If a large volume of concrete is poured, then the interior of the concrete will heat up more than the edges and cause temperature differentials which can cause the concrete to crack at the surface when hardening. Temperature rise in concrete as a result of cement hydration depends on the size of the element being cast, the cement content and type, the initial temperature of the concrete when placed, the ambient temperature, what admixtures may be included and whether the formwork is insulating or not. Hence reducing or controlling early thermal cracking may be complicated. The good news, however, is that early thermal cracking usually only affects thick elements such as raft foundations or slabs over 500 mm in depth and most concrete walls. Reducing the impact of cracking can be achieved using mixes with less cement in them and by including anti-crack reinforcement in the elements.

Another problem which arises with hardening concrete is the development of plastic cracking. Two types of plastic crack may develop, both related to the bleeding of concrete. Concrete bleeding is a process whereby water migrates up through concrete and emerges on the concrete surface. When concrete is placed and compacted, there is a tendency for the heavier and larger particles to settle, which displaces water in the lower regions of an element. The water rises through the concrete and appears on the surface. This process is not instantaneous and can take up to six hours. The rising water leaves voids which result in settlement or slumping of the remaining concrete, which produces plastic cracking. If the weather is hot and windy the effects of bleeding are disguised as no water will settle on the concrete surface; if evaporation is significant it can draw water out of the concrete, speeding up bleeding. The process is further complicated as warmer concrete hardens quicker so limiting the time over which bleeding may occur, but with cold weather the concrete will obviously be exposed to bleeding for longer.

Plastic settlement cracks occur in deeper elements and occur along the lines of top reinforcement. During bleeding the concrete settles but the reinforcement prevents that from occurring in the locality of the bars. Hence concrete settles either side of a bar and in effect hogs over where the bar is, causing longitudinal cracks.

Plastic shrinkage cracks occur in thinner elements such as slabs. Again the mechanism initiating the cracks is bleeding, but this time no reinforcement is present to cause plastic settlement cracks. As the concrete settles and reduces in volume, horizontal tensile forces will develop in the upper zone which can cause diagonal shrinkage cracks to develop. These cracks are usually not significant from a structural point of view.

With both types of shrinkage cracks, their effect can be reduced by limiting bleeding, which is undertaken by increasing the cohesiveness of the mix (reducing the possibility of larger aggregate stones from settling) and by ensuring the effects of evaporation are minimised through protecting the concrete before, during and after hardening.

12.4 The properties of hardened concrete

12.4.1 Strength development of concrete

After hardening, concrete gains strength with time, and it usually has sufficient strength at 1–2 days to be self-supporting. At 28 days it achieves its 'design strength'. This is the strength designers use to size elements in a structure. In reality concrete continues to gain strength throughout its life so any increase after 28 days is a reserve. Achieving its design strength depends to a large extent on the care the concrete is given in the first few days of its life. In particular, the concrete needs to be 'cured' when young. This is an important aspect of concrete management and requires further explanation. When concrete hardens this occurs as a result of the hydration reaction, and for cement to hydrate water must be present. If concrete is taken at an early age and dried in an oven, the strength development is arrested. When wetted again there is some further increase in strength but never to the extent with an equivalent unheated concrete. Curing is the process which ensures the concrete has sufficient moisture to hydrate continuously, especially during the first week or so of a concrete's life, which is the critical period. Concrete can be gently sprayed with water, wrapped in damp hessian or wrapped in polythene to prevent evaporation or sprayed with a curing compound, this latter material again preventing evaporation from the concrete. The curing compound is designed to degrade after six or eight weeks. Figure 12.2 indicates the strength gain of a typical concrete. The precise relationship depends on many factors, including the type of cement used and the environmental conditions at the time of and immediately after manufacture of the concrete.

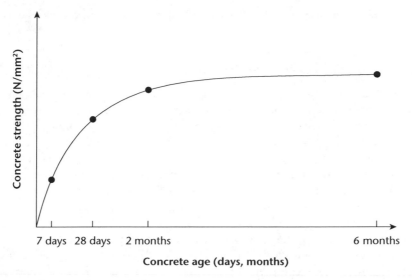

Figure 12.2 Strength vs age for a typical concrete

The relationship between concrete strength and age depends on the type of cement used. For example (Figure 12.3), if rapid-hardening Portland cement (RHPC) (CEM I 42.5R) is used, the finer grains and different raw materials used in this cement enable water to access a larger surface area of cement and so react more quickly than if a coarser-grained cement with different constituents like OPC (CEM I 42.5N) were used. The result is that the RHPC cement gains strength more rapidly in the early days than the OPC. Environmental conditions also affect the rate of hardening. The relationships indicated in Figure 12.3 are for well-cured concretes. If, for example, the concrete dries out and the hydration reaction is arrested, then a different relationship (see dotted line) will result. Clearly, even limited reductions in curing, especially in

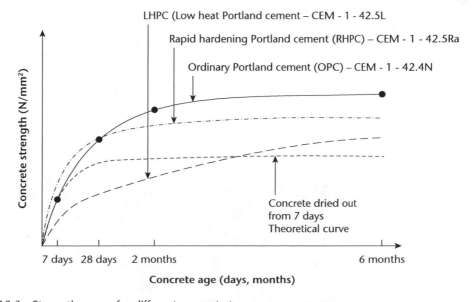

Figure 12.3 Strength vs age for different concrete types

the early days, can affect the later strength of the concrete. Any strength gained after 28 days is ignored in the design.

12.4.2 The design strength of concrete in practice

The design strength of structural concrete is related to its characteristic value. To obtain the characteristic strength of concrete, tests are carried out on small concrete samples at an age of 28 days. Traditionally in the UK concrete cubes were used. On site most cubes are 150 mm in size, although 100 mm cubes are permitted. The designer then uses these results to evaluate the characteristic compressive strength of concrete. Importantly, however, in the European Union the characteristic strength of small cylinders is used in design but cubes may be used to control quality or be converted to equivalent cylinder values for design. The characteristic strength is then reduced by a partial safety factor, usually 1.5, to give the design strength of the concrete. Materials engineers usually provide designers with a characteristic value and the designers then decide on a partial safety factor.

To evaluate the characteristic strength of a particular concrete, many small samples of the concrete are made and tested at 28 days, from which a mean strength is determined. This average strength is then reduced by subtracting what is termed the margin, which takes account of the variability in production and testing and the risk to the producer of having concrete rejected which does not conform. The spread of test results usually conforms to a normal distribution curve as shown in Figure 12.4, where $\bar{x}$ is the mean strength and the spread of the results is measured by the standard deviation, σ. Good control gives low values of σ and vice versa. In design the characteristic strength is taken as:

$$\begin{aligned}
\text{Characteristic strength} \quad &= \text{target mean strength} - \text{margin} \\
&= \text{target mean strength } 1.64x \text{ standard deviation} \\
&= 1.64x\sigma.
\end{aligned}$$

As shown in Figure 12.4, this gives a value of strength such that 95 per cent of small specimens tested exceed it. This, combined with the partial safety factor, is considered to provide adequate safety in most

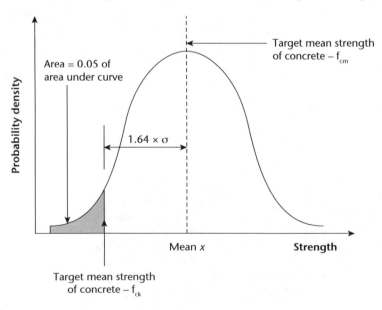

Figure 12.4   Characteristic and mean strengths of concrete

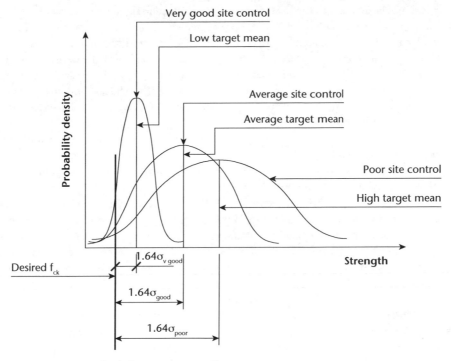

Figure 12.5 Impact of standard deviation on quality

structures. Companies which manufacture concrete have standard procedures to 'design' their concrete without having to go through the tedious procedure of a series of trial mixes and testing numerous small specimens over and over.

12.4.3 The impact of quality control on concrete strength

Poor quality control of concrete on site will produce a wider spread of results, which leads to a higher value of standard deviation σ. Very good quality control produces a very steep curve and it can be seen that the better the quality control, the lower the target mean strength needs to be to achieve the same 'characteristic strength'. In practice, though, a standard deviation of less than 3 MPa is rarely specified as it is difficult to achieve. Figure 12.5 illustrates the influence of various standard deviations.

12.4.4 Tensile strength of concrete

The tensile strength of concrete is not as important as its compressive resistance. Nevertheless, it is required in evaluating crack spacing and deflections. The tensile strength of concrete is determined in two ways: (1) the splitting tensile strength in which a cylinder of concrete is split along its length – this test determines the true tensile flexural strength of concrete; or (2) small unreinforced beam specimens subjected to a bending stress are used, in which case the flexural tensile strength is determined.

12.5 Mixing, transporting, placing and compacting concrete

About 75 per cent of concrete in the UK is manufactured at ready-mix factories. The concrete is either dry or centrally mixed. With dry mixing the ingredients, including water and admixtures, are placed in the ready-mix truck, which then mixes the ingredients thoroughly. Centrally mixed concrete is mixed by the ready-mix company and then discharged into the trucks, which agitate the mix to ensure it is acceptable when delivered to site. The management of concrete when using ready-mix factories is important, especially if significant quantities of concrete are required. Consideration of traffic delay, care with the very heavy trucks on site, and careful sampling of concrete are essential. Sometimes, though, site batching and mixing provide a more economical solution. In these instances the storage and maintenance of the equipment is important. Site batching plants need to be regularly cleaned and the scales for weighing materials must be regularly checked.

Much concrete in the UK is now placed using pumps. Pump distances of up to 500 m are common and rises of 50 m are not difficult to achieve. When using pumping, economic advantages occur when the procedure is planned into a job from the outset. This means all parties are aware that pumping will occur and the site will be prepared from the outset. In particular, the formwork, labour and concrete mix design need to be tailored specifically to pumping concrete. With formwork, the designers need to allow for the significant movement that arises in the pipe when pumping. Labour needs are usually reduced but staff must be available to deal with the high delivery rates, and the subsequent compaction and levelling of the concrete required. Finally, to pump concrete long distances, the mix needs to be right and ready-mix factories must have the right aggregate and admixtures in stock. Modern concrete pumps are reliable and capable of delivering up to 200 m^3 per hour, and even modestly sized pumps deliver 50 m^3, which is significantly more than is possible with cranes with skips.

Cranes and skips, however, are still widely used in the UK for concrete delivery, and in many instances when a site requires a crane to enable formwork and materials to be moved around it can economically double-up to place the concrete. Skips range in size from 0.2 to 1.0 m^3 and there are two basic designs available. These are the roll-over skip and the constant attitude skip. When using skips they need to be well maintained and care needs to be taken when discharging concrete onto reinforcement as the concrete can sometimes displace the bars.

The placing of concrete and its compaction are interlinked. For this reason, they need to be studied as a single entity. Designers need to consider the rate of placing and whether compaction can comfortably match this. Concrete should be placed as near as possible to where it will finally harden. Shifting concrete along forms using shovels or poker vibrators should be avoided. In most instances concrete should be placed along forms in regular layers not exceeding 500 mm in depth. If the concrete is placed in deeper layers it is possible compaction will not be uniform and air will remain trapped in the concrete. Further, if several layers of concrete are required to achieve a full form, each layer should be intermixed with the one below by using the poker vibrator to merge the two layers. With very deep lifts, placing in layers is advisable, but the rate of placement should exceed 2.0 m per hour to avoid colour variations. When dropping concrete into forms, it is acceptable to allow the mix to drop several metres, but it is ill-advised if the concrete spatters against reinforcement or other obstructions while being dropped. Self-compacting concrete should be placed in a similar manner to ordinary concrete. It is tempting to think the rate of placement of this material can be speeded up, but time is still required for the air to migrate up and out of the concrete. In general, self-compacting concrete should be placed at about the same rate as conventional concrete.

Once concrete has been placed it needs to be compacted. If left uncompacted, air trapped in the concrete matrix will form voids resulting in a weak material. Vibration of concrete effectively liquefies the material, enabling air to rise. Most concrete in the UK is compacted using poker vibrators. These vary in size from 25 to 75 mm in diameter and the size determines the spacings at which one places the vibrator into the wet concrete when vibrating. Poker vibrators are inserted rapidly into the concrete and should penetrate into the previously compacted layer below by about 75–100 mm in order to knit the two layers together. When air has ceased to rise, the vibrator is slowly removed to ensure the concrete fills the void left as the vibrator is raised. External vibrators are occasionally used on construction sites, but in the precast concrete industry they are more common. Well-proportioned concrete should not segregate when vibrated, even if the vibration time is longer than recommended. Hence the risks to concrete of over vibration are much less than under vibrating the material. Re-vibration of concrete provided the mix is still wet has no harmful effects and if the top 75–100 mm of the mix is re-vibrated while still wet, the risk of settlement cracks will be reduced.

12.5.1 Durability of concrete

This section on the durability of concrete could profitably be read alongside Section 20.4 of this book. The science underlying the various modes of degradation of concrete are covered in that section.

Concrete as a building material needs to have excellent strength properties, but in addition it must be durable and survive for the design life of a structure. For bridges, this is often in excess of 100 years, so a very durable material is required. Concrete on its own (without any metal included) is a very durable material in most situations, but when used structurally it usually has steel reinforcement included. The steel is very strong in tension, whereas concrete is strong in compression, so the two materials complement each other. It is, however, the inclusion of steel reinforcement in the concrete which usually results in durability problems.

For a concrete element to be durable it must be designed and constructed to protect the embedded reinforcement from corrosion and generally to perform in a satisfactory manner in the environment in which it is placed for the design life of the structure. Durability mainly affects the reinforcement embedded within the concrete, but there are instances in which the concrete itself is affected and degrades. For example, certain types of ground sulphates (or even pollutants with particular sulphates in them) can cause expansive reactions in concrete, significantly weakening it. Further, certain alkalis which are sometimes found in the aggregate used to manufacture concrete have similar effects. Fortunately, both of these problems are relatively rare.

12.5.2 The concrete environment

The environment in which concrete finds itself refers to any chemical or physical reactions to which the structure as a whole, individual elements of the structure, or the concrete itself is exposed and which may result in effects not included in the loading conditions considered at the structural design stage. Inadequate attention to the durability at the design and construction stages may subsequently lead to considerable expenditure on inspection, maintenance and repair. Consequently, durability has gained in importance in all codes over the years.

12.5.3 Corrosion of reinforcement

Steel reinforcement embedded in concrete is surrounded by a highly alkaline pore solution with a pH value in excess of 12.5. Such an alkaline environment causes the steel to be passivated, i.e. a highly impermeable oxide layer forms on the surface of the steel which protects it from corrosion. Corrosion of steel reinforcement is likely to occur when loss of passivity takes place, which is usually due to either carbonation or chloride ingress.

12.5.4 Carbonation ingress

Acidic gases like carbon dioxide combine with water to form weak solutions of carbonic acid which is washed over the surface of the concrete, usually as a result of rain. This acid reacts with parts of the concrete to de-passivate it and turn it acidic. Over time the acidic front penetrates deeper into the concrete, and if

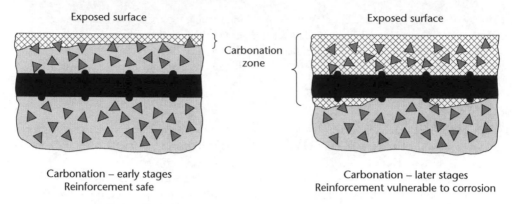

Figure 12.6   Carbonation of concrete

it reaches the reinforcement can result in corrosion if moisture and oxygen are present. The penetration of the carbon front has been observed to be proportional to $(time)^{0.5}$. Figure 12.6 indicates early and later stages of carbonation.

12.5.5 Chloride attack

Chlorides diffuse into the concrete from the surface, the concentration decreasing with depth. When chlorides reach the steel surface in sufficient concentration, the passive layer is broken down and the protective alkaline environment is degraded, so corrosion can occur if water and oxygen are present.

In the past calcium chloride was used as an accelerating admixture. It caused the concrete to gain strength more rapidly so that high-rise buildings could be built more quickly as the supporting formwork could be removed quickly. One serious side-effect of calcium chloride was that if included in sufficient quantities and if oxygen and water were present, reinforcement corrosion commenced. It is now banned, but many buildings still exist with calcium chloride included. Figure 12.7 indicates chloride profiles when salts have entered concrete from the surface and when chlorides were included in concrete during construction.

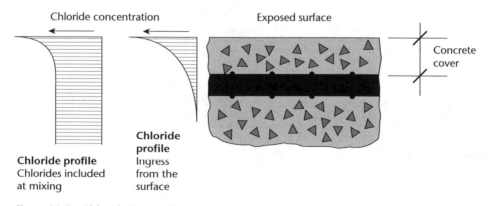

Figure 12.7   Chloride ingress into concrete

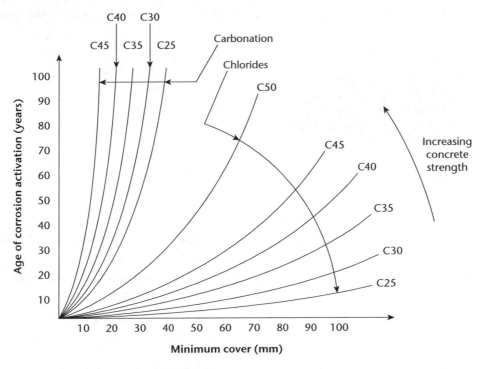

Figure 12.8   Effect of carbonation and chlorides on concrete corrosion

12.5.6 Reinforcement corrosion, cover, concrete strength

Figure 12.8 attempts to relate the age for corrosion activity to start, concrete cover and concrete grade to carbonation and chloride attack. In conclusion, chlorides are a more severe risk than carbonation. For a given age for corrosion activation, a series of concrete strengths and covers will work. For example, stronger concrete and less cover has the same effect as weaker concrete and more cover. This is an over-simplification of the situation as cover affects the effective depth (a parameter used in the design of concrete elements) and hence the overall size of elements.

12.5.7 Alkali aggregate reaction (AAR)

This reaction is an expansive reaction that can cause cracking and disruption of the concrete. AAR is a reaction within the concrete, unlike most durability problems which are associated with reinforcement deterioration. For AAR to occur all three of the following conditions must be met.

1.   There is sufficient moisture within the concrete.
2.   The concrete must have a high alkali content – alkali sources can be internal arising from cement, water, chemical admixtures and some aggregates, or external through exposure to sea water.
3.   The aggregate must contain an alkali reactive constituent. Some aggregates containing particular varieties of silica are susceptible to attack from alkalis.

AAR is relatively rare in the UK and where past experience with particular cement/aggregate combinations indicates no tendency to the reaction, precautions need not be taken. If, however, unfamiliar materials are being used, there may be a risk and additional testing or access to the national database on aggregates should be sought.

To minimise risk of AAR occurrence:

1. Limit the alkali content of the mix.
2. Use a cement with a low effective alkali content.
3. Change the aggregate to one which is known to be low risk.
4. Limit the degree of saturation or moisture content of the concrete when hardened by using, for example, impermeable membranes.

12.5.8 Sulphate attack

Normally concrete is only at risk from sulphates if it is buried, as it is the sulphates present in groundwater which cause degradation to the concrete. Foundations and retaining walls, for example, are at risk. In addition, in areas with high atmospheric pollution airborne sulphates can cause degradation to the concrete if they are washed over the concrete by rain over a long period of time.

Sulphates are, however, also present in most cements and in some aggregates, and excessive amounts of water-soluble sulphates from these sources can be deleterious to concrete. Sulphate attack in concrete is expansive and again affects the concrete itself.

Sulphate attack can often be offset by using sulphate-resisting cement. Even using this cement, which has the ingredients likely to react with sulphates removed, is not 100 per cent reliable. In association with sulphate-resisting cements, good compaction and quality control are necessary. Binary cements – those which contain PFA or GGBS, are more sulphate-resistant than conventional Portland cements.

12.5.9 Durability and design

Designing for strength is relatively straightforward, but designers need to be aware of the long-term requirements of the building, which requires consideration of durability. To assess the degree of durability required, designers should consider at least the following:

- Intended use of the structure. Designers need to view a nuclear power station differently from a garden path. Consider a situation where many music fans are dancing on a balcony in a purpose-built venue and a car-parking garage. Different needs exist in the different situations and design will be influenced by these.
- Required performance criteria. Using the nuclear power station and the garden path example again, clearly very high performance criteria are consistently needed in the former as a failure will be catastrophic, while in the garden path failure will be unlikely to affect anyone's life.
- Expected environmental conditions. Concrete protection will vary, depending on the environment. A sea wall exposed to splash will need greater protection than, say, the internal beam in a department store.
- Composition, properties and performance of materials. Durability is affected by the aggregate and cement type and in some instances by the water used in the concrete. These factors need to be considered at the design stage.
- Shape of members and structural detailing. Designers have the ability to influence the architectural details to some extent. Good detailing is essential to reduce maintenance costs.
- Quality of workmanship and level of supervision. This is obvious but sometimes difficult to implement. The construction phase is very pressured and quality control is important. A well-built structure will always be more durable than a poorly built one.

- Particular protective measures. Designers can reduce degradation of reinforced concrete by including targeted protective measures. For example, water-proof membranes can be included to prevent ground-water from saturating concrete. Good detailing can prevent concrete from being periodically wetted.

- Likely maintenance during the intended life. Clients will want the best of both worlds. Low build and zero maintenance costs. There is always a cost implication in the long term if construction costs are cut.

12.5.10 Cover as a means of influencing durability.

Cover to the reinforcement is generally regarded as the most important line of defence against reinforcement corrosion and is defined as the distance between the outer surface of the reinforcement and the nearest concrete surface. It is always the least distance as indicated in Figure 12.9. Concrete cover is required for three main reasons. First, to ensure the reinforcement is not affected by deleterious materials from the environment; second, to ensure the encased reinforcement is fully bonded; and third, to provide resistance to fire.

12.6 Critical thinking and concept review

1. What are the main constituents of concrete?
2. What are the different types of cements?
3. What is the hydration of cement?
4. What is the difference between fine and course aggregates?
5. How is the consistency of fresh concrete mix determined?
6. What is curing?
7. How does the water:cement ratio affect the compressive strength of concrete?
8. Why is the compressive strength of concrete sometimes referred to as the 28-day strength?

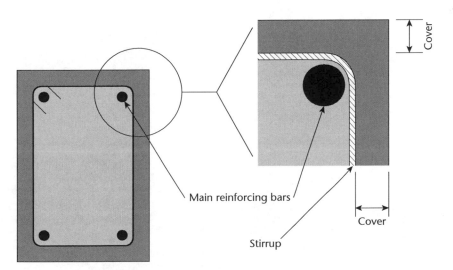

Figure 12.9 Concrete cover

12.7 References and further reading

BRITISH STANDARDS INSTITUTION, *BS EN 197-1:2011, Cement: Composition, Specifications and Conformity Criteria for Common Cements*, British Standards Institution

BRITISH STANDARDS INSTITUTION, *BS EN 15167-1:2006, Ground Granulated Blast Furnace Slag for Use in Concrete, Mortar and Grout. Definitions, Specifications and Conformity Criteria*, British Standards Institution.

BRITISH STANDARDS INSTITUTION, *BS EN 12620:2002+A1:2008, Aggregates for Concrete, Latest Amendment*, British Standards Institution.

BRITISH STANDARDS INSTITUTION, *BS 8500-2:2006, Concrete. Complementary British Standard to BS EN 206-1. Specification for Constituent Materials and Concrete*, British Standards Institution.

BRITISH STANDARDS INSTITUTION, *BS EN 933-1:1997, Tests for Geometrical Properties of Aggregates. Determination of Particle Size Distribution. Sieving Method*, British Standards Institution.

BRITISH STANDARDS INSTITUTION, *BS EN 1008:2002, Mixing Water for Concrete. Specification for Sampling, Testing and Assessing the Suitability of Water, Including Water Recovered from Processes in the Concrete Industry, as Mixing Water for Concrete*, British Standards Institution.

BRITISH STANDARDS INSTITUTION, *BS EN 934-2:2009, Admixtures for Concrete, Mortar and Grout. Concrete Admixtures. Definitions, Requirements, Conformity, Marking and Labelling*, British Standards Institution.

BRITISH STANDARDS INSTITUTION, *BS EN 12350-2:2009, Testing Fresh Concrete. Slump-Test*, British Standards Institution.

BRITISH STANDARDS INSTITUTION, *BS EN 12350-3:2009, Testing Fresh Concrete. Vebe Test*, British Standards Institution.

BRITISH STANDARDS INSTITUTION, BS EN 12350-4:2009, *Testing Fresh Concrete. Degree of Compactability*, British Standards Institution.

CONCRETE SOCIETY (2008*), Concrete Practice: Guidance on the Practical Aspects of Concreting*, British Cement Association, Camberley.

NEVILLE, A.M. (1995), *Properties of Concrete*, fourth edition, Longman Group Limited, Harlow.

13

Autoclaved aerated concrete

Contents

13.1 Introduction

Autoclaved aerated concrete (AAC or aircrete) may not sound as familiar as some of the other construction materials such as steel, concrete, bricks, etc., but it is one of the most widely used material in construction, due especially to its excellent thermal insulating properties, which can be used to keep in the heat in cold climates, or keep out the heat in warm climates.

This chapter will look into aircrete in more detail and why it is such an important and widely utilised construction material, especially in the UK, Europe and North America.

13.2 History and general overview

Aircrete was developed in Sweden in the 1920s and first used in the UK in the late 1950s as an alternative to building with timber. Currently over 30 million m^3 of aircrete (of different densities) is produced annually worldwide. Of this, 3 million m^3 are produced in the UK – a third of all concrete blocks used in the UK construction/building industry.

AAC, now known commercially in the UK as aircrete, is produced by H+H Celcon Ltd, Tarmac Topblock Ltd, Hanson-Thermalite and Quinn plc. Aircrete blocks perform well as the load-bearing inner leaf to cavity walls, as well as providing the requisite thermal resistance. They may also be used as the outer leaves of masonry walls, other external walls, internal load-bearing and non-load-bearing walls and walls below ground level. Aircrete's superior thermal performance in most cases eliminates the need for expensive cavity insulation.

The lightweight and faster build speed with aircrete blocks in comparison to other building materials mean foundations and other walls can be constructed quickly, easily and cost-effectively as units are easy to handle and quick to lay, where they create an effective moisture barrier with significant thermal insulation properties. Their cellular structure and durability make them a recognised alternative in most below-ground situations. Further, it is now more popular than timber or solid concrete floors in new housing at ground level, and is increasingly the preferred solution for upper floors too.

Aircrete is a versatile material with exceptionally good workability and is easy to cut, shape and chase using ordinary woodworking tools. This makes it ideal for closing the cavity at reveals and for cutting around and over joists, or for special shapes such as infills. Blocks are easily cut using a hammer and bolster or wood saw. A straight cut ensures less wastage and reduces the need to make good. Hammer and wood chisels can be used for chasing out, which cannot be done with concrete and other dense materials.

As we shall see later, dwellings and buildings constructed from aircrete comply with all the stringent building regulations.

13.2.2 Sizes and classification of blocks

Aircrete blocks are produced in a range of thickness from 60 mm to 355 mm, with a range of face dimensions. The most common work face dimensions (especially in the UK) are 440 mm or 620 mm long by 215 mm high. Other work face dimensions are also available ranging from coursing bricks 215 mm long by 65 mm high to large-format blocks 620 mm long by 440 mm high.

Aircrete blocks are usually classified according to their compressive strength, as shown in Table 13.1. There are three categories: low-, medium- and high-density blocks. These are based on density, but also correspond to the compressive strengths. The higher the density (strength), the lower the porosity – porosity implies how much air is in the material; thus a 100 per cent dense material theoretically has 0 per cent

Table 13.1 Physical properties and general classification of aircrete blocks

Aircrete density	Compressive Strength $(N\ mm^{-2})$	Density $(kg\ m^{-3})$	Thermal conductivity $(W\ m\ K^{-1})$
Low	2.0–3.5	350–450	0.09–0.11
Medium	4.0–4.5	620	0.15–0.17
High	7.0–8.5	750	0.19–0.20

porosity. Porosity confers both favourable and undesirable properties. First, as air is a poor conductor of heat, the higher the porosity the lower the thermal conductivity of the material, hence the better it is for thermal insulation. Conversely, the higher the air content the lower the strength, thus as porosity increases the strength of the material decreases.

13.3 Production of aircrete

Cement or cement and lime with pulverised fuel ash (PFA) and/or ground sand form the calcareous and siliceous ingredients, which are mixed together with water to form a low-viscosity slurry. The mix receives a predetermined small dose of finely divided aluminium powder just before it is poured into pre-oiled large steel moulds. The mould is of a size and dimensions suitable for the volume required to make a specific number of blocks or slabs. Approximately one-third of the mould is filled with the slurry. Therefore, the mould is kept in a warm and humid environment and the slurry is left to hydrate. The period of time depends on the process used. The aluminium powder soon reacts with the alkaline environment in the mix to generate tiny bubbles, which stabilise to form the aircrete cellular structure. During this process the mix rises in a similar way to cake in an oven, hence giving the material the porous structure. After the initial set has taken place it is cut into block size. The final process involves autoclaving for approximately 9–12 hours at 180–200 °C, which provides the product with its final strength. It is ready for despatch usually after cooling. In the UK most aircrete is manufactured using PFA, which confers environmental benefits as PFA is a waste product from coal power stations.

Aircrete consists of 60–85 per cent air by volume, depending on density. The solid material part is a crystalline binder, which is called tobermorite. Besides the binding phase, tobermorite, grains of quartz and small quantities of other minerals also exist. The chemical composition of tobermorite comprises of silicium dioxide, calcium oxide and H_2O (water). It is tobermorite that provides the relatively high compressive strength and stability of aircrete in spite of the high proportion of pores and lack of course aggregate in this material.

During manufacturing, gas-forming tiny aluminium flakes within the powder produce millions of small bubbles of hydrogen in the slurry. The mixture rises in its moulds until all aluminium has reacted and the desired volume is reached. Bubbles formed are mostly about 1 mm in diameter, which gives the porous structure. The production process is modified accordingly to attain the appropriate level of porosity. The production process is simplified and illustrated in Figure 13.1.

13.4 Properties of aircrete

This section looks at the physical/mechanical properties of aircrete, which will be reviewed as follows:

- Strength: compressive, tensile, shear, modulus of elasticity and fracture mechanics parameters
- Thermal
- Sound/acoustic
- Moisture movement
- Fire resistance
- Durability

It should be noted that this section looks at the generic material, while Section 13.5 looks into more detail at how buildings and dwellings constructed using aircrete conform to the building regulations.

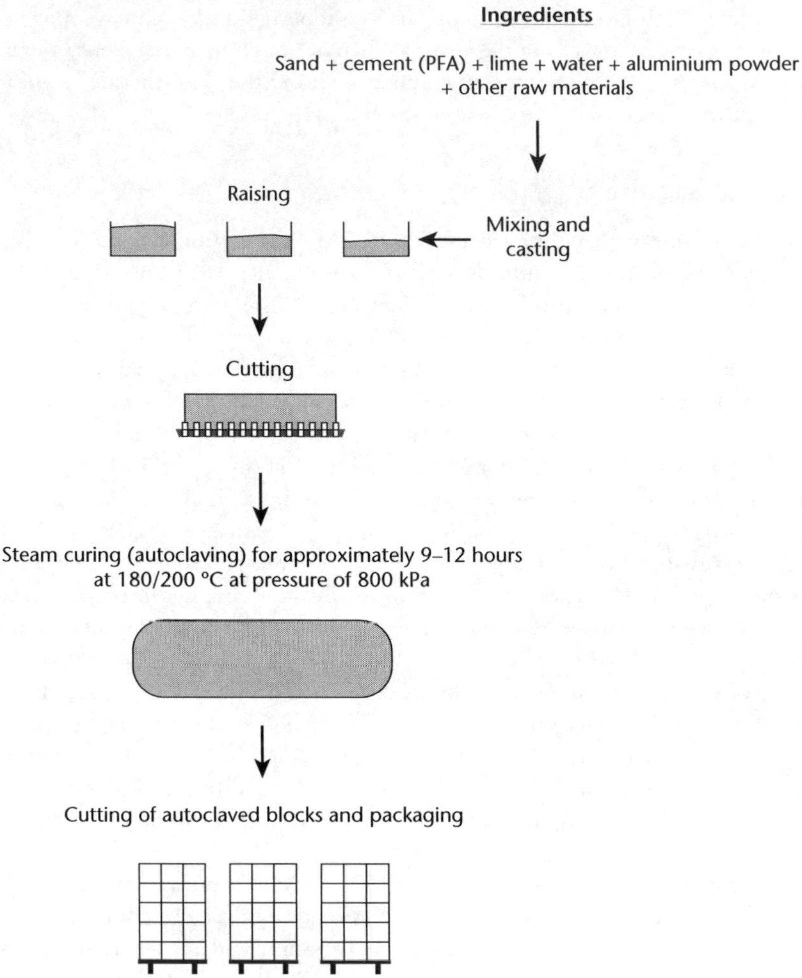

Ingredients

Sand + cement (PFA) + lime + water + aluminium powder
+ other raw materials

Raising

Mixing and
casting

Cutting

Steam curing (autoclaving) for approximately 9–12 hours
at 180/200 °C at pressure of 800 kPa

Cutting of autoclaved blocks and packaging

Figure 13.1   Manufacturing process of aircrete

13.4.1 Strength/mechanical properties

Compressive strength

The compressive strength of aircrete is related to its density and increases with increasing density. Commonly produced compressive strengths are 2.0, 2.8, 3.5, 4.0, 7.0 and 8.4 N mm^{-2} (MPa). The compressive strength of aircrete is nearly independent of specimen size due to its homogeneity. Aircrete achieves its final strength during the autoclaving process without further curing. Drying leads to a strength increase when the moisture content falls under a value of about 10 per cent by mass, as indicated in Figure 13.2.

Tensile strength

In general the direct tensile strength is about 15–35 per cent of the compressive strength of aircrete. The measurement of the tensile strength is more sensitive to the test conditions than the measurement of the compressive strength.

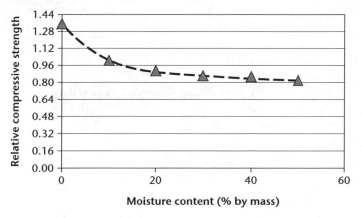

Figure 13.2 Compressive strength variation with moisture content for aircrete

Shear strength

A shear force is one applied so as to cause or tend to cause two adjacent parts of the same body to slide relative to each other, in a direction parallel to their plane of contact. The shear strength of aircrete is approximately 20–30 per cent of the compressive strength.

Modulus of elasticity

Like most other materials, the instantaneous deformation of aircrete can be described approximately by Hooke's law:

$$\frac{\varepsilon}{E} = \sigma$$

where

ε is the elastic strain
σ is the applied stress
E is the Young's modulus.

The modulus depends both on the density and the humidity content of the material. Linear regression analysis of experimental data leads to the following equation:

$$E = (-520 + 4.7_{\rho dry}) \pm 500 \ N/mm^2$$

This equation holds true approximately for aircrete with a dry density (ρ_{dry}) ranging between 300 and 800 kg m^{-3}. The elastic modulus decreases proportionately (linearly) with an increase in humidity (inverse relationship):

$$E_{cm} = 5\,(\rho m - 150)$$

where

E_{cm} is the mean value of the modulus of elasticity of AAC (MPa)
ρm is the mean value of the dry density of AAC (kg m^{-3})

The Young's modulus (modulus of elasticity) of aircrete is generally assumed to be $1.5 - 3.2 \times 10^3$ N mm^{-2} (1.5–3.2 GPa).

13.4.2 Thermal

Thermal conductivity

Thermal conduction is the phenomenon by which heat is transported from high to low temperature regions of a substance. The property that characterises the ability of a material to transfer heat is the thermal conductivity. It is best defined in terms of the expression:

$$q = \frac{-k\ dt}{dx}$$

where

q is the heat flux, or heat flow, per unit time per unit area (area being taken as that perpendicular to the flow direction)
k is the thermal conductivity
dt/dx is the temperature gradient through the conducting medium.

The units of q and k are W m^{-2} and W m K, respectively. This equation is valid only for steady state heat flow, that is, for situations in which the heat flux does not change with time. Also, the minus sign in the expression indicates that the direction of heat flow is from hot to cold, or down the temperature gradient.

The high degree of porosity of aircrete has a significant influence on thermal conductivity. Internal pores normally contain still air, which has extremely low thermal conductivity – approximately 0.02 W m k. Furthermore, gaseous convection within the pores is also comparatively ineffective.

Aircrete has excellent thermal insulation properties (low thermal conductivity) and the thermal conductivity is in the range 0.1–0.2 W m K for 350–750 kg m^{-3} density aircrete, as shown in Table 13.1.

Specific heat capacity

A solid material, when heated, experiences an increase in temperature signifying that some energy has been absorbed. Heat capacity is a property that is indicative of a material's ability to absorb heat from the external surroundings; it represents the amount of energy required to produce a unit temperature rise. In mathematical terms, the heat capacity C is expressed as follows:

$$C = \frac{dQ}{dT}$$

where

dQ is the energy required to produce a dT temperature change.

Specific heat (often denoted by a lower case c) is sometimes used; this represents the heat capacity per unit mass (J kg K).

The specific heat of AAC is approximately 1,050 J kg K or 1.1 kJ kg K.

13.4.3 Sound/acoustic resistance

Acoustic

The combination of the internal structure (porous) and the stiffness characteristics of aircrete enable sound reduction performance for walls and partitions superior to the results predicted by the general mass law curve for masonry. As the pores are not interconnected, this enables aircrete to impart satisfactory sound insulation properties.

The mass law curve for aircrete is Sound Reduction Index,

$$R = 22.9 \log (m) - 4.2 \text{ dB}$$

average for the range 100–3150 Hz)

where

m is the mass in kilograms per square metre.

Sound absorption

Sound absorption is a property relevant to particular applications. When it is exposed, the aerated internal structure of aircrete provides good sound absorption properties. However, painting reduces the sound absorption. Specialist finishes, which are outside the scope of this text, may provide superior sound absorption. Table 13.2 shows typical values of sound absorption coefficient for aircrete.

Sound insulation

Airborne sound can be transmitted from one part of a structure to another. The degree to which a wall may reduce the passage of airborne sound depends on its 'mass law'. The aircrete blockwork mass law curve can be used to calculate the sound reduction index of single-leaf internal walls and partitions. The parameter

Table 13.2 Typical values of sound absorption coefficient for aircrete

Mean frequency (Hz)	Unpainted aircrete blockwork density (kg m^{-3})		Painted aircrete units (density 480 kg m^{-3})
	700	480	
125	0.16	0.05	0.05
250	0.22	0.10	0.10
500	0.28	0.15	0.10
1000	0.20	0.15	0.10
2000	0.20	0.20	0.10

m is the superficial density in kilograms per square metre. The superficial density is the mass per unit area of the wall and its finishes and may be calculated from the formula:

$$\frac{M_B + \rho_m \, Td \, (L + H - d)}{LH} + \text{the weight of the finishes kg/m}^2$$

where

M_B is block mass (kg)
ρ_m is mortar density (kg m^{-3})
T is block thickness (m)
d is mortar joint thickness (m)
L is coordinating length (m)
H is coordinating height (m)

13.4.4 Movement

Moisture content and moisture transfer

Moisture movement through porous building materials is a very complex process. Therefore, for practical predictions simplifying assumptions have to be introduced.

There are at least three different origins of water in aircrete. Immediately after autoclaving, aircrete contains typically about 30 per cent water by weight of the dry material. Most of this excess water is lost under normal conditions to the surrounding air after a few years. If the relative humidity of the surrounding air increases temporarily, aircrete will take up water again by absorption and capillary condensation. If the surface of a structural element is in contact with liquid water the material absorbs water quickly by capillary suction.

Water vapour permeability

Assuming that two half spaces have a water vapour pressure $P1$ and $P2$, respectively, and are separated by a plate of aircrete of thickness ΔX, then a water vapour flux towards the space with lower partial vapour pressure will take place. The mass flux per unit area j is given by:

$$j = \frac{\lambda_h \, \Delta P}{\Delta X}$$

where

j is the mass flux per unit area (kg m^2 h)
λ_h is the hygral water vapour permeability (kg m h torr)
ΔP is the difference $P1 - P2$ of the vapour pressure (torr)
ΔX is the thickness of the specimen (m).

λ_h depends on the humidity range. At 100 per cent relative humidity the partial vapour pressure at 20 °C is about 17 torr. If vapour permeability is determined between 100 per cent relative humidity and 50 per cent, typical values around 8.1×10^{-6} kg h torr are found. λ_h also depends on the dry density of the material. Figure 13.3 shows how λ_h varies as a function of the relative humidity for medium–density aircrete (500 kg m^{-3}).

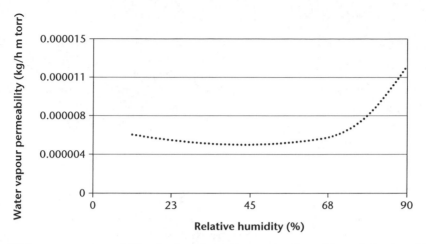

Figure 13.3 Water vapour permeability of general aircrete as a function of the relative humidity

Liquid water transfer: capillary suction

If the surface of a porous material is brought in contact with liquid water, water is absorbed by capillary suction into the material. Under simplifying assumptions the water absorption per unit area can be deduced:

$$W = a_w \, (t_w)^{0.5}$$

where

W is the absorbed water per unit area (kg m^{-2})
a_w is the water absorption coefficient (kg m^{-2} h$^{-0.5}$)
t_w is the duration of contact with water (h).

The water absorption coefficient of aircrete depends on the dry density of the material and on the pore size distribution. Typically a_w is found to be between 4 and 8 kg m^{-2} h$^{-0.5}$. The water front penetrates into the material as a function of time. This process can be described in a simplified way with the following equation:

$$x_{cs} = B_{cs} \, (t_w)^{0.5}$$

where

x_{cs} is the penetration depth (m)
B_{cs} is the coefficient of water penetration (m h$^{-0.5}$).

Typical values of B_{cs} for aircrete are found to be between 0.01 and 0.03 m h$^{-0.5}$. The ratio a_w/B_{cs} indicates the percentage of the available pore space that is filled by capillary suction. This value corresponds to the water capacity of the porous material.

13.4.5 Fire

Aircrete is combustible and has Class O surface spread of flame. All aircrete products provide excellent fire protection. Subsequently, the requirements of the building regulations can be easily satisfied, as elaborated in the next section. Aircrete blocks are classified as non-combustible in accordance with the UK building regulations; thus, in many structures aircrete blocks are used to encase and protect steel structures in buildings. Steel columns completely encased in at least 60 mm aircrete will give a fire resistance of two hours, which is critically important in the event of a fire and can save lives.

13.4.6 Durability

Aircrete products have been in use for nearly 70 years. Since that time, the material has proved its durability under extremely different climatic and chemical conditions. Aircrete is mainly attacked by acids, solutions of acid salts and acid-forming gases. The degree of attack depends on the acid concentration, relative humidity and temperature. Alkalinity of aircrete is between pH 9 and 10.5.

Destruction of aircrete can be caused by the formation of ice or salt crystals. In some countries, possible damage caused by freeze–thaw action is very important. Precaution should, however, be taken in the presence of non-structural steel so that no damage occurs due to corrosion.

Resistance to freezing

The resistance to freezing and thawing action of a construction material is determined by its pore size distribution and, in particular, the percentage of capillary pores, size and shape of the pores and the mechanical strength of the inner pore walls. In this case porosity is a very useful property as the free internal area can accommodate the expansion of ice (when water freezes this is accompanied by an increase in volume of ~11 per cent).

Aircrete usually possesses good resistance to freezing, which is proved by unrendered buildings situated in areas where frequent freeze–thaw cycles occur remaining undamaged. The reason for the good resistance is that the introduced big spherical pores are almost closed, the material has comparatively very little capillary suction and therefore the moisture content does not normally reach the critical degree.

Where aircrete is used below ground level, water-proof coatings, finishes or foils may be necessary. The resistance to frost is superior to that of many stronger, denser masonry materials, although the degree of resistance is to some extent dependent on strength and density. The high freeze–thaw resistance as discussed above is due to the aerated internal structure of the material.

Resistance to sulphate attack

The resistance to attack from sulphates likely to be found in soils and groundwater is high. The degree of resistance is dependent on the strength and density of aircrete. Aircrete has excellent resistance to sulphate attack for exposure to sulphate concentrations >600 mg l^{-1} in the water. Possible measures for protecting subsoil aircrete is to apply protective coatings, usually based on bitumen.

Resistance to attack by liquids

Aircrete is normally unaffected by all alkaline solutions. Because of its alkalinity (pH is 9–10.5), solutions such as soda, potash or ammonia cannot form soluble compounds capable of destroying the matrix. Substances such as alcohol, benzene or fuel oil do not damage aircrete.

Resistance to attack by gases

Aircrete is not affected by gases which do not form acidic solutions with water.

High carbon dioxide concentrations may occur, for example, in rooms where fermentation is taking place, or in fruit stores. The stability and strength of the material is not reduced when subjected to high carbon dioxide concentrations. Cracking does not occur if exposure to carbon dioxide is not excessive. Attack from carbon dioxide occurs very slowly if the aircrete material has reached equilibrium moisture content or if it is saturated with water.

Resistance to attack from biological sources

Aircrete has proved to be resistant to attack by living organisms.

13.5 Conformability to the Building Regulations

So far we have looked at the general properties of aircrete, which are quite favourable for construction. However, how does aircrete conform to the building regulations? Although, many regulations exist, the most important ones are considered to be Part L (heat loss), Part E (sound) and fire. This section will look into how construction using aircrete helps conform to the aforementioned regulations.

13.5.1 Part L

Since the 1990s it has been suggested that the globe has been undergoing an increase in its temperature as a result of man-made pollution — emission of carbon (greenhouse) gases. Thus, there has been a drive to reduce greenhouse gas emissions. As a result, house builders are under pressure to design and build homes that not only conserve energy, but more specifically reduce CO_2 emissions. The construction industry in the UK now has to comply with Part L of the building regulations. Part L relates to the heat loss from dwellings measured in U-values; therefore, a house with a low U-value is very well thermally insulated (there is little heat loss). It follows that if a dwelling has a low U-value, it will need less energy, thus reducing CO_2 emissions.

Over the last 17 years, following the Kyoto Protocol agreement in 1997, where countries agreed to reduce greenhouse gas emissions, there has been a drive by the UK government to gradually reduce U-values. This is imperative as nearly 50 per cent of the UK's CO_2 emissions are produced by the energy consumption of buildings. To help comply with the stringent building regulations, aircrete is the preferred option for builders.

Due to aircrete's excellent thermal insulation characteristics, it reduces the extremes of internal temperature within a building, keeping it at a more consistent and comfortable level — this also applies to warm and tropical climates where, conversely, less air-conditioning (energy) is required. Unlike other materials, such as timber and metal frame, aircrete does not require large amounts of additional insulation to achieve good thermal performance. The cellular structure of aircrete products minimises heat loss, saving vital energy and reducing the need for costly insulation.

Aircrete allows for both full or partial fill insulation without necessarily increasing cavity widths — a primary concern for builders seeking to maximise liveable space. This allows compliance with any modifications to the Part L requirements, which will continue to be achieved more cost-effectively with aircrete masonry construction than other materials. As a result, as of 2010 aircrete masonry construction accounted for greater than 60 per cent of all new housing. Currently, Part L has put increased emphasis on air tightness, with air pressure testing now being required for new dwellings. Although aircrete construction (with general purpose mortar) complies admirably, the efficiency can be further improved using thin layer mortar (previously known as thin joint mortar), discussed in more detail in Section 13.6.

13.5.2 Part E

The Part E regulations relate to sound insulation between dwellings. The term 'noisy neighbours' has over a long period of time caused anxiety and suffering to dwellers, especially in terraced housing and apartments. Therefore the Part E regulations have been continuously modified to ensure there is adequate sound insulation between and within individual dwellings and in certain multi-occupancy buildings.

The main considerations in Part E are:

- E1: Protection against sound from other parts of the building and adjoining buildings. This deals with the performance aspects of separating walls and floors.
- E2: Protection against sound within a dwelling – walls and floors. This deals with the performance aspects of internal walls and floors.

In comprehensive tests, aircrete walls and floors comply with Part E regulations. Given the porous structure of aircrete (Figure 13.1), surprisingly, the material is an excellent sound insulator. It is widely assumed that dense materials provide superior sound insulation properties, but it is actually aircrete's cellular structure of thousands of individual cell walls that each interrupt and diminish the passage of sound.

13.5.3 Fire

One of the fundamental issues with any building is fire protection, for obvious reasons. Part B of the Building Regulations aims to safeguard householders from the risk of fire. Briefly, it requires:

- safe means of escape from the building;
- the stability of a building to be maintained in a fire, both internally and externally;
- that the internal finishes of the building will resist the spread of fire and the materials used in the building will not react in a way that contributes to the development of the fire;
- fire and smoke to be prohibited from spreading to concealed spaces in a building's structure;
- the external walls and roof to resist spread of fire to walls and roofs of adjacent buildings;
- the building to be easily accessible for fire fighters and their equipment.

In terms of fire protection, aircrete is a superior materials as discussed in Section 13.4.5. Using 100 mm aircrete masonry blocks can give up to four hours of fire resistance on non load-bearing walls and up to two hours on load-bearing walls. Aircrete blocks can also resist temperatures up to 1,200 °C and do not emit toxic smoke or gases as a result of fire.

13.6 Sustainable construction using aircrete

With an ever-increasing emphasis on sustainability, the construction industry is regularly looking for good practice (construction), to utilise greener materials and, more importantly, to reduce greenhouse gas emissions. Aircrete contributes favourably to sustainability for several reasons:

- Aircrete blocks are manufactured using environmentally responsible renewable or waste materials, e.g. PFA.
- Aircrete products present no risk of pollution to water or air.

- As aircrete is easy to cut and versatile, there is minimum waste.
- Aircrete's thin layer system is a major contributor to sustainable construction as it speeds the construction process, cuts waste, minimises the effect of construction on the environment and creates an energy-efficient home to live in (see below).
- Due to its low density, aircrete allows a greater volume to be transported thus substantially reducing its carbon footprint.
- Aircrete can be re-used or recycled at the end of a building's life.

13.6.1 Aircrete using thin layer mortar

As mentioned already, there is a lot of interest in constructing dwellings from aircrete using thin layer mortar. Thin layer mortar is a special type of mortar which comes in a pre-mixed powder form which only requires the addition of water; it is then applied with a scoop as shown in Figure 13.4. Unlike traditional mortar it has accelerated setting time of within 30 minutes, which results in outstandingly high productivity, with one storey being constructed within a day. For conventional cement-based mortar, the mortar joints are typically 10 mm thick – however, for thin layer mortar the mortar joints are only 3 mm thick. As mortar has a higher thermal conductivity than aircrete, the use of thin layer mortar can reduce the U-value by 10 per cent. Thin layer mortar also increases the bond strength between blocks, resulting in a stronger wall; this is illustrated in Figures 13.5 and 13.6, which are aircrete walls tested to failure (flexural four-point bend load) using conventional and thin layer mortar, respectively.

For the conventional mortar wall, failure is typically along the mortar joint with a general strength value of 0.2 N mm^{-2} (approximate values for low-density aircrete). For the thin layer wall, however, the wall behaves as a plate with very little de-bonding of the blocks (Figure 13.6); the typical strength is about 0.3 N mm^{-2} (for low-density aircrete), which is 50 per cent stronger.

Figure 13.4 Application of thin layer mortar

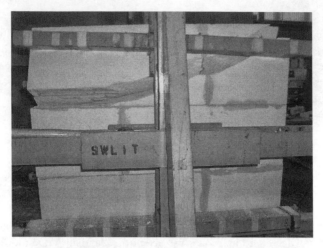

Figure 13.5 Failure of an aircrete wall using conventional mortar under four-point loading

Figure 13.6 Failure of an aircrete wall using thin layer mortar under four-point loading

Thus, in addition to help reduce U-values and increase productivity, aircrete wall construction using thin layer mortar also creates walls that are stronger.

13.7 Critical thinking and concept review

1. What does AAC stand for?
2. What other name does AAC have?
3. How is AAC different to concrete?
4. What are the main advantages of using AAC in construction?
5. What makes AAC such a sustainable material?
6. How does AAC perform in the event of a fire?
7. How durable is AAC?
8. How is AAC classified?
9. What are the advantages when used with thin layer mortar.

Polymers: properties, structure and characteristics

Contents

14.1 Introduction

Polymers are one of the most widely utilised materials in all industries, including aerospace (aeroplane windows, seats, tables), automobile (dashboard, seats, tyres), food and beverage (bottles, packaging) and, of course, construction. The polymers used by the construction industry are generally low-density materials and they are used mainly in non-loadbearing applications. Unlike metals, they are not subject to corrosion, but are rather degraded by the action of the ultraviolet (UV) radiation that comes with sunlight (we shall look at this in more detail later). As a group, polymers perform badly in fire. They are made from hydrocarbons, and are therefore fuel for a fire. They soften and melt at low temperatures, and often give off noxious fumes when they burn.

Approximately 20 per cent of plastics produced in the UK go into construction. The most commonly used polymer is polyvinyl chloride (PVC), and this material finds use as pipe materials for rain water, waste and sewage systems, electrical cable sheathing, cladding, window frames and doors and flooring applications.

Two chapters are dedicated to polymers: this one will look at the chemistry, structure and generic properties, whereas the following chapter looks at all the common and different polymers used in construction and the rationale for its selection.

14.2 What is a polymer?

A polymer is a material usually composed of hydrocarbon compounds with extensive applications and usage. Polymers have a wide range of applications in construction; the polymers used by the construction industry are generally low-density materials and they are used mainly in non-load bearing applications. The most commonly used polymer is polyvinyl chloride (PVC), and this material finds use as pipe materials for rainwater, waste and sewage systems, electrical cable sheathing, cladding, window frames and doors and flooring applications (see Chapter 15).

Polymers are generally classified into three categories based on their mechanical properties: *thermoplastics*, *thermosets* and *elastomers* (as discussed later). Typical synthetic polymers are neoprene, nylon, PVC and polystyrene. The structural properties of a polymer relate to the physical arrangement of monomers along the backbone of the chain. Structure has a strong influence on the other properties of a polymer. The identity of the monomers comprising the polymer is generally the first and most important attribute of a polymer. Polymers that contain only a single type of monomer are known as homopolymers, while polymers containing a mixture of monomers are known as copolymers. Polystyrene, for example, is composed only of styrene monomers, and is therefore classified as a homopolymer. Ethylene–vinyl acetate, on the other hand, contains more than one variety of monomer and is thus a copolymer. In polymers the molecular mass (related to density) may be expressed in terms of degree of polymerisation – essentially the number of monomer units which comprise the polymer.

It is vitally important to understand the polymer chemistry as this directly impacts upon the polymer and more importantly its mechanical and physical properties. Although civil engineers, architects, surveyors and construction professionals need not to familiarise themselves too intimately with the chemistry, nevertheless, it is essential to grasp the basic fundamentals.

14.2.1 How polymers are formed

Consider a polymer with which we are all familiar – polyethylene. This is the polymer from which washing-up bowls and buckets are made. Polyethylene is made from ethylene (C_2H_4), specifically, it is made by *polymerising* ethylene, which has the molecular shown in Figure 14.1.

If the ethylene gas is subject to elevated temperature and pressure in the presence of a catalyst (R•), it will transform to polyethylene, which is a solid polymeric material. The polymerisation process begins

Figure 14.1   Molecular structure of ethylene

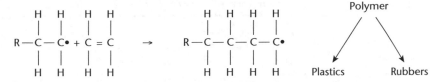

Figure 14.2 Schematic of the polymerisation process for polyethylene

when an active monomer is formed by the reaction between the catalyst (R•) and the ethylene unit, as shown below.

The polymer chain then forms by the sequential addition of ethylene monomers to this active initial monomer. The active site, i.e. available bond electron (denoted by •), is transferred to each successive end monomer as it is linked to the chain. This may be represented as shown in Figure 14.2.

The resulting polyethylene chain can be very long indeed. In high–density polyethylene (HDPE) it can be 20,000 or 30,000 carbon atoms long ($C_{30,000}H_{60,000}$). There is little or no side branching or cross-linking. A lump of solid polyethylene consists of a tangle of tens of thousands of these long-chain molecules all intertwined and tangled in a completely random way. Because these chains are flexible, the polymer itself is flexible to a degree. Thus, polymerisation is a process of reacting monomer molecules together in a chemical reaction to form polymer chains – hence the final polymer product. Furthermore, *mer* represents the group of atoms that constitutes a polymer chain repeat unit and *monomer* is a molecule consisting of a single *mer*.

The degree of polymerisation determines the density and properties of the polymer, e.g. typically in low (or very low) density polyethylene (LDPE) the chain has only a few thousand carbon atoms, thus the final product is very light and not very strong – e.g. supermarket polythene bags. Conversely, in ultra high molecular weight polyethylene (UHMWPE) the polymer chain can contain 50,000 carbon atoms, the resultant material is extremely tough and can be used as bulletproof vests – something a supermarket bag couldn't be used for, although the chemical formula for both materials is the same! Table 14.1 is a list of the monomer structures for some of the most commonly utilised polymers.

With so many polymer chains forming, simplistically we can assume that several polymers have a spaghetti-like (micro)structure – this is only visible under very high magnifications (electron microscope).

Polyethylene is an example of a thermoplastic polymer. There are two other types, *thermosets* and *elastomers*. Before we look into the three categories it is important at this stage to clarify that both rubbers and plastics are polymers as illustrated below.

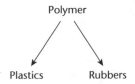

Table 14.1 A listing of mer structures for some of the common polymeric materials

Polymer	Repeating (mer) structure
Polyethylene	$\begin{array}{cc} H & H \\ | & | \\ -C-C- \\ | & | \\ H & H \end{array}$
Polyvinyl chloride (PVC)	$\begin{array}{cc} H & H \\ | & | \\ -C-C- \\ | & | \\ H & Cl \end{array}$
Polytetrafluroethylene (PTFE)	$\begin{array}{cc} F & F \\ | & | \\ -C-C- \\ | & | \\ F & F \end{array}$
Polypropylene (PP)	$\begin{array}{cc} H & H \\ | & | \\ -C-C- \\ | & | \\ H & CH_3 \end{array}$
Polystyrene (PS)	$\begin{array}{cc} H & H \\ | & | \\ -C-C- \\ | & | \\ H & \bigcirc \end{array}$
Polymethyl methacrylate	$\begin{array}{cc} H & CH_3 \\ | & | \\ -C-C- \\ | & | \\ H & C-O-CH_3 \\ & \| \\ & O \end{array}$

As we shall see later, the classification of polymers is more intricate. Primarily, lets look at the main grouping of polymers.

14.3 Classification of polymers

Polymers may be classed according to their mechanical properties into *three* groups:

1. thermoplastics
2. thermosetting polymers (thermosets)
3. elastomeric polymers (elastomers).

These polymer types have very different properties from each other, and so are used for quite different applications.

14.3.1 Thermoplastics

These polymers soften when heated, and stiffen when cooled. As a result they are very ductile and workable, and they can be recycled without too much difficulty. Structurally, they consist of long carbon chains tangled together and intertwined, without any cross-linking (tied up in knots), as illustrated in Figure 14.3. This explains their flexibility and resilience as a group. Examples include polyethylene, PVC, nylon, polystyrene, polycarbonate. Uses in construction include: polyethylene sheet for DPC and vapour barrier systems, uPVC window frames and doors, etc.

Thermoplastics can deform elastically. They also yield and can deform plastically to a considerable extent (nylon will give over 300 per cent plastic strain). Before they yield they also show non-linear visco-elasticity, a form of behaviour which is partly elastic and partly viscous flow. This means that thermoplastics are very ductile as a group. They can easily be shaped into complex shapes to make products such as window frame sections, waste pipe systems, water cisterns and simpler things such as polyethylene sheet, etc.

14.3.2 Thermosetting polymers

These polymers do not soften when heated. Their molecular structure is heavily cross-linked, as illustrated in Figure 14.4. Cross-linking can be loosely imagined as if the polymer chains (spaghetti strands) are tied

Figure 14.3   Long-chain structures of thermoplastic polymers (tangled carbon chains)

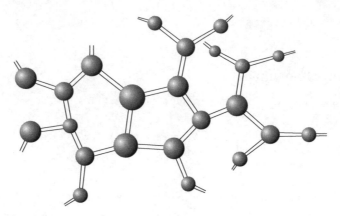

Figure 14.4 Heavily cross-linked structure of thermosetting polymers

in knots; thus, if we imagine a plate of spaghetti tied in knots it would make the dish rather inflexible and almost impossible to move freely with a fork. This cross-linking gives them great rigidity, hardness and wear resistance. Examples include melamine and urea formaldehyde. The thermosets are hard, wear-resistant, stiff (high elastic modulus, E) and have no ductility. They tend to be strong and brittle. Thermosets are not ductile, and so they have to be shaped and polymerised in one operation. Once polymerisation has taken place no further shaping is possible. Thermosets lend themselves to the production of accurately moulded components such as switch and socket plates, boxes, etc. for electrical products. They are also used for kitchen surfaces. Since thermosets have no ductility once they are polymerised, they must be shaped and polymerised in a single operation.

14.3.3 Elastomers

Classification of polymers which are completely elastic and cannot be plastically deformed. The molecular structure of elastomers consist of long, helically wound carbon chains, as shown in Figure 14.5, thus explaining their remarkable ability to extend elastically by over 300 per cent, and then to spring back to their original length.

Elastomers possess excellent elasticity (hence their name), with very low modulus (they can be stretched elastically with a very low load). They have no capacity for even minimal plastic (permanent) deformation. In the construction industry elastomers can be moulded to form sealing rings for waste pipe and rain-water systems and sheet materials and gaskets for other seals.

14.3.4 Summary of properties of the three classes of polymers

The properties of these three types of polymers can be understood in terms of their molecular structures. Because the structures are different, they have quite different properties. The thermosets

Figure 14.5 Helical structures of elastomeric polymers

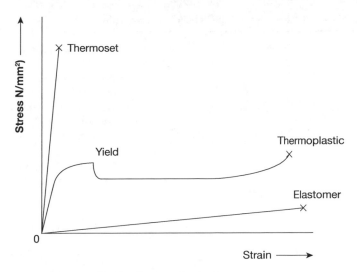

Figure 14.6 The stress–strain curves for the three polymer types

are hard, wear-resistant, stiff (high elastic modulus, *E*) and have no ductility. They tend to be strong and brittle.

Elastomers possess excellent elasticity (hence their name), with very low modulus (they can be stretched elastically with a very low load). They have no capacity for plastic (permanent) deformation.

Thermoplastics have the most interesting properties. They can deform elastically, they also yield and can deform plastically to a considerable extent (nylon will give over 300 per cent plastic strain). Before they yield they also show non-linear visco-elasticity, a form of behaviour which is partly elastic and partly viscous flow.

The stress–strain graphs for the three types of polymer are shown in Figure 14.6.

Thermoplastics are very ductile as a group. They can easily be shaped into complex shapes to make products such as window frame sections, waste pipe systems, water cisterns and simpler things such as polyethylene sheet, etc.

Thermosets are not ductile – they have to be shaped and polymerised in one operation. Once polymerisation has taken place no further shaping is possible. Thermosets lend themselves to the production of accurately moulded components such as switch and socket plates, boxes, etc. for electrical products. They are also used for kitchen surfaces.

Elastomers can be moulded into sealing rings, O-rings, gaskets, etc. which are the usual products. Table 14.2 shows the three polymer types with examples of the use of each in building construction.

The main range of thermoplastics, thermosets and elastomers used in construction are divided into their families and shown in Figure 14.7.

Recyclability (environmental implications)

Nowadays one of the most important factors regarding material selection is sustainability or environmental implications. With polymers one of the perceived main disadvantages is how degradable a waste polymer is. Indeed, polymers can be considered as a green material, particularly thermoplastics. As thermoplastics soften when heated, and stiffen when cooled (this can be repeated over and over again), they are recyclable. Thus, when a thermoplastic material is no longer considered to be useful it can be recycled and used again – a common everyday example is supermarket polythene bags. The area of sustainable materials is comprehensively addressed in Chapter 22.

Table 14.2 Examples of uses of each polymer type in building construction

Polymer type	Polymer	Application in construction
Thermoplastic	uPVC	Window frames and rain-water goods
	Polyethylene	Water-proof membranes
Thermosets	Melamine	Kitchen work surfaces
	Urea formaldehyde	Electrical switch plates and socket outlets
Elastomer	Neoprene	O-ring seals for waste pipes and rain-water goods

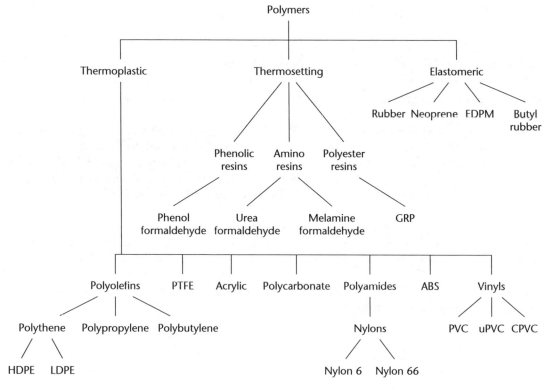

Figure 14.7   Polymer family tree

14.4 Shaping of polymers

14.4.1 Thermoplastics

Thermoplastics are often supplied in the form of small pellets of material, which can be fed into either a moulding or an extrusion process. Moulding is used to produce individual items and components, and extrusion is used to make long lengths of water pipe or window frame sections, which can subsequently be cut to length after shaping. In moulding, the pellets are fed into a mould under heat and pressure; in extrusion the pellets are put into a container, and then put under heat and pressure as they are squeezed through a die. Extrusion can be a continuous process. The processes of moulding and extrusion are illustrated in Figures 14.8 and 14.9, respectively.

Thermoplastics can be produced in sheet form by the process of film blowing. A tube is first extruded, and then air is blown into it to form a continuous, cylindrical plastic sheet. This is then rolled flat and trimmed to produce a folded sheet. By adjusting the air pressure, the film thickness can be controlled.

Thicker sheet plastics for gaskets, etc. can be produced from plastic granules by compression and fusion between a series of heated rollers, a process called clandering. Laminates may be produced by heating together two or more thermoplastic sheets, and during this process sheet reinforcement can be incorporated, if required.

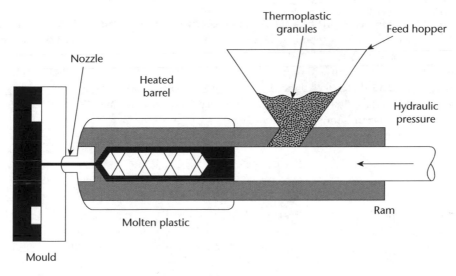

Figure 14.8   Shaping of thermoplastics by injection moulding

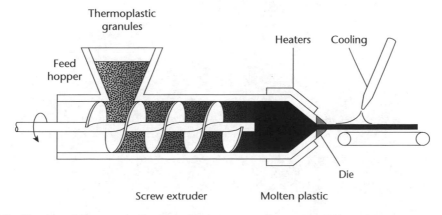

Figure 14.9   Shaping of thermoplastics by extrusion

14.4.2 Thermosets

Since thermosets have no ductility once they are polymerised, they must be shaped and polymerised in one operation. In practice, they are compression moulded, i.e. moulded and polymerised together. Once the items are moulded, no further shaping is possible. So a weighed charge of thermosetting monomer in granular form would be placed into a die, compressed and heated, and the polymerised component ejected once the die is opened, and allowed to cool.

14.4.3 Elastomers

Elastomers can be moulded to form sealing rings for waste pipe and rain-water systems. A clandering process can also be used to form sheet materials and gaskets for other seals.

14.5 Degradation of polymers

Polymers do not corrode, nor are they attacked by liquids occurring naturally in the environment. However, they are degraded by the UV component of sunlight. As a result of the action of UV, they lose their mechanical strength and become embrittled.

A photon of short-wavelength UV has enough energy to sever a C–C (carbon–carbon) bond or a C–H (carbon–hydrogen) bond if it strikes them. Because the long-chain molecules are progressively cut into shorter and shorter lengths, the plastic loses its resilience and flexibility and becomes embrittled. The process of cutting or scission is shown in Figure 14.10.

During the scission process the cut ends (Figure 14.10b) then usually oxidise – this process is called photo–oxidation.

14.5.1 Protection of polymers against ultraviolet radiation

Without protection, polymers such as uPVC would be seriously degraded (embrittled) within 3–6 months when exposed to sunlight. They are protected by mixing UV absorbing materials into the polymer during manufacture in powder form, either carbon black (C; black in colour) or titanium dioxide (TiO_2; brilliant white in colour).

The carbon black and the titanium dioxide work by absorbing the photons of UV before they can strike the C–C and C–H bonds and sever them. Obviously, some UV will eventually get through, but with

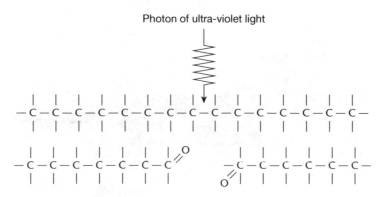

Figure 14.10 The cutting or scission process in polymers

the correct amount of UV absorber added in a well-mixed polymer, the service life of the uPVC items will be extended from a few months to 25–30 years at least.

The carbon black and titanium dioxide therefore serve more than one purpose: they protect the polymer, colour it and also act as a filler. Rain-water goods are frequently required to be black in colour, while window frames are usually brilliant white. Another example is a white Apple laptop made from polycarbonate This laptop contains titanium dioxide both to colour and for protection.

14.6 Polymer structures

Many polymers are composed of molecules that consist of many thousands of atoms arranged into long linear chains. But they don't have to be long, straight chains. Polymers can be made in a lot of other arrangements, too.

Polymers have a wide range of properties that are dependent on their molecular structure. We know that polymers are essentially a long chain of hydrocarbons. However, how are the chains arranged/aligned? This section looks into the different polymer structures.

14.6.1 Linear polymers

A linear polymer is a polymer molecule in which the atoms are more or less arranged in a long chain (Figure 14.11). This chain is called the backbone. Normally, some of these atoms in the chain will have small chains of atoms attached to them. These small chains are called pendant groups (Figure 14.12). The chains of pendant groups are much smaller than the backbone chain. Pendant chains normally have just a few atoms, but the backbone chain usually has hundreds of thousands of atoms.

Figure 14.11   A linear polymer chain

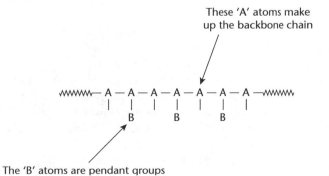

Figure 14.12   Illustration of a linear polymer with pendant groups

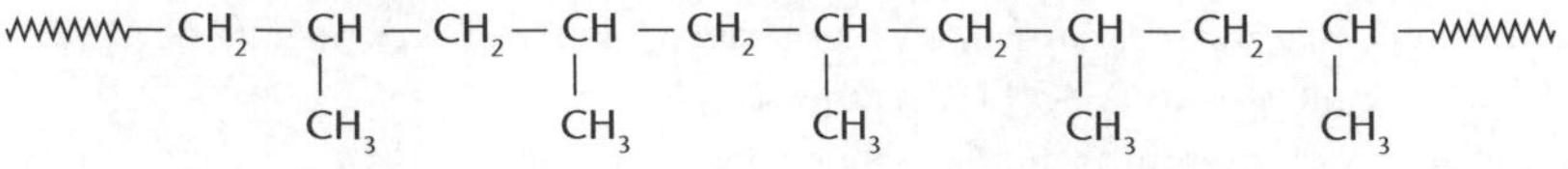

$$\text{\small wwwww— } CH_2 - CH - CH_2 - CH - CH_2 - CH - CH_2 - CH - CH_2 - CH \text{ —wwwww}$$

$$| \qquad\qquad | \qquad\qquad | \qquad\qquad | \qquad\qquad |$$

$$CH_3 \qquad CH_3 \qquad CH_3 \qquad CH_3 \qquad CH_3$$

Figure 14.13   The repeat structure for polypropylene (linear polymer)

To recap, earlier it was mentioned that the *mer* unit is repeated over and over again along the backbone chain; hence, in polypropylene the repeat structure is as shown in Figure 14.13

At this point it is important to note that it is possible to have different variations for the repeat unit (or *mer*) for polymers, e.g. it is possible for polypropylene to have its repeat unit expressed in either of the following ways:

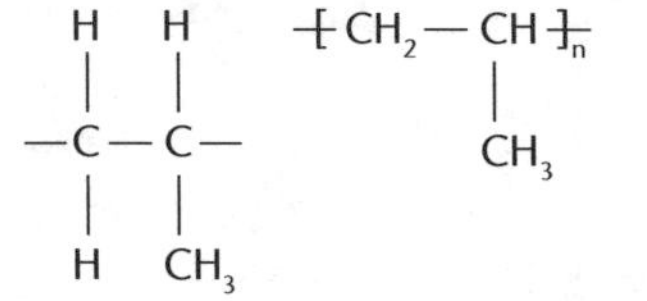

$$\begin{array}{cc} H & H \\ | & | \\ -C - C - \\ | & | \\ H & CH_3 \end{array} \qquad \begin{array}{c} \text{—}(CH_2 - CH)_n \\ | \\ CH_3 \end{array}$$

Please note the chemistry for both structures is exactly the same but expressed in different ways.

14.6.2 Cross-linked polymers

One structural quality of a polymer that can drastically affect its behaviour is cross-linking. Cross-links are covalent bonds linking one polymer chain to another. They are the characteristic property of thermosetting plastic materials. Cross-linking inhibits close-packing of the polymer chains, preventing the formation of crystalline regions (we shall look at crystalline polymers later in this chapter). The restricted molecular mobility of a cross-linked structure limits the extension of the polymer material under loading (e.g. tensile). In simplistic terms it is a bit like tying knots in spaghetti so that they are held together better. A typical cross-linked structure is illustrated in Figure 14.14.

Cross-links are formed by chemical reactions that are initiated by heat and/or pressure, or by the mixing of an unpolymerised or partially polymerised resin with various chemicals; cross-linking can be induced in materials that are normally thermoplastic through exposure to radiation.

In the same manner, polymers that are composed of cross-linked polymer chains are often more rigid than similar, non-cross-linked polymers, as illustrated in Figure 14.15. The higher number of cross-links (shown in bold) in the polymer at the right of the illustration enable it to be more rigid and less deformed by an applied stress, which is represented in the each figure by an identical weight.

In most cases, cross-linking is irreversible and the resulting thermosetting material will degrade or burn if heated, without melting. In some cases, though, if the cross-link bonds are sufficiently different, chemically, from the bonds forming the polymers, the process can be reversed.

The chemical process of vulcanisation is a type of cross-linking and it changes the property of rubber to the hard, durable material we associate with car and bike tyres. This process is often called sulphur curing, or vulcanisation after Vulcan, the Roman god of fire. However, this is a slow process, taking around eight hours. A typical car tyre is cured for 15 minutes at 150 °C. However, the time can be reduced by the addition of accelerators.

Cross-links can be made also by purely physical means. For example, electron beams are used to cross-link polyethylene. Other types of cross-linked polyethylene are made by addition of peroxide during extruding or by addition of a cross-linking agent (e.g. vinylsilane).

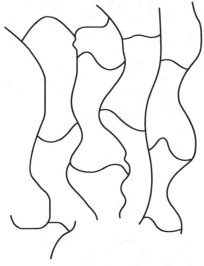

Figure 14.14 Polymer cross-linked structure (a little bit like spaghetti with knots)

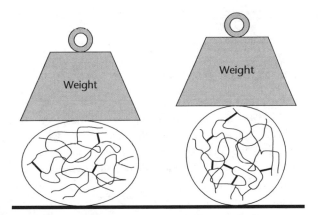

Figure 14.15 Illustration of how cross-linking can make a polymer much stiffer

14.6.3 Branching and graft copolymers

Sometimes there are chains attached to the backbone chain which are comparable in length to that backbone chain. This is called a branched polymer. Some polymers, like polyethylene, can be made in linear or branched versions.

In polymer chemistry, branching occurs by the replacement of a substituent, e.g. a hydrogen atom, on a monomer subunit, by another covalently bonded chain of that polymer; or, in the case of a graft copolymer, by a chain of another type. Branching may result from the formation of carbon–carbon or various other types of covalent bonds. Polymers which are branched but not cross-linked are generally thermoplastic. Branching sometimes occurs spontaneously during synthesis (manufacture) of polymers. In fact, preventing branching to produce linear polyethylene requires special methods. The ultimate in branching is a completely cross-linked network such as found in Bakelite, a phenol–formaldehyde thermoset resin.

To facilitate understanding, a branched polymer has a structure like branching in trees, as illustrated in Figure 14.16.

14.6.4 Block copolymers

A block copolymer is a linear copolymer in which identical *mer* units are clustered in blocks along the molecular chain. A special type of copolymer is called a block copolymer. Block copolymers are made up of blocks of different polymerised monomers. For example, PS-b-PMMA is short for polystyrene-b-polymethylmethacrylate and is made by first polymerising styrene, and then subsequently polymerising MMA (methylmethacrylate). This polymer is a diblock copolymer because it contains two different chemical blocks. It is possible to also make triblocks, tetrablocks, pentablocks, etc.

Types of copolymers

Since a copolymer consists of at least two types of repeating units (not structural units), copolymers can be classified based on how these units are arranged along the chain. These include

- Random copolymer: –A–A–B–B–A–A–A–B–A–A–B–B–A–B–A–A–A–B–A–A–A–B–B–
- Alternating copolymer: –A–B–A–B–A–B–A–B–A–B–, or –(–A–B–)$_n$–
- Block copolymer: –A–A–A–A–A–A–A–B–B–B–B–B–B–A–A–A–A–A–A–B–B–
- Graft copolymer: –A–A–A–A–A–A–A–A–A–A–A–A–A–A–A–A–A–A–A–A–

$$|$$

B–B–B–B–B–B–B

- Star copolymers
- Brush copolymers.

14.7 Polymer crystallinity

Polymer crystallinity is one of the important properties of all polymers. A polymer exists both in crystalline and amorphous form. Thus far we have assumed that polymers have a random structure (spaghetti), as illustrated in Figure 14.3. Although a large number of polymers do indeed have this structure, others also exist. First, when polymers have this random structure, they are considered to be amorphous.

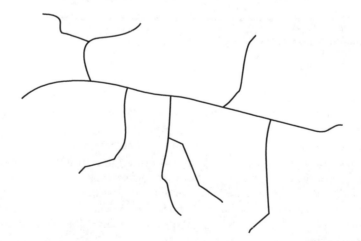

Figure 14.16 A branched polymer

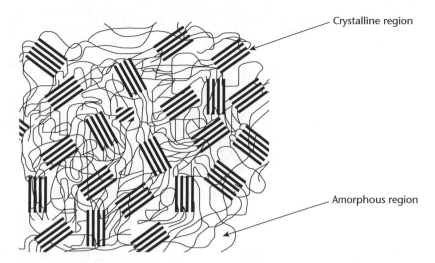

Crystalline region

Amorphous region

Figure 14.17 A polymer with amorphous and crystalline regions

However, if the polymer chains are arranged in an ordered manner or array, the polymer is said to be crystalline. It is also possible for a polymer to have a mixture of amorphous and crystalline regions, as illustrated in Figure 14.17.

Figure 14.17 shows the arrangement of a polymer chain forming crystalline and amorphous regions. It can be seen that part of the molecules are arranged in a regular order – these regions are called crystalline regions. In between these ordered regions molecules are arranged in a random, disorganised state and these are called amorphous regions.

The difference between amorphous and crystalline structures can be simplistically summarised as follows:

- Amorphous polymers have a structure which can be described as being untidy (spaghetti-like).
- Crystalline polymers have a structure which is very tidy and ordered.

Crystallinity is an indication of the amount of crystalline region in a polymer with respect to amorphous content; in other words, how tidy the structure is. Crystallinity influences many of the polymer properties, such as:

- hardness
- modulus
- tensile strength
- stiffness
- crease
- melting point.

Therefore, when selecting the correct polymer for the required application its crystallinity plays a very important role. Thermoplastic polymers are generally either amorphous or semi–crystalline. Table 14.3 lists a selection of amorphous and semi–crystalline polymers.

Table 14.3 Amorphous and semi-crystalline polymers

Amorphous	Semi-crystalline
Polyamideimide	Polyetheretherketone
Polyethersulphone	Polytetrafluoroethylene
Polyetherimide	Polyamide 6,6
Polyarylate	Polyamide 11
Polysulphone	Polyphenylene sulphide
Polyamide (amorphous)	Polyethylene terephthalate (PET)
Polymethylmethacrylate	Polyoxymethylene
Polyvinyl chloride (PVC)	Polypropylene (PP)
Acrylonitrile butadiene styrene (ABS)	High-density polyethylene (HDPE)
Polystyrene (PS)	Low-density polythene (LDPE)

14.8 The glass transition temperature

Have you ever left a plastic bucket or some other plastic object outside during the winter, and found that it cracks or breaks more easily than it would in the summer? What you experienced was the phenomenon known as the glass transition temperature.

The glass transition temperature (Tg) is the temperature at which, upon cooling, a non-crystalline polymer transforms from a supercooled liquid (rubber-like behaviour) to a rigid glass. This transition is unique to polymer materials. When the polymer is cooled below this temperature, it becomes hard and brittle, like glass. Conversely, when a polymer is heated above its Tg, it becomes ductile and flexible. The glass transition phenomenon is present only in amorphous polymers; crystalline polymers do not have a Tg, instead they have a melting point only. However, semi-crystalline polymers (polymers having amorphous and crystalline regions) have both a Tg and melting point – the amorphous portion undergoes the glass transition only, and the crystalline portion undergoes melting only. Some polymers are used above their Tg and some are used below. Rubber elastomers like polyisoprene and polyisobutylene are used above their Tg, that is, in the rubbery state, where they are soft and flexible. Hard plastics like PVC, polystyrene and polymethylmethacrylate (Perspex) are used below their Tg, especially in construction applications.

Thus, to summarise for any amorphous or semi-crystalline polymer:

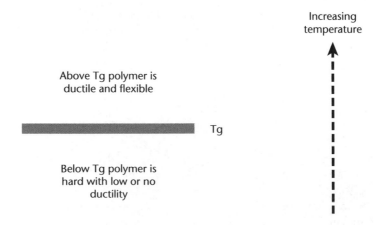

It is important to understand some of the fundamental science behind this behaviour. If we assume a fully amorphous polymer, it will have a totally random structure as illustrated in Figure 14.18.

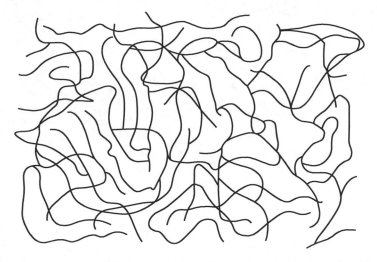

Figure 14.18 Long-chain structures of thermoplastic polymers (tangled carbon chains)

We will look at what happens to the chains below Tg, at Tg and above Tg. Below Tg the chains are in effect frozen. As energy is supplied in the form of heat the chains start to vibrate (without changing position). This vibration increases with an increase in temperature until there comes a point when the chains actually have enough to move positions – this point (or temperature) is the Tg. As there is a further increase in temperature the chains move much more freely. Thus, the flexibility of the chains below and above Tg explains why the ductility and other properties of the polymer changes with temperature. Above the Tg, when you take a piece of the polymer and bend it, the molecules, being in motion already, have no trouble moving into new positions to relieve the stress you have placed on them. But if you try to bend a polymer below its Tg, the polymer chains won't be able to move into new positions to relieve the stress placed on them.

With rising temperature and above the Tg, solid polymers become softer and progress through the rubbery state to finally become a viscous melt capable of flow. The term 'rubbery' refers to the ability to deform sluggishly, but the deformations recover when the load is removed. The term 'glassy' relates to the hardness, stiffness and brittleness of the polymer at low temperatures.

14.8.1 Measuring the glass transition temperature (Tg)

To determine the Tg and/or melting point of a polymer, usually a differential scanning calorimetry (DSC) is utilised. In a DSC the thermal properties of a sample are compared against a standard reference material, which has no transition in the temperature range of interest, such as powdered alumina. Each is contained in a small holder within an adiabatic enclosure, as illustrated in Figure 14.19.

The temperature of each holder is monitored by a thermocouple and heat can be supplied electrically to each holder to keep the temperature of the two equal. A plot of the difference in energy supplied to the sample against the average temperature – as the latter is slowly increased through one or more thermal transitions of the sample – yields important information about the transition, such as latent heat or a relatively abrupt change in heat capacity. The glass transition process is illustrated in Figure 14.20 for a glassy polymer which does not crystallise and is being slowly heated from below Tg.

In Figure 14.20, the drop marked Tg at its midpoint represents the Tg. A melting process is also illustrated in Figure 14.21 for the case of a highly crystalline polymer which is slowly heated through its melting temperature:

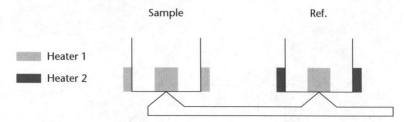

Figure 14.19   A differential scanning calorimetry (DSC) measuring the thermal properties of a polymer and a reference sample to determine the Tg and/or melting point

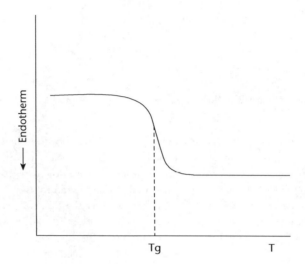

Figure 14.20   How Tg is determined from a DSC curve (endotherm is a measure of the energy absorbed by the sample, T is temperature)

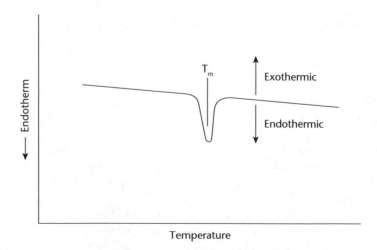

Figure 14.21   How the melting point (Tm) is determined from a DSC curve

Again, as the melting temperature is reached, an endothermal peak appears because heat must be preferentially added to the sample to continue this essentially constant temperature process.

14.8.2 Difference between glass transition and melting

It's tempting to think of the glass transition as a kind of melting of the polymer. But this is a totally inaccurate way of looking at things. There are a lot of important differences between the glass transition and melting. As mentioned earlier, melting is something that happens to a crystalline polymer, while the glass transition happens only to polymers in the amorphous state. A given polymer will often have both amorphous and crystalline domains within it, so the same sample can often show a melting point and a Tg. But the chains that melt are not the chains that undergo the glass transition.

There is another big difference between melting and the glass transition. When a crystalline polymer is heated at a constant rate, the temperature will increase at a constant rate. The amount of heat required to raise the temperature of 1 g of the polymer by 1 °C is called the *heat capacity*. The temperature will continue to increase until the polymer reaches its melting point. When this happens, the temperature will hold steady for a while, even though heat is being added to the polymer. It will hold steady until the polymer has completely melted. Then the temperature of the polymer will begin to increase once again. The temperature rising stops because melting requires energy. All the energy added to a crystalline polymer at its melting point goes into melting, and none of it goes into raising the temperature. This heat is called the latent heat of melting, the word latent means hidden. Once the polymer has melted, the temperature begins to rise again, but now it rises at a slower rate. The molten polymer has a higher heat capacity than the solid crystalline polymer, so it can absorb more heat with a smaller increase in temperature. But when an amorphous polymer is heated to its Tg, something different happens. First, when the material is heated the temperature goes up. It goes up at a rate determined by the polymer's heat capacity. Once the Tg is reached the temperature doesn't stop rising as there is no latent heat of glass transition. The temperature keeps going up; however, the temperature doesn't go up at the same rate above the Tg as below it. The polymer does undergo an increase in its heat capacity when it undergoes the glass transition. Figure 14.22 helps explain this. The plots show the amount of heat added to the polymer on the *y*-axis against heat on the *x*-axis.

The plot on the left shows what happens when a 100 per cent crystalline polymer is heated. The discontinuity represents the melting temperature. At that break, a lot of heat is added without any temperature increase at all; that's the latent heat of melting. The slope gets steeper on the high side of the

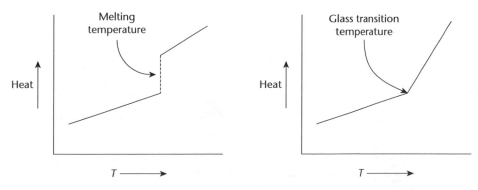

Figure 14.22   A heat vs temperature plot for a crystalline polymer (left) and an amorphous polymer (right)

break; the slope of this kind of plot is equal to the heat capacity, so this increase in steepness corresponds to the increase in heat capacity above the melting point, but in the plot on the right, which shows what happens to a 100 per cent amorphous polymer when it is heated, there is no discontinuity; the only change is at the Tg corresponding to an increase in the slope, which means there is an increase in heat capacity. There is a heat capacity change at the Tg, but no break, as in the plot for the crystalline polymer. For a semi-crystalline polymer (which is typically 40–70 per cent amorphous), the polymer sample experiences both a glass transition temperature and a melting temperature. The amorphous portion undergoes the glass transition only, and the crystalline portion undergoes melting only.

Table 14.4 shows the Tg and melting points for various polymers.

It must be noted that the Tg of a specific polymer is not fixed, i.e. it is possible to modify it accordingly, e.g. adding a small amount of plasticiser to PVC can lower the Tg to −40 °C. This addition renders the PVC a soft, flexible material at room temperature, ideal for applications such as garden hoses and clothing (e.g. jackets, skirts, ladies' boots, etc). A plasticised PVC hose can, however, become stiff and brittle in winter. In this case, as in any other, the relation of the Tg to the ambient temperature is what determines the choice of a given material in a particular application.

There are many factors which influence the Tg; however, it is beyond the scope of this book to elaborate on this. Furthermore, it is highly unlikely the intended range of reader for this text will require this information in their academic or professional careers. Nevertheless, very simplistically the Tg of any polymer is determined by how easily its chains move. A polymer chain that can move around fairly easily will have a very low Tg, while one that doesn't move so well will have a high one. The more easily a polymer can move, the less heat it takes for the chains to commence movement and break out of the rigid, glassy state and into the soft, rubbery state. The most important factor is the backbone flexibility – the more flexible the backbone chain is, the better the polymer will move and the lower its Tg will be.

Table 14.4 Glass transition and melting temperatures for various thermoplastics and elastomers

Polymer	Tg (°C)	Tm (°C)
Low-density polyethylene (LDPE)	−120	115
High-density polyethylene (HDPE)	−120	130
Polypropylene (PP)	−15	170
Polystyrene (PS)	90–120	240
Polymethylmethacrylate (PMMA)	100	–
Polyvinyl chloride (PVC)	90	200
Polyacrylonitrile	105	320
ABS	90–120	–
Polycarbonate (PC)	150	230
Polyethyleneterephthalate (PET)	70	265
Acetal	−90	180
Nylon 6 (PA6)	50	215
Polybutadiene	−90	120
Polyester	75	250
Polyisoprene	−70	30
Silicone	−120	–
Polytetrafluoroethylene (PTFE)	20	325

14.9 Critical thinking and concept review

1. What is a polymer material?
2. What is its chemistry?
3. What is the polymerisation process?
4. What are the three groups of polymers and how are they categorised?
5. What are the unique properties of the three groups?
6. How are plastics and rubbers related to polymers?
7. What are the relative environmental advantages with polymers?
8. How do polymers degrade and how can this be prevented?
9. Why are most polymer products either coloured white or black?
10. What are cross-linked polymers? How do they affect the mechanical properties?
11. What is the difference between an amorphous and crystalline polymer?

15

Polymers utilised in construction

Contents

15.1 Introduction

In the previous chapter we looked at the chemistry and physical properties of polymer materials. In this chapter we look at the main polymer materials and their application in civil engineering and construction applications. The following are the most commonly utilised polymer materials in architecture, construction or civil engineering applications; the fundamental chemical properties, main mechanical properties and practical applications are discussed. In a few examples the chemical structure is illustrated though this is not totally relevant for the intended reading audience.

15.2 Polymers used in a variety of applications

15.2.1 Polyethylene (PE)

This is the most widely utilised polymeric material in the world. A molecule of polyethylene is a long chain of carbon atoms with two hydrogen atoms attached to each carbon atom; the chemical formula is CH_2, as illustrated in Figure 15.1. In construction, applications include interior plumbing pipes (as cross-linked polyethylene   see below) and water-proof sheets for damp-proof applications as polyethylene is impermeable to the passage of water. Polyethylene is available in many different densities which determine its application.

Polyethylene is classified into several different categories based mostly on its density and branching. The mechanical properties of PE depend significantly on variables such as the extent and type of branching, the crystal structure and the molecular weight:

- UHMWPE (ultra high molecular weight PE)
- HMWPE (high molecular weight polyethylene)
- HDPE (high-density PE)
- HDXLPE (high-density cross-linked PE)
- PEX (cross-linked PE)
- MDPE (medium-density PE)
- LDPE (low-density PE)
- LLDPE (linear low-density PE)
- VLDPE (very low-density PE)

Figure 15.1   Ethylene monomer

Polyethylene possibly has the widest range of applications of all polymers; non–construction examples include grocery bags (VLDPE), shampoo bottles (MDPE) and even bullet-proof vests (UHMWPE).

Cross-linked polyethylene

Commonly abbreviated to PEX or XLPE, this is a form of polyethylene with cross-links. It is formed into tubing and is used predominantly in radiant heating systems, domestic water piping and insulation for high-tension electrical cables. Recently, it has become a viable alternative to polyvinyl chloride (PVC), chlorinated polyvinyl chloride (CPVC) or copper pipe for use as residential water pipes. The advantageous properties of PEX also make it a candidate for progressive replacement of metal and thermoplastic pipes, especially in long-life applications, because the expected lifetime of PEX pipes reaches 50–200 years. PEX has become a contender for use in residential water plumbing because of its flexibility. It can be turned through 90 degrees either by a wide turn or using an adapter; PVC, CPVC and copper all require elbow joints. It also has the capacity to run tubing directly from a distribution point continuously to the desired outlet fixture without cutting or splicing; this reduces the need for potentially weak and costly joints. The cost of material can also be approximately 20 per cent less than alternatives, and installation is much less labour intensive. Almost all PEX is made from high-density polyethylene (HDPE). PEX contains cross-link bonds that change the material from a thermoplastic into an elastomer. The cross-linking imparts excellent properties: first, the high-temperature properties of the polymer are improved with the strength maintained to 120–150 °C; second, its impact and tensile strength, scratch resistance and resistance to brittle fracture are enhanced.

15.2.2 Unplasticised polyvinyl chloride (uPVC)

This is a common thermoplastic vinyl polymer which is widely utilised in the construction industry. PVC is similar to polyethylene, but on every other carbon in the backbone chain, one of the hydrogen atoms is replaced with a chlorine atom, as illustrated in Figure 15.2. PVC is probably the most widely utilised polymer in the construction industry; globally, over 50 per cent of PVC manufactured is used in construction. As a building material, PVC is cheap and easy to assemble. Since the 1980s PVC has been replacing traditional building materials such as wood, concrete and clay in many areas. The type of PVC used in construction is called uPVC – unplasticised polyvinyl chloride; 'unplasticised' refers to the fact that the material contains no plasticisers. uPVC is primarily used in the building industry as window frames, doors, waste pipes, drainpipes, gutters and down pipes, as illustrated in Figure 15.3. The material comes in a range of colours and finishes, including a photo-effect wood finish, and is also used as a substitute for painted wood. Furthermore, it is also used to make linoleum for the floor. Another advantage of PVC is that it has flame resistance because it contains chlorine; when PVC burns, chlorine atoms are released, which inhibit combustion.

15.2.3 Ethylene propylene diene rubber

This is an elastomeric polymer also called EPDM and EPM. EDPM is one of the most widely used and fastest-growing synthetic rubbers, having both specialty and general-purpose applications. In construction,

Figure 15.2 Vinylchloride monomer

Figure 15.3   Front door of a house made from uPVC

EPDM is used in radiators, garden and appliance hose, tubing, belts, electrical insulation and roofing membranes. Ethylene propylene rubbers are valuable for their excellent resistance to heat (up to 160 °C), oxidation, ozone and weather ageing due to their stable chemical properties. They have good electrical resistivity, as well as resistance to certain potentially corrosive liquids, such as water, acids, alkalis, phosphate esters and many ketones and alcohols. Amorphous EDPM has excellent low temperature flexibility with glass transition points of about −60 °C. EPDM polymers have high tensile and tear properties, excellent abrasion resistance, as well as improved oil swell resistance and flame retardance.

15.2.4 Melamine

Melamine is a thermosetting polymeric material with a wide range of applications. Melamine is a strong organic base with chemical formula $C_3H_6N_6$, with the name 1,3,5-triazine-2,4,6-triamine. The melamine monomer is illustrated in Figure 15.4. Melamine is primarily used to produce melamine resin, which when combined with formaldehyde produces a very durable thermoset plastic. This plastic is often used as kitchen worktops (Figure 15.5), kitchen utensils or plates. Melamine is also used to make decorative wall panels and is often used as a laminate.

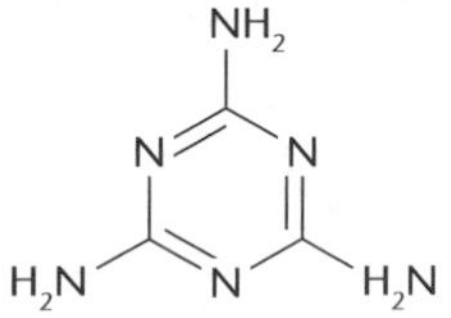

Figure 15.4   Melamine monomer

Figure 15.5 Melamine kitchen worktop

Melamine foam has an interlinking bubble format which produces a structure more like a block of microscopic fibreglass than normal foam. It can also be used for soundproofing and as a fire-retardant material, but not as insulation because it allows air to pass through its structure.

15.2.5 Polypropylene (PP)

This thermoplastic polymer has a level of crystallinity intermediate between that of low-density polyethylene (LDPE) and high-density polyethylene (HDPE). Although it is less tough and flexible than LDPE, it is much less brittle than HDPE. This allows polypropylene to be used as a replacement for engineering plastics such as ABS. Thus, in construction PP is occasionally used as sewer pipes or lavatory seats. Polypropylene is rugged, often somewhat stiffer than some other plastics, reasonably economical and can be made translucent when uncoloured, but not completely transparent like other polymers. It can also be made opaque and/or have many kinds of colours. PP has very good resistance to fatigue. It is a versatile polymer as it serves double duty, both as a plastic and as a fibre. As a fibre, polypropylene is used to make indoor–outdoor carpeting, such as that used around swimming pools. PP is also used as outdoor carpets as a variety of coloured polypropylene is available, and also because PP doesn't absorb water. Structurally, it's a vinyl polymer, and is similar to polyethylene, except that every other carbon atom in the backbone chain has a methyl group attached to it, as illustrated in Figure 15.6.

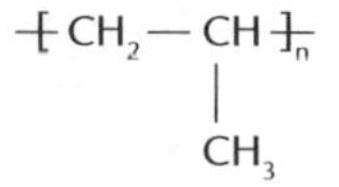

$$\left[CH_2 - CH \right]_n$$
$$|$$
$$CH_3$$

Figure 15.6 Propylene monomer

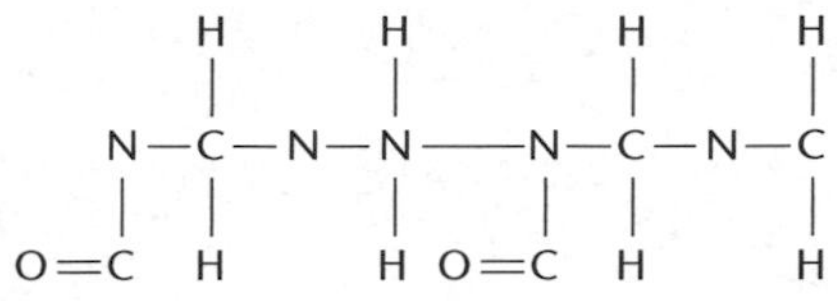

$$N-C-N-N \quad N-C-N-C$$

Figure 15.7 Formaldehyde monomer

15.2.6 Urea formaldehyde (UF)

This thermosetting plastic (polymer) is made from urea and formaldehyde. Applications in construction include electrical switch plates and electrical sockets (white plastic plates), highly prevalent in dwellings and municipal buildings. UF has a high tensile strength, flexural modulus, low water absorption and mould shrinkage and high surface hardness, elongation at break and volume resistance. Previously, UF foam insulation, or UFFI, was used as cavity-wall insulation, beginning in the 1950s. UF is typically made at a construction site from a mixture of urea–formaldehyde resin, a foaming agent and compressed air. When the mixture is injected into the wall, urea and formaldehyde unite and cure into an insulating foam plastic. During the 1970s, when concerns about energy efficiency led to efforts to improve home insulation worldwide, UFFI became an important insulation product for existing houses. Most installations occurred in the 1970s. In the insulating process, a slight excess of formaldehyde was often added to ensure complete curing with the urea to produce the urea–formaldehyde foam. That excess was given off during the curing, almost entirely within a day or two of injection. Properly installed, UFFI might not have resulted in any problems. Unfortunately, however, UFFI was sometimes improperly installed or used in locations where it should not have been. In the 1980s concerns began to develop about the toxic formaldehyde vapour emitted in the curing process, as well as from the breakdown of old foam; consequently, its use was discontinued. However, UF is still almost exclusively used as electrical switch plates and sockets, as illustrated in Figures 15.8–15.9.

15.2.7 ABS (acrylonitrile butadiene styrene)

This common thermoplastic polymer is used to make light, rigid, moulded products. Applications in construction include lavatory seats, highway safety devices and pipes. ABS is a copolymer made by

Figure 15.8 UF polymer light switch

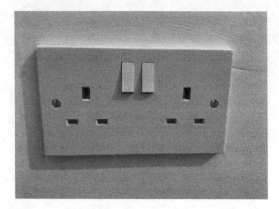

Figure 15.9   Electrical socket made from UF

polymerising styrene and acrylonitrile in the presence of polybutadiene. The proportions can vary from 15 to 35 per cent acrylonitrile, 5 to 30 per cent butadiene and 40 to 60 per cent styrene. The result is a long chain of polybutadiene criss-crossed with shorter chains of poly(styrene-co-acrylonitrile). The styrene gives the plastic a shiny, impervious surface; the butadiene, a rubbery substance, provides resilience even at sub-ambient temperatures. ABS can be used between –25 °C and 60 °C and is used for its good shock absorbance. The structure for ABS is illustrated in Figure 15.10.

The advantage of ABS is that this material combines the strength and rigidity of the acrylonitrile and styrene polymers with the toughness of the polybutadiene rubber. The impact resistance can be amplified by increasing the proportions of polybutadiene in relation to styrene and acrylonitrile, although this causes changes in other properties. Impact resistance does not fall off rapidly at lower temperatures and its stability under load is excellent. The final properties are influenced to some extent by the conditions under which the material is processed to the final product; for example, moulding at a high temperature improves the gloss and heat resistance of the product, whereas the highest impact resistance and strength are obtained by moulding at low temperature.

15.2.8 Polyvinylidene fluoride (PVDF)

This is a highly non-reactive and pure thermoplastic fluoropolymer (a polymer containing fluorine). PVDF is used generally in applications requiring the highest purity, strength and resistance to solvents, acids, bases and heat and low smoke generation during a fire event. Compared to other fluoropolymers it has an easier melt process because of its relatively low melting point. PVDF has a relatively low density and low cost compared to the other fluoropolymers. It is available as piping products, sheet, tubing, films, plate and an insulator for premium wire. As a fine powder grade, PVDF is also used as the principal ingredient

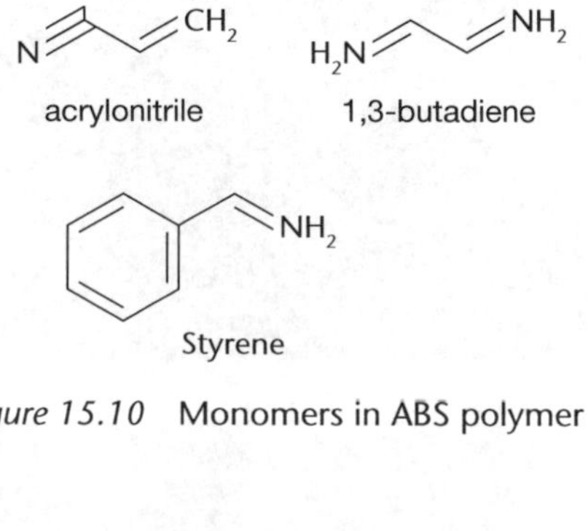

acrylonitrile   1,3-butadiene

Styrene

Figure 15.10   Monomers in ABS polymer

of high-end paints for metals. These PVDF paints have extremely good gloss and colour retention, and they are in use on many prominent buildings around the world, e.g. the Petronas Towers in Malaysia, as well as on commercial and residential metal roofing.

15.2.9 Polystyrene

Polystyrene is a material made from the monomer styrene, as illustrated in Figure 15.11. Pure solid polystyrene is a colourless, hard plastic with limited flexibility. It can be cast into moulds with fine detail. Polystyrene can be transparent or can be made to take on various colours. Polystyrene's most common use, however, is as expanded polystyrene or polystyrene foam. Expanded polystyrene is produced from a mixture of about 5–10 per cent gaseous blowing agent and 90–95 per cent polystyrene by weight, most commonly pentane or carbon dioxide. The solid plastic is expanded into a foam through the use of heat, usually steam. In polystyrene foam the voids filled with trapped air give it low thermal conductivity. This makes it ideal as a construction material and it is used in structural insulated panel building systems. As a result this can lead to improved/lower U-values for dwellings. Like most polymers, polystyrene is highly flammable or easily ignited; consequently, though it is an efficient insulator at low temperatures, it is prohibited from being used in any exposed installations in building construction. It must be concealed behind drywall, sheet metal or concrete. Foamed plastic materials have been accidentally ignited in buildings and caused huge fires and losses. Polystyrene foam is also predominantly used in packaging.

15.3 Polymers used primarily in sealants, adhesives and paints

15.3.1 Butyl rubber

Also known as polyisobutylene and PIB $(C_4H_8)_n$, butyl rubber is a synthetic rubber, a homopolymer of 2-methyl-1-propene. Polyisobutylene is produced by polymerisation of about 98 per cent isobutylene with about 2 per cent isoprene. Structurally, polyisobutylene resembles polypropylene; it has excellent impermeability and the long polyisobutylene segments of its polymer chains give it good flex properties. As polyisobutylene is impermeable to air it is used in many applications requiring an airtight seal. Butyl rubber is commonly utilised in construction; it is particularly useful as a sealant used for rubber roof repair and for maintenance of roof membranes, especially around the edges. Rubber roof is typically referred to as a specific type of roofing material that is made from ethylene propylene diene monomer (EPDM).

15.3.2 Neoprene

This polymeric material is the DuPont chemical trade name for a family of synthetic rubbers based on polychloroprene; the monomer is illustrated in Figure 15.12. Neoprene is chemically inert, which makes it well suited for industrial applications such as gaskets, hoses and corrosion-resistant coatings. Neoprene is widely used in the construction industry as a mechanical seal in a device which helps join systems or

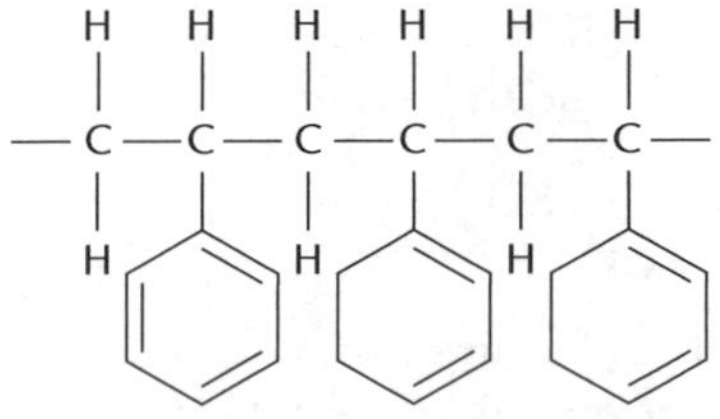

Figure 15.11 A styrene monomer

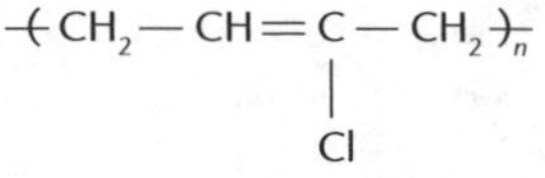

$$-\!\!\left(CH_2 - CH = C - CH_2\right)_{\!\!n}$$
$$|$$
$$Cl$$

Figure 15.12   Neoprene monomer

mechanisms together by preventing leakage, especially in plumbing systems containing pressure or excluding contamination. Neoprene (originally called duprene) was the first mass-produced synthetic rubber compound. It can also be used as a base for adhesives and for noise isolation in power transformer installations.

15.3.3 Elastomeric sealants

A type of sealant made from elastomers. Sealants are a type of material used to seal gaps or joints between materials or building elements, usually to prevent the ingress of water. Elastomeric polymer sealants are considerably more expensive than other types of sealants; however, the extra cost is offset by the superior durability. These sealants have the advantage of being very resilient, together with the ability to withstand great joint movements (in the region of 20 per cent). The anticipated life is about 25 years in the case of polysulphide sealants. Elastomeric sealants bond well to metals, brick, glass and to many plastics. Elastomeric sealants do not change in volume when stressed.

15.2.4 Enamel paint

This is a special type of paint based on polyurethane or alkyd resins to give very durable, impact-resistant, easily cleaned, hard, gloss surfaces. Colours are usually bright, and these are suitable for machinery and plant in interior and exterior locations.

15.3.5 Epoxy

Epoxy is a thermosetting epoxide polymer that cures (polymerises and cross-links) when mixed with a catalysing agent or hardener. Epoxies are used as powder coatings for a variety of applications. Fusion-bonded epoxy powder coatings are extensively used for corrosion protection of steel pipes and fittings used in the oil and gas industry, portable water transmission pipelines (steel), concrete reinforcing rebar, etc. Epoxy coatings are also widely used as primers to improve the adhesion of paints, especially on metal surfaces where corrosion (rusting) resistance is important. Metal cans and containers are often coated with epoxy to prevent rusting. Epoxy resins are also used for high-performance and decorative flooring applications, especially terrazzo flooring, chip flooring and coloured aggregate flooring. Epoxy adhesives are a major part of the class of adhesives called structural adhesives or engineering adhesives (which also includes polyurethane, acrylic and cyanoacrylate). These high-performance adhesives are used in the construction industry as well as many others where high-strength bonds are required. Epoxy adhesives can be developed to suit almost any application. They are exceptional adhesives for wood, metal, glass, stone and some plastics. They can be made flexible or rigid, transparent or opaque/coloured, fast setting or extremely slow setting. Epoxy adhesives are almost unmatched in heat and chemical resistance among common adhesives. In general, epoxy adhesives cured with heat will be more heat- and chemical-resistant than those cured at room temperature.

Epoxy paints

These are a special type of paint which are highly resistant to abrasion and spillages of oils, detergents or dilute aqueous chemicals. They are often applied to concrete, stone, metal or wood in heavily-trafficked workshops and factories.

15.3.6 Ethylene vinyl acetate copolymers (EVA)

This polymer is based on copolymerisation products of ethylene with vinyl acetate. EVA copolymers are useful for coatings, hot melt adhesives, etc., while high-density EVA can be used for tougher applications. EVA is primarily used in construction as hot melt and heat seal adhesives. As the level of vinyl acetate in the copolymer increases, so the level of crystallinity found in polythene alone reduces from about 60 per cent to 10 per cent. This yields products ranging from materials similar to low–density polythene to flexible rubbers. Common grades can contain from 2 per cent to 50 per cent vinyl acetate. Clarity, flexibility, toughness and solvent solubility increase with increasing vinyl acetate content. A major advantage is the retention of flexibility of EVA rubber grades down to −70 °C. Good resistance to water, salt and other environments can be obtained, but solvent resistance decreases with increasing vinyl acetate content.

15.3.7 Polyvinyl acetate (PVA or PVAc)

This is a rubbery synthetic polymer prepared by polymerisation of vinyl acetate monomer, also referred to as VAM. As an emulsion in water, PVA is utilised as an adhesive for porous materials, particularly wood, paper and cloth. It is the most commonly used wood glue, both as white glue and yellow carpenter's glue. PVA is a common copolymer with more expensive acrylics, used extensively in paper, paint and industrial coatings, referred to as vinyl acrylics.

15.3.8 Polyvinyl fluoride (PVF)

This polymer is mainly used in flammability lowering coating applications in both the construction and aeronautical industries. It is commonly applied to coat metal sheeting. PVF is a thermoplastic fluoropolymer with the repeating vinyl fluoride unit; it is structurally very similar to PVC. PVF has low permeability for vapours, burns very slowly and has excellent resistance to weathering and staining. It is also resistant to most chemicals, except ketones and esters. It is available as a film in a variety of colours and formulations for various end uses, and as a resin for specialty coatings.

15.4 Polymers used in glass or as glass replacements

15.4.1 Fibreglass

Fibreglass is made from extremely fine fibres of glass. It was developed in the late 1940s. It was the first modern composite and is still the most common. It makes up about 65 per cent of all the composite materials produced today. Glass reinforced plastic (GRP) or glass fibre reinforced plastic consists of two distinct materials, fibres of glass (inorganic), which is the reinforcement, and a polymer resin called polyester, which serves as the matrix. The polyester resin polymer is brittle and has low strength, but when fibres of glass are embedded in the polymer the resultant composite becomes strong, tough, resilient and flexible. It is an ideal material for roofing and furniture. The other advantage of GRP is that it is very light with a very good strength–weight ratio.

15.4.2 Laminated glass

This is a special type of glass which is a combination of two or more glass sheets with one or more interlayers of plastic polyvinyl butyral (PVB) or resin. In case of breakage, the interlayer holds the fragments together and continues to provide resistance to the passage of persons or objects. This glass is particularly suitable where it is important to ensure the resistance of the whole sheet after breakage, such as: shop-fronts, balconies, stair-railings and roof glazing. There are two types of laminated glass: PVB and resin laminated glass. PVB laminated glass is two or more sheets of glass which are bonded together with one or more layers (PVB) under heat and pressure to form a single piece. Resin-laminated glass is manufactured by pouring liquid resin into the cavity between two sheets of glass, which are held together until the resin cures.

15.4.3 Polyvinyl butyral (PVB)

This is a resin, usually used for applications that require strong binding, optical clarity, adhesion to many surfaces, toughness and flexibility. It is prepared from polyvinyl alcohol by reaction with butyraldehyde. The major application is laminated safety glass for a variety of applications. Laminated glass, commonly used in the automotive and architectural fields, comprises a protective interlayer, usually polyvinyl butyral (PVB), bonded between two panes of glass. The bonding process takes place under heat and pressure. When laminated under these conditions, the PVB interlayer becomes optically clear and binds the two panes of glass together. Once sealed together, the glass 'sandwich' (i.e. laminate) behaves as a single unit and looks like normal glass. The polymer interlayer of PVB is tough and ductile, so brittle cracks will not pass from one side of the glass to the other.

15.4.4 Polymethylmethacrylate

This polymer is also known as Perspex, PMMA or acrylic resin. PMMA is a transparent plastic, used as a shatterproof replacement for glass. Due to its excellent transmission properties and toughness, PMMA has been used as a substitute for glass, especially in municipal buildings and windows in bus stops. Polymethylmethacrylate is the synthetic polymer of methylmethacrylate, illustrated in Figure 15.13. This thermoplastic and transparent plastic is sold under the trade names Plexiglas, Perspex, Acrylite, Acrylplast, Altuglas and Lucite, and is commonly called acrylic glass or simply acrylic.

Differences in the properties of glass and PMMA include:

- PMMA is lighter: its density is 1,190 kg m^{-3}, about half that of glass.
- PMMA is softer and more easily scratched than glass. This can be overcome with scratch-resistant coatings.
- PMMA can be easily formed by heating it to 100 °C.
- PMMA transmits more light (92 per cent of visible light) than glass.
- Unlike glass, PMMA does not filter UV (ultraviolet) light. PMMA transmits UV light, though coating with protective films can alleviate this problem.

In comparison with other polymers, PMMA is generally cheaper. When it comes to making windows, PMMA has another advantage over glass, which is that it is more transparent than glass. When glass windows are made too thick, they become difficult to see through. But PMMA windows can be made as much as 33 cm thick and remain perfectly transparent. Other non-construction applications of PMMA include aeroplane windows. PMMA is also found in acrylic paints.

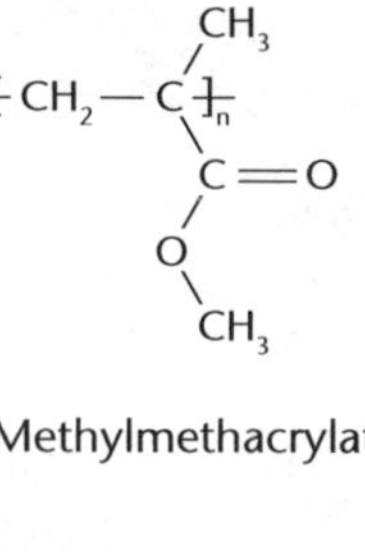

Figure 15.13 Methylmethacrylate monomer

15.4.5 Polycarbonate (PC)

Like PMMA, PC is a material commonly used as an unbreakable glass substitute. Polycarbonates are a particular group of thermoplastics. They are easily worked, moulded, and thermoformed; as such, these plastics are very widely used in modern manufacturing. PC glass is very popular as bullet-proof or vandal proof windows utilised in banks, public transport stations, high-risk buildings, etc. Polycarbonate is becoming more common in housewares, as well as laboratories and in industry. It is also used to create protective features for lighting lenses for many buildings. They are called polycarbonates because they are polymers having functional groups linked together by carbonate groups (–O–(C=O)–O–) in a long molecular chain.

Differences in the properties of glass and PC include:

- PC is lighter: its density is 1,200 kg m^{-3}, about half that of glass.
- PC transmits 91 per cent of visible light, which is significantly more than glass.
- PC can be easily formed by heating it to 100 °C.
- PC has poor weathering in an ultraviolet (UV) light environment in comparison to glass; however, this can be rectified by coating the PC.

15.4.6 Polychlorinated biphenyls (PCBs)

In construction PCBs are used as stabilising additives in flexible PVC coatings of electrical wiring and electronic components, flame retardants, sealants, adhesives, wood floor finishes, paints, etc. Commercial PCB mixtures are clear to pale yellow viscous liquids; the more highly chlorinated mixtures are more viscous and more yellow. PCBs have very high thermal conductivity and are chemically almost inert, being extremely resistant to oxidation.

15.4.7 Polytetrafluoroethylene (PTFE), Teflon

This is a polymeric material, also commercially known as Teflon, with the lowest coefficient of friction (against polished steel) of any known solid material. PTFE is very non-reactive, and so is often used in containers and pipework for reactive chemicals. Its melting point is 385 °C, but its properties degrade above 260 °C. An early advanced use was as a material to coat valves and seals in the pipes holding highly reactive uranium hexafluoride in uranium enrichment plants. PTFE is also used to coat petroleum pipelines due to its excellent non-reactive properties. Other well known non-construction related applications include non-stick coating for pans and other cookware. The fluorinated ethylene monomer is illustrated in Figure 15.14.

15.5 Other polymers not widely used in construction

15.5.1 Polyamide (nylon)

Polyamides comprise the largest family of engineering plastics with a very wide range of applications. Polyamides are often formed into fibres and are used for monofilaments and yarns. Characteristically

Figure 15.14   Teflon, a polymer of fluorinated ethylene

polyamides are very resistant to wear and abrasion, have good mechanical properties even at elevated temperatures, have low permeability to gases and have good chemical resistance. They can occur both naturally, examples being proteins, such as wool and silk, and can be made artificially, examples being nylons, aramids and sodium poly(aspartate). Polyamides currently have limited application in the construction industry.

15.5.2 Kevlar

Kevlar is a fibre based on polyamide. The fibres have good strength, but are weak in compression. Kevlar 49 is currently the most widely used aramid fibre for reinforcement of plastics. Kevlar fibres have high stiffness and tensile strength, superior resistance to heat and wear and strength far greater than any other reinforcing material. It has been used for bullet-proof vests and composite body parts for the racing and aerospace industries.

15.5.3 Polyvinylidene chloride (PVdC)

This polymer is derived from vinylidene chloride. PVdC is a heavy fibre providing a remarkable barrier against water, oxygen and aromas, has superior chemical resistance to alkalis and acids, is insoluble in oil and organic solvents, has very low moisture regain and is impervious to mould, bacteria and insects. PVdC fibre has a high elastic recovery and resists wrinkling and creasing. As PVdC is pigment dyed before fibrespinning, it has excellent colour-fastness and high light permeability. PVdC is also flame-retardant and self extinguishing – it may soften or char in flame and decomposes in moderate heat. The major disadvantage of PVdC is that it will undergo thermally induced dehydrochlorination at temperatures very near to processing temperatures. This degradation easily propagates, leaving polyene sequences long enough to absorb visible light, and change the colour of the material from colourless to an undesirable transparent brown. Therefore, there is a significant amount of product loss in the manufacturing process, which increases production and consumer costs.

15.5.4 Styrene/butadiene copolymer

This is a thermoplastic copolymer comprising styrene and butadiene. One of the outstanding characteristics of styrene/butadiene thermoplastic copolymers is their combination of high transparency, brilliance and impact resistance. The good miscibility allows adjustment to the desired toughness, while at the same time reducing material costs. Styrene/butadiene can be extruded, thermoformed and injection moulded into a variety of high-quality products.

15.6 Critical thinking and concept review

1. What are different types of polyethylenes and what are their applications in construction?
2. What are the applications of uPVC?
3. What is melamine used for?
4. What is UF?
5. What is unique about ABS?
6. Which polymers are used in sealants, adhesives and paints?

16

Timber

This chapter covers one of the oldest and most important construction materials – timber. The cell structures of hardwoods and softwoods are dealt with first, explaining how they are formed during the growth process. The shape and distribution of the cells leads to the anisotropy of timber and the non-uniform strength properties and dimensional response to changes in moisture content and humidity. The seasoning and conversion of timber and the manufacture of timber products are dealt with, and the chapter concludes with a discussion of how the use of timber might contribute to a more sustainable future for the construction industry.

Contents

16.1 Introduction

Timber has been used as a construction material since the earliest times. It is still an important construction material today, because of its versatile properties, its diversity and its aesthetic qualities. In the UK, about two-thirds of all sawn softwoods and over one-third of sawn hardwoods go into construction – the rest goes into paper production, as fuel or is wasted during the logging process. Worldwide, perhaps half of all timber harvested is burned for fuel.

It is important to emphasise at the outset that timber is not just a versatile and important construction material, it is also a vital component of the Earth's ecosystems, and it plays a vital role in maintaining life on this planet. We ignore this fact at our peril. Animals, including humans, need oxygen and food to live, and these are produced by photosynthesis in green plants. Trees are the largest forms of plant life, and so they play a pre-eminent part in providing the oxygen necessary for life, and in the sequestration (removal) of CO_2 from the Earth's atmosphere. So the Earth's forests, besides being a source of timber, also act as a carbon sink.

Timber is also a renewable resource, and so can make an important contribution towards the achievement of *sustainable* building. Timber is called renewable, because if we cut down a mature tree for timber we can replace it by replanting with seedlings which will eventually grow to maturity. So while we can say that timber is renewable in principle, there is more to it than that. A mature forest, whether a northern temperate forest with deciduous and coniferous trees or a tropical rainforest containing various hardwoods, will contain a whole variety of other plants besides, together with numerous insects, birds, mammals, reptiles, etc.; in other words, the forest will contain a complete ecosystem – it will contain *biodiversity*. So if we harvest trees, we shall damage the forest if we do not manage to preserve this biodiversity. Loss of biodiversity will lead to extinction of species, and runs against the principles of sustainability. Forests provide habitats for a whole range of plants and animals, and these must be preserved.

Forests also play a part in climate control. Rain that falls on a forest is held and immediate evaporation and run-off is prevented, i.e. they play a part in the hydrological cycle, and help regulate the rate at which water circulates. The presence of the tree cover enables all the species living in the forest to find food and water. In some cases, such as the Amazon rainforest, the trees are growing in very poor soil; if the trees are cut down this poor soil will be eroded and lost within a decade. Although the vegetation looks very lush, it is quite fragile; it is only the tree cover which enables it to survive.

In conclusion, it is very important that forests are not regarded just as sources of timber; their vital contribution to the maintenance of life must always be borne in mind. There is evidence that societies that cut all their trees down do not survive (Diamond, 2005).

16.2 Structure of timber

Timber is obtained from trees, the largest type of plant life on Earth. Trees grow with a single main stem, commonly called the trunk or bole, and it is the trunk that provides the timber. A tree trunk serves a number of purposes:

- to support the crown;
- to conduct dilute mineral solutions from the soil to the crown;
- to store organic substances manufactured by photosynthesis.

Timber is used as a structural material, and also for its aesthetic qualities, among others. Both of these important properties depend on its cell structure or microstructure, and so we need to understand the structure of timber at the level of the individual cells. Figure 16.1 shows a wedge-shaped section cut from a five-year-old tree, and it shows the way its cell structure is arranged. The cell structure has to meet the three functions given above. Wood can be classified into two types – hardwood and softwood – and the differences between them lie in their cell structures. This classification is based solely on the cell structures and not on the actual hardness or strength of individual species. Balsa wood, for example, is classed as a hardwood despite being weaker and softer than most softwoods.

We can distinguish between them by examining their structures under a microscope; typical examples are shown in Figure 16.2.

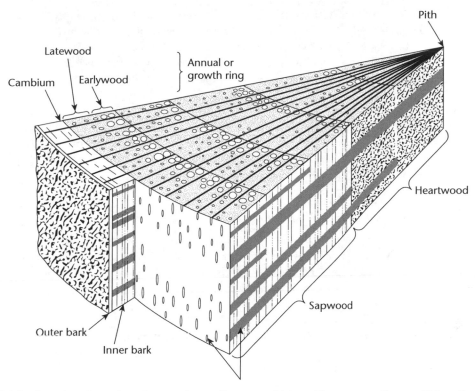

Figure 16.1   A wedge-shaped section cut from a five-year-old tree. The cross-section, radial section and tangential section faces are all shown.

(Princes Risborough Laboratory, BRE © Crown copyright)

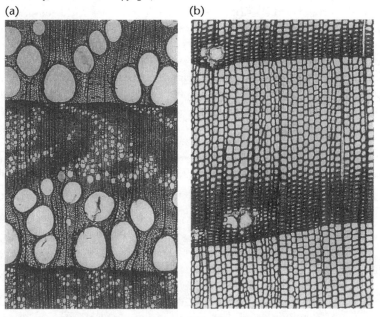

Figure 16.2   (a) Sweet chestnut; (b) Scots pine

(BRE © Crown copyright)

In each case it is possible to differentiate between the early wood – very porous – and the late wood – more dense. The early wood is so-called because it grows early in the annual growing season, when the tree needs huge volumes of water to open its leaves to the sun. The buds are opened and the leaves spread by hydraulic (water) pressure. Once the leaves are open and synthesising food, the water requirement reduces dramatically; the later growth is of higher density, providing more strength and less conduction.

The vessel cells themselves are long, tubular cells composed of cellulose, and they are closed at each end. They are very long compared to their diameters, but they are connected to each other by small circular valves called pits. These pits allow the transmission of moisture and nutrients from one vessel to the next, in a long vertical chain of cells. Alternatively, vertical vessel cells may be connected to those aligned radially, to allow moisture and nutrient movement in the radial direction too.

16.2.1 Anisotropy of timber

Because of the way timber grows, with most of the vessel cells running along the length of the tree, the timber has different properties in different directions. In other words, the properties are different along the grain from those measured across the grain – this is called *anisotropy*. In contrast, materials where the properties are the same in all three directions are called *isotropic*, and the condition of having equal properties in all three directions is called *isotropy*. So timber is an *anisotropic* material, and examination of the cell structures illustrated in Figure 16.4 confirms this.

A tree trunk section is approximately circular, and so we can measure properties in *three* directions:

1. longitudinally (direction from top to bottom of the trunk);
2. radially (direction from the centre to the outer bark);
3. tangentially (circumferentially around the trunk).

Earlier we saw the cross-sectional microstructures of a hardwood and a softwood. Because timber is anisotropic, the microstructures of the radial and tangential sections will look quite different. Figure 16.4 shows (a) the transverse or cross-section of Scots pine, (b) the radial-longitudinal section of Scots pine and (c) the tangential-longitudinal section of Scots pine. Despite their differing appearances, they are all sections of the same wood.

We are all familiar with the fact that we can split wood along the grain, but not across the grain. All the other properties, such as stiffness (E), compressive strength and tensile strength will also vary depending upon the direction (parallel to the grain or perpendicular to the grain). Table 16.1 gives values of stiffness (E) for five hardwoods and four softwoods in the longitudinal (E_L), radial (E_R) and tangential (E_T) directions.

Inspection of the data above shows that the properties of timber in the three directions differ to a high degree; this material is extremely anisotropic. Besides causing the mechanical properties to vary widely, the timber's response to moisture is also heavily direction-dependent, which will be examined in Section 16.2.3.

16.2.2 Variability of timber

We have seen that timber is anisotropic and this is because of the way that the annual growth rings form. However, cut timber shows other features arising from its growth that can seriously impact upon its strength properties. These include defects such as juvenile wood, sloping grain, resin pockets and knots. The presence of any of these defects can cause the strength of an individual length of wood to be seriously reduced.

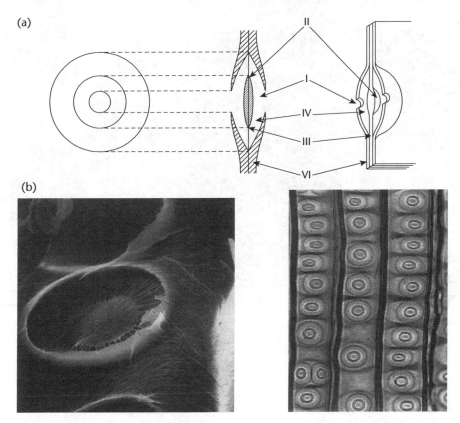

Figure 16.3 (a) Section through a bordered pit in conducting cells: (I) pit opening, (II) torus, (III) margo strands formed from the primary wall, (IV) pit cavity, (V) secondary wall. (b) Bordered pit on the radial wall of a softwood. (c) Array of pits of neighbouring cells

(After Dinwoodie, 1996; BRE © Crown copyright)

Table 16.1 Stiffness (*E*) values for five hardwoods and four soft woods in the three directions longitudinal, radial and tangential

Species	Density (kg/m³)	Moisture content (%)	E_L	E_R	E_T
Balsa	200	9	6,300	300	106
African mahogany	440	11	10,200	1,130	510
Birch	620	9	16,300	1,110	620
Ash	670	9	15,800	1,510	800
Beech	750	11	13,700	2,240	1,140
European whitewood	390	12	10,700	710	430
Sitka spruce	390	12	11,600	900	500
European redwood	550	10	16,300	1,100	570
Douglas fir	590	9	16,400	1,300	900

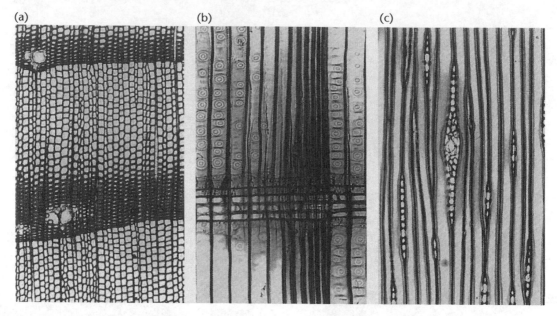

Figure 16.4 (a) Transverse section of Scots pine; (b) radial-longitudinal section of Scots pine; (c) tangential-longitudinal section of Scots Pine

(After Dinwoodie, 1996; BRE © Crown copyright)

Figure 16.5 shows a picture of a so-called 'live' knot, and Figure 16.6 a 'dead' knot. Both are discontinuities in the mass of vertical fibres, but the live knots are attached to the surrounding wood, whereas the dead knots are physically separated from the surrounding wood by a layer of bark. When thin sections of wood are cut, dead knots will often fall out, leaving a hole and seriously weakening the length of timber, so dead knots are more seriously weakening than live ones.

So timber is both anisotropic and variable. It is still a very attractive material for many applications in construction, and over recent decades the timber industry has developed many ways to overcome these variations and make the material more useful, and we shall examine some of these later in Section 16.4.

16.2.3 Moisture movement effects

We all know that timber can swell in damp weather and shrinks in dry weather, and this swelling and shrinkage is called *moisture movement*. It is important that we understand how this movement occurs, and what results from it.

When the tree is growing, the vessel cells are full of moisture. When the timber is cut and felled, the vessel cells are cut through, and the cut ends are exposed to the atmosphere. During seasoning, all the moisture inside the vessels is lost and the moisture content of the timber falls to about 28 per cent, and no movement occurs. However, if we dry the timber further, the vessel cell walls (cellulose) begin to dry out, and they shrink as they do so. *So it is the drying out of the vessel cell walls (drying below 28 per cent moisture) which causes the movement.* Because the vessel cells are arranged in rings, the moisture movement will be greatest in the circumferential direction. There will be no movement in the longitudinal direction. In the radial direction, we have alternate rings of porous vessel cells and denser, non-porous wood. Therefore, in this direction we shall have an intermediate amount of movement as shown in Figure 16.7. Specifically, Figure 16.7 explains why timber which is initially machined square can finish up with a diamond-shaped

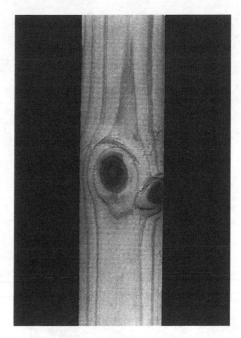

Figure 16.5 A 'live' knot. The knot is continuous with the wood of the trunk
(BRE © Crown copyright)

Figure 16.6 A 'dead' knot. The knot is separated from the wood of the trunk by a layer of bark
(BRE © Crown copyright)

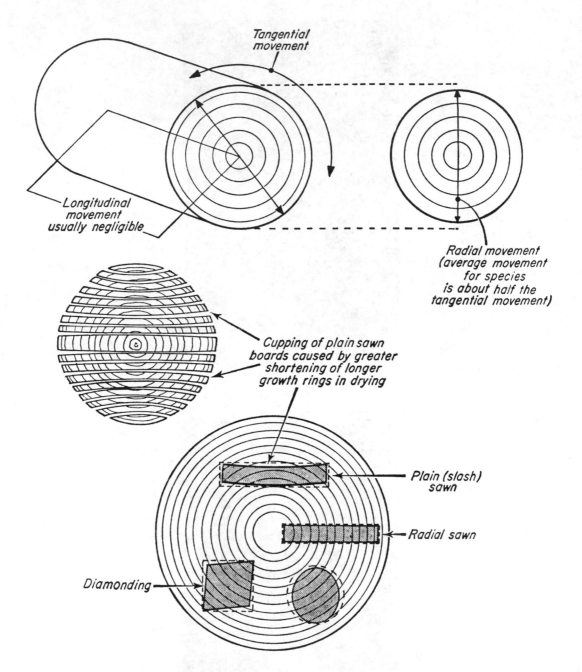

Figure 16.7 Movements in drying of timber

(After Everett, 1970)

cross-section, why round dowels can become oval and why some initially flat floorboards can become 'cupped'. All these effects are caused by the fact that movement in the circumferential (or tangential) direction is twice that in the radial direction. The diagram shows movement during drying, and obviously the movement would be in the reverse direction if the moisture content of the timber increased.

These movements are an intrinsic result of the ring structure, which is a result of the growing process, and there is nothing we can do to stop them in a piece of solid timber. Hardwoods tend to be denser than the softwoods, and so the amount of movement will depend on the species of wood; dense hardwoods such as teak will show little movement, whereas yellow Canadian birch shows high movement. Other species such as English elm give medium movement effects.

In Figure 16.8 we can see the relationship between moisture content and various uses of timber, susceptibility to decay, etc. We can see that 28 per cent moisture is the limit of fibre saturation. In timber containing more than 28 per cent moisture there will be liquid water inside the vessels. Below 28 per cent

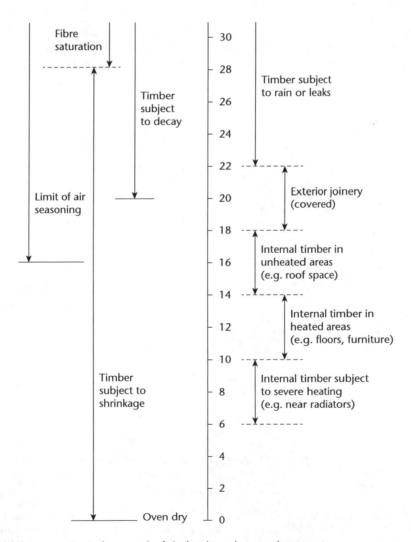

Figure 16.8 Moisture contents (per cent) of timber in various environments

moisture the cell walls begin to dry out by losing their combined water, and so will begin to shrink. Below 28 per cent moisture is where movement problems begin. We can also see that when using air drying about 16 per cent is the minimum moisture content we can achieve. Using kiln drying we can reduce the moisture content to whatever level we choose, and this will depend upon the use to which the timber is to be put.

Very importantly, Figure 16.8 shows that 20 per cent is the critical moisture content for decay by fungal attack. Fungus spores will not germinate if they land on dry timber. The timber must contain above 20 per cent moisture to be liable to attack. The best way to avoid fungal attack is by good building design in the first instance. By this is meant design that ensures that all timber is kept dry and well ventilated. Then the building must be thoroughly maintained during its service life, so that any leaks that develop are spotted and corrected as soon as possible.

16.2.4 Effect of moisture on strength

While moisture has major effects on dimensional stability of timber, as we have seen, it also has an effect on strength. Figure 16.9 shows a plot of longitudinal compressive strength against percentage moisture content. From this graph we can see that as the moisture content of the timber reduces, the strength

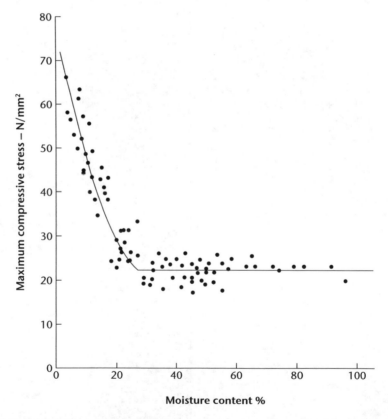

Figure 16.9 Relationship between longitudinal compressive strength and moisture content. The point of inflexion corresponds to the state where there is no free water inside the vessels and there is maximum bound water in the vessel cell walls

(BRE © Crown Copyright)

remains constant until it gets down to around 28 per cent. At this point there is an inflexion and the strength then rises rapidly as the moisture content is further reduced, reaching a maximum at zero moisture. Why is this so?

We saw in Figure 16.8 that 28 per cent moisture is the lower limit of fibre saturation. Above 28 per cent moisture there is liquid water inside the cellulose vessel cells. At 28 per cent the vessel cells are completely emptied of liquid, and further drying begins to dry out the cellulose vessel cell walls, which contain chemically combined water. As this drying out process proceeds the vessel cell walls shrink in volume due to the loss of this combined water. Of course, it is this loss that causes the dimensional changes discussed in Section 16.2.2. As long as there is liquid water inside the vessels themselves, no movement occurs.

When the timber dries below 28 per cent it becomes harder and stronger, but it also becomes more brittle. Two hundred years ago, when the battleships of the Royal Navy were made of wood, the gun crews suffered terrible injuries during sea battles, and many of their injuries were caused by flying shards and splinters of wood when enemy fire impacted on the hulls of their ships. The timber in the ship's walls would typically have dried to 16–18 per cent moisture content, and so would be nowhere near as tough as timber containing more than 28 per cent moisture. To protect themselves against this the gunners used to line the internal walls around the gun ports with quilting and other fabrics, measures which were often only partially effective.

16.3 Conversion and seasoning

Conversion is the process of cutting up the tree trunks into sections prior to seasoning.

Seasoning refers to the process of removing moisture from the timber before it is used and put into service. In a living tree, the weight of water in the tree's vessel cells will frequently be greater than the dry weight of the tree itself. Seasoning is the removal of most of this moisture, and the stabilisation of the moisture content before putting the timber into service.

16.3.1 Conversion of timber

Converting a log into a set of planks may appear to be a simple process, but in practice it is not. The aim is to maximise output in financial rather than volume terms; one batten of large cross-section will be worth considerably more than two battens each of half the cross-section. Similarly, one long batten is worth more than two half as long.

The aim of maximising the financial return from sawn timber is frequently limited by a number of factors about which quick decisions may be needed. These factors include:

- the shape of the log, including
    - taper of the log – there will be more waste where there is a high degree of taper;
    - bowed logs – is it better to cut one long length of small cross-section, or two short lengths of large cross-section?
    - large knots – this is important where structural softwoods are being cut. How much will a knot degrade the timber's worth?
- growth stresses,
- possible exclusion of sapwood and juvenile wood.

The principal equipment used for conversion is a set of saws, of the following types: circular saws, bandsaws and framesaws. Many modern sawmills now have a chipper canter as well. The purpose of a

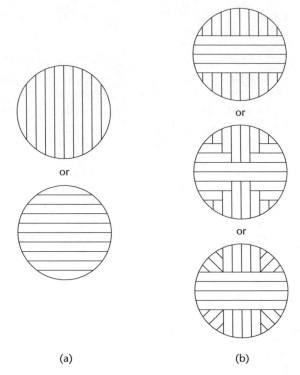

(a)   (b)

Figure 16.10   Conversion of hardwood logs

(After Dinwoodie, 1996; BRE © Crown Copyright)

chipper canter is to convert the portions of round timber cut from the edge of a log into wood chips (for chipboard, etc.).

Conversion of hardwood logs. Most hardwoods are sawn on a 'through and through' basis to produce a series of flat-sawn planks of equal thickness (Figure 16.10(a)). Where it is desired to obtain as many radial or near-radial faces as possible, the logs may be quarter-sawn. This is a rather expensive operation, requiring a great deal of manipulation of the log on the saw table (Figure 16.10(b)).

Conversion of softwood logs. Figure 16.11 shows the sequence of operations in the conversion of softwood logs using a non-automated saw-bench operated by two workers. The waste material (shown black in the diagram) is converted to chips (for making chipboard) or used as firewood.

These diagrams show the traditional sawing sequences used in the conversion of timber. However, these have been superseded by computer-aided methods. A tree trunk can rapidly be scanned from several viewpoints to give a three-dimensional model, and the computer can equally rapidly decide on an optimal cutting sequence. The tree trunk is automatically manipulated and cut to generate minimum waste. Because timber is such a precious resource, otherwise unusable off-cuts can be turned into chips for chipboard manufacture, etc.

Two other processes are *peeling* (used in making plywood) and *slicing* (used to make decorative veneers). Peeling is shown in Figure 16.12.

The lamellae produced by peeling are glued together as shown in Figure 16.13 to produce plywood. Notice that there is always an odd number of lamellae.

Slicing to produce decorative veneers is shown schematically in Figure 16.14.

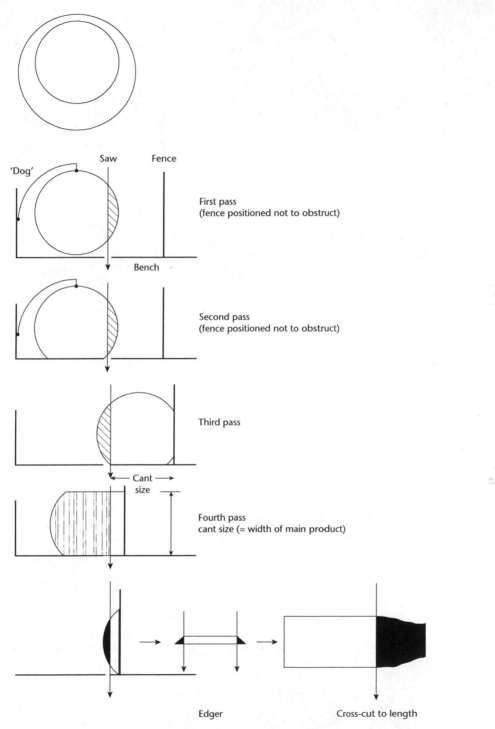

Figure 16.11 Sawing sequence for softwood log conversion using a non-automated sawbench with two operators

(After Dinwoodie, 1996; BRE © Crown Copyright)

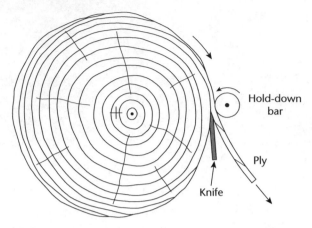

Figure 16.12 Cutting ply veneer by peeling (non-decorative)
(BRE © Crown Copyright)

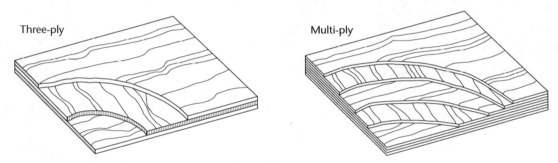

Figure 16.13 Typical three-ply and multi-ply plywood boards
(BRE © Crown Copyright)

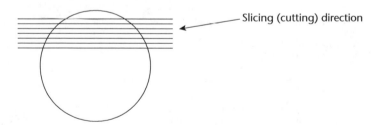

Figure 16.14 Slicing direction. Sliced sections show the wood grain to the best effect, whereas peeled veneers (cut circumferentially) do not

16.3.2 Seasoning of timber

The primary aim of seasoning is to render the timber as dimensionally stable as possible. This ensures that once put into service as flooring, furniture, doors, etc., movement will be minimal. *Seasoning involves removing most (but not all) of the moisture from the timber*, and when this is done other advantages accrue. Most wood-rotting fungi can grow in timber only if the moisture content is above 20–22 per cent.

Drying of woods occurs because of differences in vapour pressure from the centre of a piece of wood outwards. As the surface layers dry, the vapour pressure in these layers falls below the vapour pressure in the wetter wood further in, and a vapour pressure gradient is built up which results in the movement of moisture from centre to surface; further drying is dependent on maintaining this vapour pressure gradient. The steeper the gradient, the more rapidly the drying (seasoning) progresses, but in practice too steep a gradient must be avoided to prevent splitting of the wood.

Two main methods of seasoning are used; *air seasoning* (natural seasoning) and *kiln seasoning* (artificial seasoning). For large-sized timbers, a combination of the two methods is often used.

Air seasoning. Air seasoning is still practised in some countries where the cost of kiln drying is high, and it is still used to at least partially dry timber of large cross-sectional area. Where such timber would take more than 4–5 weeks to kiln dry, it is often air dried to a moisture content of 25–30 per cent first. In the UK hardwood planks are often air-dried for 18–24 months before being kiln dried.

Very large-section hardwood is often only air dried because of the cost involved. Beam-quality oak measuring 350 mm × 300 mm is usually air dried for at least eight years prior to sale. Softwood 25 mm thick, if air dried in spring, should dry to about 20 per cent moisture content in 1.5–3 months. Hardwood 25 mm thick, if stacked in autumn, will take 7–9 months to dry to 20 per cent, and 50 mm hardwood will take about one year. The extent of drying will depend on the local weather conditions; for example, in the UK it is not usually possible to dry below 22 per cent in winter and about 16–1 per cent during summer.

Properly constructed, well-ventilated drying sheds are essential for satisfactory air seasoning. The stacking technique is also important, as is end-protection for the lengths of wood. End-protection prevents the ends of the timber pieces from drying out quicker than the rest and splitting.

Kiln seasoning. Kiln drying is carried out in a closed chamber, providing maximum control of air circulation, humidity and temperature. Therefore drying can be carefully regulated so that shrinkage occurs with the minimum of degradation problems (cracking and splitting, etc.). One advantage of kiln seasoning is that lower moisture contents are possible than can be achieved using air seasoning. Other advantages include rapidity of turnaround, adaptability and precision. It is the only way to season timber intended for interior use, where required moisture contents may be as low as 10 per cent or less.

It is necessary to regulate kiln drying to suit the particular circumstances, i.e. different species and different sizes of timber require drying at different rates. As a general rule, softwoods can withstand more rapid drying than hardwoods, thin boards than thick planks and partially dry stock than green timber. Air circulation is necessary to ensure that humidity and temperature are maintained as uniform throughout. The progress of drying can be checked by testing sample boards for moisture. It is also possible to check to see whether drying stresses are being induced in the timber.

16.4 Timber products

We have seen that timber, while being valuable and versatile, is also a variable material. Therefore, while solid timber is still in widespread use, the industry has increasingly developed products that obviate this variability and its effects. As well as this, converting an item of approximately circular cross-section into lengths of material with square or rectangular section inevitably leads to waste. So the industry is driven by economic as well as technical considerations. What the construction industry requires are timber sections and items with uniform, reliable properties. In particular, the need is for a material with reduced susceptibility to moisture movement and uniform strength properties, i.e. a material with the variability removed. At the same time, the industry seeks to produce less waste than before.

There is a well-developed market for sheet and board materials of a size that cannot be obtained from round timber with normal cutting. The timber industry developed a range of products including *plywood*, *block-board* and *chipboard* many years ago in response to these requirements. These products have economic advantages for they allow the use of smaller-diameter, immature trees obtained during the thinning out of properly managed forests, off-cuts and thin or odd sections produced during conversion, etc. The thin laminae required for plywood are made by peeling, and block-board will utilise the thin strips of timber left over during the milling process. Other timber oddments can be chipped, bonded together with glue, pressed into sheets and cured to produce chipboard. Block-board is dimensionally more stable than solid wood, and chipboard has excellent sound-proofing and acoustic properties. In these ways the industry makes maximum use of the timber coming into the mill, and leaves very little waste.

These products have been in use for many years; plywood was invented before the Second World War and was used by the aircraft company de Havilland during the war to make the airframes of its very successful Mosquito aircraft. Plywood solves the anisotropy problem, as the laminae are laid with the grain direction alternately at 90 degrees to each other, and usually with an odd number of leaves, i.e. three-ply, five-ply, seven-ply, etc. Similarly with block-board, if strips of wood are glued edge-to-edge, the orientation of the vessel cells will not matter, neither will the presence of a knot in one of the strips. The strips on either side will compensate. In effect all local weaknesses will be 'averaged out'. The industry then invented the 'glulam' beam, where lengths of wood typically 10–21 mm thick are glued together. Such beams have very uniform properties due to the averaging-out effect mentioned above. Furthermore, by butting and finger jointing lengths together, a glulam beam can be made much longer than any individual piece of solid timber. Curved beams can be made, tapering beams can be made and designed to carry the specific loads and bending moments called for in the building structure. In other words, the items can be engineered to suit, knowing that the material has uniform predictable properties.

Of course, not only do these products possess uniform strength properties but they are also not prone to the moisture movement problems outlined in Section 16.2.3. Having devised these materials with uniform strength and dimensional stability, the industry has gone on to develop various other products based upon plywood, such as I-beams, box beams and stiffened and stressed panel materials. It is immediately apparent that these products are all very economical in their use of timber, and they are illustrated in Figure 16.15.

Chipboard was introduced before the Second World War as a way of utilising the small chips of wood produced during the milling process. However, in the decades after the Second World War it was recognised that it might be advantageous to manufacture board material from larger pieces of wood, to obtain greater strength and stiffness properties. This idea was first tried many years ago in North America with the production of *waferboard*. Large, square wafers of wood were bonded together using phenol formaldehyde resins. No fine material or chips were used so the volume of resin used was minimal.

Waferboard was superseded by *oriented strand board* (OSB). Here the wood particles are narrower, with a width about half their length, and most importantly, their fibre directions are aligned either in each of three layers or only on the outer two layers of the board. In the second case, the surface and centre layers are oriented at right angles to each other to obtain properties similar to those of plywood. Having produced this material, it was realised that an almost infinite variety of board and section properties could be obtained by designing in the degree of fibre alignment required. Phenol formaldehyde resin was also replaced by improved adhesives, and this technology led directly to the production of engineered timber.

In the production of engineered timber, lamellae cut parallel to the grain or bundles of parallel fibres are mixed with adhesive, and then compressed into regular sections and then cured. Because it consists of lengths of uncut fibres, engineered timber possesses high strength and uniform properties. It contains no knots, sloping grain, resin pockets or any of the irregular and random defects that can weaken solid timber. The use of engineered timber allows designers to create structures with predictable and uniform behaviour. It also allows the use of random strips and odd lengths of timber left over from the initial sawing operations, thereby avoiding waste.

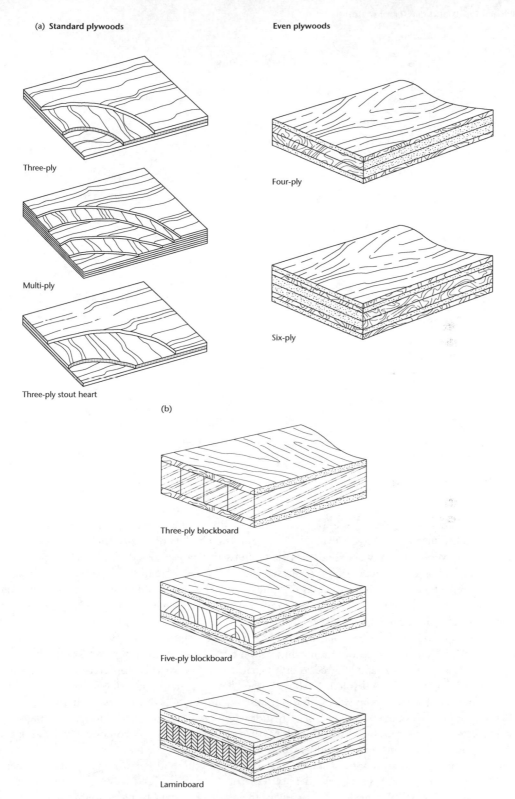

Three-ply

Multi-ply

Three-ply stout heart

Four-ply

Six-ply

(b)

Three-ply blockboard

Five-ply blockboard

Laminboard

Figure 16.15   (a) Plywoods; (b) block-boards; (c) beams and panels

(BRE © Crown Copyright)

(c)

Plywood beams

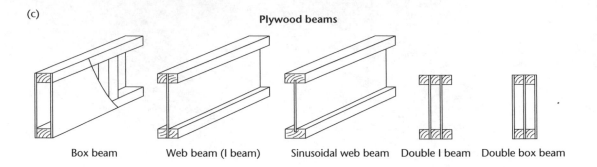

Box beam   Web beam (I beam)   Sinusoidal web beam   Double I beam   Double box beam

Plywood panels

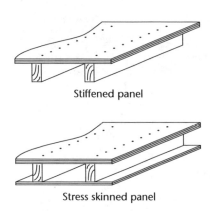

Stiffened panel

Stress skinned panel

Figure 16.15   (Continued)

Another development is the class of products known as fibreboards. These types of board offer different densities and hardness values, and are manufactured from wood in the fibrous state. Fibreboard can be made from both hardwoods and softwoods, either by the original 'wet' process or by the newer 'dry' process. The raw material for fibreboard production is small-diameter timber or waste material from timber processing. First, this is reduced to small chips (typically 25 mm × 25 mm × 4 mm), and then these are defibrated to small bundles of timber fibres, and sometimes to individual fibres, by mechanical grinding, or explosively by the 'Masonite' process.

In the wet process the fibres are mixed with water, giving a pulp to which any necessary chemicals are added, and the pulp is deposited onto a moving wire mesh. The fibres interlock and form a sheet. Low-density boards are cut to size and allowed to dry, higher density boards are compacted at a temperature of 180 °C. The lignin in the wood softens and bonds the fibres together, so no adhesive is required. Boards produced this way have a smooth side, with the reverse side imprinted with a 'mesh' pattern.

With the dry process, the fibre bundles are first dried to a low moisture content before being sprayed with an adhesive and formed into a mat. This mat is then hot-pressed to produce a board with two smooth faces. Medium-density fibreboard (MDF) is the main type of board produced by the dry process. The various types of fibreboard produced include hardboards (density 800 kg/m^3), mediumboards (density 350–800 kg/m^3 and above), softboards (density less than 350 kg/m^3) and medium-density fibreboard (MDF). Much more MDF is produced than any of the other fibreboard materials, with some going into furniture production. However, an increasing amount goes into construction in the form of window boards, skirting boards and architraves. Each new house in the UK now contains perhaps 150–250 m of MDF skirting and architrave (Dinwoodie, 1996). It is not difficult to understand why this is; solid timber skirtings and architrave gave rise to cutting and aesthetic problems because of the presence of knots, and other defects.

In all the foregoing discussion it can be seen that timber technologists have been developing useful timber products (plywood, block–board, chipboard, glulam, OSB, MDF, etc.) that eliminate the sometimes extreme variability and moisture movement problems that characterise solid wood, while at the same time reducing waste.

The processes of decay and degradation of timber, and their prevention, will be dealt with in Chapter 20.

16.5 Timber as a sustainable construction material

In every field of engineering and technology, including construction, sustainability is now an important topic. The drive is on for renewable and sustainable materials, as well as the construction of minimum-energy buildings. How can timber help?

At the present time steel and concrete are widely used, especially for high-rise buildings. However, very recently timber has been used to construct a nine-storey residential block in London, approximately 100 feet high – the Stadthaus Building. This success could presage the construction of more such buildings, perhaps even 25 storeys or more in height. What difficulties stand in the way of doing this? What advantages would the use of timber confer?

It will be useful to examine some of the basic properties of the three structural materials – steel, concrete and timber. Some of the key properties are given in Table 16.2. We can see immediately that timber has the lowest density of the three materials. It also has the lowest stiffness, the lowest values of thermal conductivity and thermal diffusivity and a low value of embodied energy.

The first benefit of using timber instead of steel and/or concrete would be the lightness of the building and a saving in weight of concrete for foundations. Another very desirable advantage of timber is in speed of erection. Timber lends itself to machining and pre-cutting of the pieces of timber, and of doing some pre-fabrication under factory conditions. On-site work is then limited to assembly of factory-made units and items, and this leads to more accurate construction and to a considerable reduction in the time spent on site. In the construction of the Stadhaus, construction time using timber was 49 weeks, as against the 72 weeks that would have been needed for construction with concrete. The weight of the timber building was 300 tonnes compared with 1,200 tonnes for concrete. In terms of carbon footprint, the timber in the Stadhaus stores 185 tonnes of carbon, whereas 125 tonnes of carbon would have been emitted in producing a corresponding concrete and steel building.

The Stadhaus project has been so successful that the building of even taller structures from timber is being planned. The factor that immediately militates against this is the low stiffness of timber (see Table 16.2). However, the problem could be solved by building a steel-reinforced protected lift and stairwell

Table 16.2 Showing values of stiffness, thermal conductivity and thermal diffusivity for steel, concrete and timber

Material	Density kg/m³	Stiffness – Young's modulus GN/m²	Thermal conductivity W/m K	Thermal diffusivity m²/s × 10⁻⁶	Embodied energy GJ/tonne
Steel	7.87	200–207	60.0	15.96	22.5
Concrete	2.4	45–50	1.4	0.663	1.6
Timber	0.6–1.0	9–16	0.11	0.074	1.6

shaft, and then assembling the timber modules around this stiff core. The shaft would solve two problems; it would stiffen the tall structure, and it would be of great benefit from the fire safety point of view. The overall build time and building weight would be reduced, and the outer timber walls would ensure the energy efficiency of the building.

Durability is always an important consideration in the choice of construction materials, and timber is excellent in this respect. There are many timber-framed houses in England that are hundreds of years old; England is a northern country and so these wooden buildings are not subject to attack by termites, as they might be in warmer climates. One of the reasons that these timber-framed buildings have lasted so well is that they were constructed of English oak, which was almost ubiquitous in earlier times. Oak is a hardwood. Nevertheless, the longevity of these buildings is testament to the durability of timber, making it an attractive material for continued use in the UK.

Finally, in our energy-conscious world, we need our buildings to be well-insulated against heat loss, and here again, timber proves to be an excellent material. The values for thermal conductivity and thermal diffusivity given in Table 16.2 indicate that timber is excellent in this respect. To summarise, timber is a remarkable, versatile material, being renewable and, in principle, sustainable. It possesses a unique combination of properties which make it just as attractive and important a construction material in the twenty-first century as it has ever been in the past.

16.6 Critical thinking and concept review

1. Give *three* reasons why forests should not be regarded as mere sources of timber.
2. List the types of cells in: (a) hardwoods and (b) softwoods.
3. Timber is an *anisotropic* material. Explain what this means.
4. Explain what is meant by moisture movement in timber.
5. Explain (a) why moisture movement in timber is such a non-uniform process, and (b) the ways in which timber can change shape in response to variations in humidity.
6. What is meant by the *seasoning* of timber, and why is it necessary?
7. Outline the relative merits and drawbacks of air seasoning and kiln seasoning.
8. Give the reason why drying shrinkage in timber only begins once the moisture content falls below about 28 per cent.
9. Briefly explain why moisture movement in timber is greatest in the circumferential direction, zero in the longitudinal direction and intermediate in the radial direction.
10. When used for load-bearing beams, what advantage would a glulam beam have compared with a solid timber beam (a) from a technical point of view, and (b) from an economic point of view?
11. In general, why would timber intended for internal fittings and furniture be dried to a lower moisture content than timber for roof joists?
12. What advantages would an engineered timber I-beam offer over a solid timber joist?

16.7 References and further reading

CLIFTON-TAYLOR, A. (1987), *The Pattern of English Building*, Faber and Faber, London.
DESCH, H.E. & DINWOODIE, J.M. (1996), *Timber, Structure, Properties, Conversion and Use*, MacMillan Press, London.
DIAMOND, J. (2005), *Collapse: How Societies Choose to Fail or Survive*, Allen Lane, London.
DINWOODIE, J.M. (1996), *Timber Structure, Properties, Conversion and Use*, Macmillan, London.
EVERETT, A. (1970), *Materials*, Batsford, London.
GIBSON, L.J. & ASHBY, M.F. (1988), *Cellular Solids: Structure and Properties*, Pergamon Press, Oxford.
LYONS, A.R. (1997), *Materials for Architects and Builders*, Arnold, London.

17

Soil as a material

Contents

17.1 Introduction

Soil is one of the most important materials in construction. However, it is often neglected in other construction material textbooks. It is one of the most variable materials to consider as it is naturally occurring. Also, it will be, almost certainly, inconsistent from one site to the next. The most important elements to consider are: the range of soils, their properties/parameters and what causes variations in these values.

17.2 Soil, rock and mineral types

From an engineering perspective soil can be defined as a material that can be worked without drilling or blasting and consist of relatively soft, loose, uncemented deposits, such as gravels, sand, silts and clays.

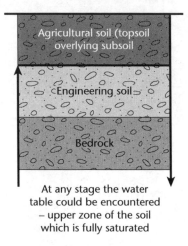

Ground level

At any stage the water
table could be encountered
– upper zone of the soil
which is fully saturated

Figure 17.1 Typical ground profile

A typical ground profile is illustrated in Figure 17.1. Soil can be classed into four main groups:

1. *Agricultural soil* (e.g. topsoil, subsoil) may be organic in nature (e.g. peat, etc.) – not considered as engineering soils.
2. *Engineering soil* (e.g. gravels, sands, silts and clays) consist of soft, loose, uncemented deposits.
3. *Organic soil* is formed from the decomposition of vegetable matter, e.g. topsoil and peat (sands, silts and clays may also contain organic matter). Soils containing organic matter are unsuitable for engineering purposes; typically they are weak and subject to large settlement.
4. *Fill or made ground* is material that has been placed in a (semi) controlled manner, e.g. embankments and controlled landfill.

Rock, is a hard, rigid cemented deposit and is grouped according to how it has been formed. There are three main types of rocks (Table 17.1):

1. *Igneous rocks* are those that solidify from the magma through either intrusive or extrusive processes. Extrusive rocks form when magma cools rapidly on the earth's crust. Intrusive rocks are formed when magma cools very slowly within the earth's crust.
2. *Metamorphic rocks* are those that have been altered by heat and/or pressure so that they have lost their original character. They have often been re-crystallised to form new rock types.
3. *Sedimentary rocks* result from external forces on the earth's crust and are formed from particles deposited by rivers, glaciers, the wind, the sea or by chemical deposition from lakes or the sea.

Rocks are formed from minerals. For example:

Granite (rock) = feldspar + quartz + mica + amphibole (a collection of different minerals)

A mineral is a naturally occurring, inorganic (non-living) substance. Mineralogy is the identification of minerals and the study of their properties, origin and classification. There are approximately 3,000 minerals, and each mineral has a particular chemical composition which can be expressed as a chemical formula. The

Table 17.1 Various mineral and rock types

Silicate minerals	Non-silicate minerals	Igneous rocks	Sedimentary rocks	Metamorphic rocks
Amphibole	Azurite	Andesite	Breccia	Amphibolite
Augite	Barite	Basalt	Chalk	Anthracite
Biotite mica	Calcite	Camptonite	Chert	Blue schist
Chiastolite slices	Chromite	Diorite	Coal	Eclogite
Garnet	Fluorite	Dolerite	Conglomerate	Gneiss
Hornblende	Galena	Gabbro	Dolomite	Granulite
Kyanite	Graphite	Granite	Flagstone	Hornfels
Labradorite	Gypsum	Keratophyre	Flint	Kyanite
Lepidolite mica	Haematite	Lamprophyre	Greywacke	Marble
Muscovite mica	Halite	Microgranite	Ironstone	Mylonite
Olivine	Magnetite	Obsidian	Limestone	Phyllite
Feldspar	Malachite	Peridotite	Mudstone	Quartzite
Quartz	Copper	Pitchstone	Sandstone	Schist
Sodalite	Pyrite	Rhyolite	Siltstone	Serpentine
Talc	Sphalerite	Volcanic tuff	Shale	Slate

majority of minerals are compounds comprising two or more elements: for example, NaCl, which comprises sodium and chlorine to form sodium chloride or halite. A small number of minerals contain just one element (e.g. sulphur, copper and gold).

Minerals can be grouped into those that contain silica (silicon dioxide) and those that do not (Table 17.1). It is the silicate minerals that tend to be the rock-forming minerals. The non-silicate minerals tend not to form rocks but are important in their own right as ore minerals.

17.3 Engineering soils

Engineering soils occur in two distinct types:

1. Granular soils:
    *   boulders, cobbles and gravels are angular to rounded rock fragments;
    *   sands are granular in nature;
    *   silts are similar to sands but have smaller grains with some plasticity and cohesion.

The above particles are formed by the mechanical weathering process of rocks. Their shape tends to be bulky and 'equi-dimensional' (Figure 17.2a). The structure arrangement of each individual particle is supported by points of contact with adjacent particles forming frictional resistance (Figure 17.2b).

2   Cohesive soils:
    *   clays are formed from rock weathering, mainly due to chemical action.

Clay particles are plate-like in shape, with varying degrees of plasticity and cohesion (but possess no frictional resistance). A small amount of clay will influence the soil behaviour. The structure arrangement tends to be, either:

*   *flocculated* from inter-particle attraction (Van der Waal's or secondary bonding) – these forces pull particles closer together (Figure 17.3a);

Figure 17.2   Granular soils

(a)

(b)

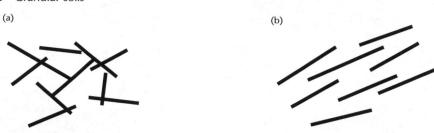

Figure 17.3   Cohesive soils

- *dispersed* from repulsive electrically negative forces from the particle surface, so particles are held apart (Figure. 17.3b).

In reality a soil may contain a mixture of particles. This can be represented by the soil triangle (Figure 17.4).

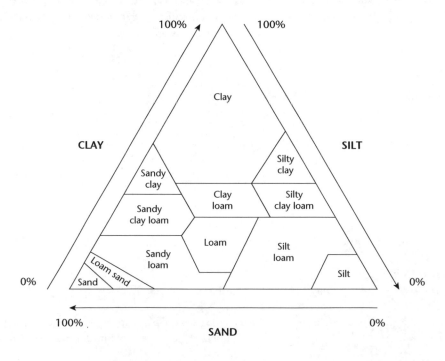

Figure 17.4   The soil triangle

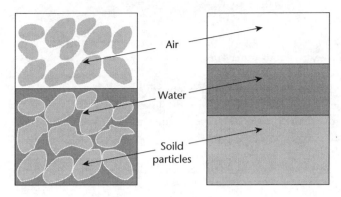

Figure 17.5 Soil structure

17.4 Soil structure

Soil is a collection of solid individual particles forming a porous structure. The pores or voids will contain air and/or water, depending on the saturation state of the soil. The individual particles can move relative to each other. However, in a rock mass the grains cannot move. Soil is therefore regarded as a 'particulate system' (Figure 17.5).

Within the soil structure air is unimportant from an engineering point of view; however, air voids should be reduced as far as possible by the process of compaction. Water is, however, very important for the engineering properties and behaviour – particularly for cohesive soils. The solid matter (i.e. the soil skeleton) varies widely in particle shape, size and mineral composition and can contain both organic and inorganic matter.

17.5 Soil classification

Soil is classified by determining a number of fundamental parameters, which are defined in BS1377: 1990, as:

Moisture content	Part 2: 3.2
Consistency limits (Atterberg limits)	
• Liquid limit	Part 2: 4.3 and 4.4
• Plastic limit	Part 2: 5.3
• Plasticity index	Part 2: 5.4
• Shrinkage limit	Part 2: 6.3
Density	
• Linear measurement	Part 2: 7.2
• Immersion in water	Part 2: 7.3
• Water displacement	Part 9: 7.4
• Sand replacement	Part 9: 2.1 and 2.2
• Core cutter	Part 9: 2.4
Particle density	
• Gas jar	Part 2: 8.2
• Pyknometer	Part 2: 8.3 and 8.4
Particle size distribution	
• Dry sieve	Part 2: 9.2
• Wet sieve	Part 9.3

17.5.1 Moisture content

The amount of water in a soil has a profound effect on its engineering properties. For example, under compaction dry density varies with water content (Figure 17.6):

- A soil compacted dry will reach a certain dry density.
- If compacted again with the same compacted effort but with water, a higher dry density will be reached. The water helps to lubricate the soil – the particles move closer on compaction. Also, air in the mixture is removed during the process.
- However, too much water reduces the density.
- Optimum moisture content (OMC) occurs when the compacted dry density is at its maximum $(\rho_{d\,(max)})$.

A point to note is that OMC varies with the nature of the soil and compaction method. Heavier compaction plant produces a higher value of dry density at a lower OMC.

To simply determine the moisture content of a soil, the following procedure can be followed:

1. Record the mass of a dry container.
2. Add the wet soil to the container and record the new mass (container + wet soil).
3. Place the container including the wet soil in an oven at a temperature of around 100 °C for about 24 hours.
4. After the duration, record the mass of the container + dry soil.
5. The moisture content is then calculated from the following formula, and is expressed as a percentage:

$$Moisture\ content\ (w) = \frac{mass\ of\ water}{mass\ of\ dry\ solid}\%$$

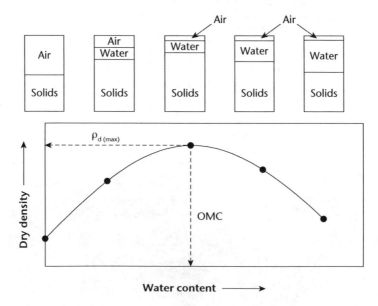

Figure 17.6   Illustration to show how water content influences dry density.

Example 1: moisture content of soil

Calculate the moisture content for the following:

Test	w_1	w_2	w_3
A	16.5	26.6	24.3

where:

w_1 = mass of container (g)
w_2 = mass of container and wet soil (g)
w_3 = mass of container and dry soil (g)

The moisture content equation can be written in the form:

$$w = \frac{w_2 - w_3}{w_3 - w_1} \times 100$$

The recorded values can be substituted in to give:

$$w = \frac{(26.6 - 24.3)}{(24.3 - 16.5)} \times 100 = 29.5\%$$

17.5.2 Consistency limits (Atterberg limits)

These stages describe the consistency of the soil, which in turn relate to its engineering properties:

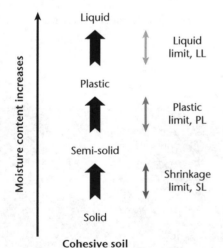

Liquid

Liquid limit, LL

LL is defined as the moisture content at which the soil is assumed to flow under its own weight – measured using a cone penetrometer or Cassegrande cup (Fig. 17.7).

Plastic

Plastic limit, PL

PL is defined as the moisture content at which a 3 mm diameter thread of soil can be rolled by hand without breaking up (Fig. 17.8).

Semi-solid

Shrinkage limit, SL

SL is defined as the moisture content at which no further reduction in volume will occur. It is measured by slowly drying a sample out and periodically measuring its volume and mass.

Solid

Moisture content increases

Cohesive soil

The range of moisture content over which the soil remains in a plastic condition is defined as the plasticity index (PI) and is calculated from: PI = LL – PL.

For fine-graded soils the 'A–line' classification chart can be used (Figure 17.9). The classification chart divides the soil according to ranges of liquid limit and plasticity index. Generally, the higher the liquid limit, the higher the plasticity of the soil. Silts and organic soils tend to have a low plasticity index (i.e. a small range of moisture content over which they are plastic) compared to their liquid limit. Clays tend to have a high plasticity index in relation to their liquid limit.

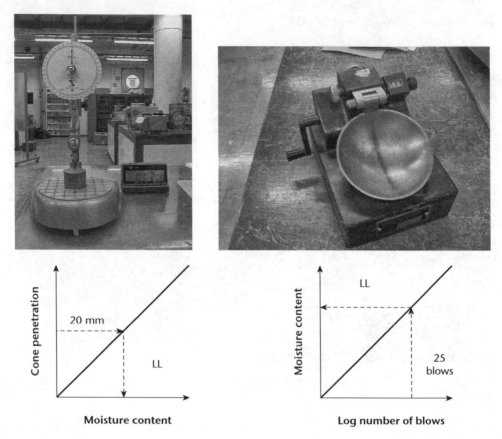

Figure 17.7   Liquid limit tests

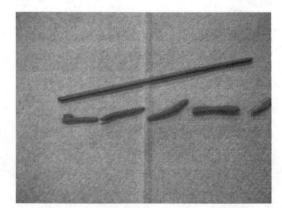

Figure 17.8   Plastic limit test

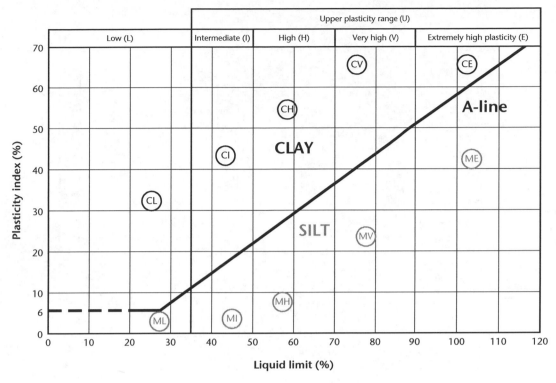

Figure 17.9 A-line classification chart

To classify a fine soil according to the 'A–line' chart:

1. Determine liquid and plastic limits; from these calculate the plasticity index.
2. Plot the liquid limit and plasticity index values on the chart.
3. Observe which segment the soil falls into (CL, CI, ML, etc).
4. Write down the soil name, CLAY for 'C' soils, SILT for 'M' soils.
5. Add the plasticity grading to the name, i.e. CI = CLAY of intermediate plasticity.

The classification of fine soils (e.g. silts and clays) is based on plasticity, which is a function of the soil's capacity to absorb water. If water is added to a coarse soil (e.g. sands and gravels), the water will fill the pores and saturate the sample. Any further water will simply drain off and not be absorbed further into the soil structure. Figure 17.10 demonstrates what will happen if water is added to a fine soil; to summarise:

1. Water will initially fill the pores.
2. When the soil is saturated it will continue to absorb water due to the properties of the clay minerals and an increase in volume of the soil mass will occur.
3. At the same time the soil is progressively softened by the water, which results in a decrease cohesion. That is, the water increases the distance between the clay minerals, hence decreases the attracting forces between them.
4. As the soil gets weaker it becomes pliable. When it is sufficiently pliable to be rolled out into a thread it is in a plastic state.
5. Eventually the soil loses all its strength. When it starts to flow under its own weight it is in a liquid state.

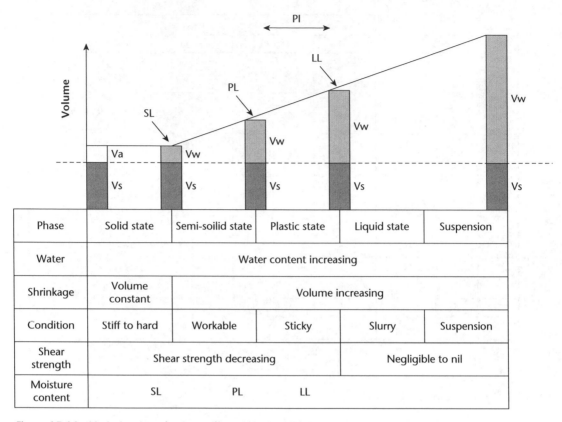

Phase	Solid state	Semi-soilid state	Plastic state	Liquid state	Suspension
Water	Water content increasing				
Shrinkage	Volume constant	Volume increasing			
Condition	Stiff to hard	Workable	Sticky	Slurry	Suspension
Shear strength	Shear strength decreasing			Negligible to nil	
Moisture content	SL	PL	LL		

Figure 17.10 Variation in cohesive soil's properties with increasing moisture content

17.5.3 Density

Density is the ratio in a mass of a material compared to its volume:

$$Density = \frac{Mass}{Volume} \text{ which can be written as: } \rho = \frac{M}{V}; \text{ the units are: } \frac{Mg}{m^3}$$

Bulk density represents the *in-situ* density of the soil at a current moment in time, e.g. the soil might also contain some water and air in the pores:

$$Bulk \ density \ (\rho_b) = \frac{Total \ mass}{Total \ volume}$$

Dry density is when water has been removed from the soil:

$$Dry \ density \ (\rho_D) = \frac{Mass \ of \ solids}{Total \ volume} \text{ or}$$

where w = the moisture content, and is expressed as a decimal in this equation:

$$\rho_d = \frac{\rho}{1+w}$$

Saturated density is when the weight of the water is included in the weight of the soil:

$$Saturated\ density\ (\rho_{sat}) = \frac{Saturated\ mass}{Total\ volume}$$

When a volume of soil is submerged (e.g. below the water table), it displaces an equal volume of water, so that its effective or submerged density is:

$$Submerged\ density\ (\rho') = \rho_{sat} - \rho_w$$

where ρ_w = density of water

Note:

- mass is when gravity is not included;
- weight includes gravity.

Therefore, a soil's unit weight = density × gravity, where gravity is equal to 9.81 m s^{-2}. For example:

- Bulk density (ρ_b) Mg m^{-3} × 9.81 m s^{-2} = Bulk unit weight (γ_b) kN m^{-3}.
- Dry density (ρ_d) Mg m^{-3} × 9.81 m s^{-2} = Dry unit weight (γ_d) kN m^{-3}.
- Saturated density (ρ_{sat}) Mg m^{-3} × 9.81 m s^{-2} = Saturated unit weight (γ_{sat}) kN m^3.
- Submerged unit weight (γ') kN m^{-3} = $\gamma_{sat} - \gamma_w$.
- If the density of fresh water (1 Mg m^{-3}) is multiplied by gravity (9.81 m s^{-2}) the unit weight of water (γ_w) is obtained, i.e. 9.81 kN m^{-3}. Note the density of seawater is 1.025 Mg m^{-3} or 1,025 kg m^{-3}.

Tests to determine the density of cohesive soils include:

1. Linear measurement

This involves extracting an undisturbed cylindrical sample of soil, measuring its mass and volume (Figure 17.11) and then comparing the ratio of the two to compute the density.

2. Immersion or displacement in water

This is based on Archimedes' principle, which states that a body immersed in water loses weight equal to the weight of the fluid it displaces. The procedure involves extracting an irregular cohesive soil sample and determining its mass. The sample is then coated in wax (to prevent water ingress) by dipping it into the wax pot. The mass of the coated sample is also recorded. In the immersion method, the sample's mass is recorded when suspended in water. In the displacement method the mass, or volume, of water displaced by the sample is recorded (Figure 17.12). In the calculation stage the density of the wax is eliminated from the sample.

17.5.4 Particle density

The particle density (ρ_s) of a material is the ratio of the mass of a soil to the mass of an equal volume of water.

For coarse-grained soils, a 500–1,000 ml density bottle (gas jar) may be used, but for fine-grained soils a special conical topped glass jar, called a pyknometer, should be used (Figure 17.13).

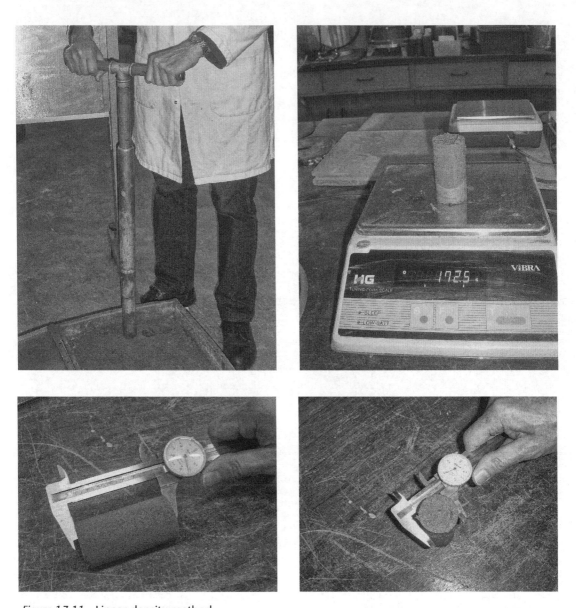

Figure 17.11 Linear density method

The procedure to determine particle density is simply illustrated in Figure 17.14. It should be noted that the range of particle density of common soil particles is very narrow, typically 2.60–2.70 Mg m^{-3}.

17.5.5 Particle size distribution (PSD)

The classification of coarse–graded granular soils involves passing a mass of soil through a series of interconnecting sieves with decreasing aperture size (Figure 17.15a). The mass retained on each sieve is then recorded, from which the percentage of material passing a given sieve can be calculated. The results

Figure 17.12   Immersion/displacement in water – density methods

are then plotted on a graph to determine the shape of the grading curve. For a given soil mass, the shape of the curve can be defined as (Figure 17.15b):

1. *Uniformly graded*, indicating that the particles are all the same size.
2. *Well graded*, indicating that there is a large range of particle sizes.
3. *Gap graded*, indicating that there are only small and large particles.

17.6 Soil's strength

The strength of soil is defined in terms of its shear strength (Figure 17.16). This is the maximum shear stress that can be applied to the soil in any direction. When this maximum has been reached the soil yields and is regarded to have failed.

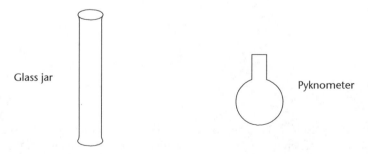

Figure 17.13   Glass jars used in particle density tests

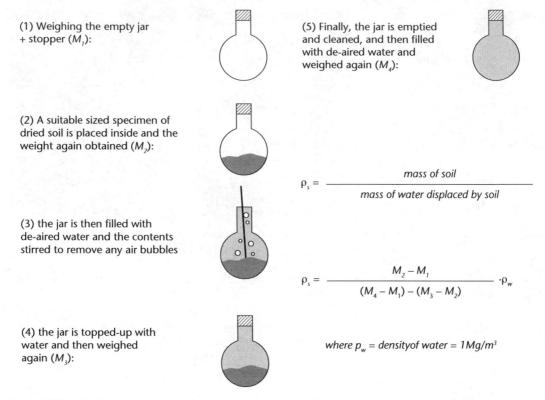

(1) Weighing the empty jar + stopper (M_1):

(2) A suitable sized specimen of dried soil is placed inside and the weight again obtained (M_2):

(3) the jar is then filled with de-aired water and the contents stirred to remove any air bubbles

(4) the jar is topped-up with water and then weighed again (M_3):

(5) Finally, the jar is emptied and cleaned, and then filled with de-aired water and weighed again (M_4):

$$\rho_s = \frac{\textit{mass of soil}}{\textit{mass of water displaced by soil}}$$

$$\rho_s = \frac{M_2 - M_1}{(M_4 - M_1) - (M_3 - M_2)} \cdot \rho_w$$

where ρ_w = density of water = $1\,Mg/m^3$

Figure 17.14   Outline of the procedure to determine particle density

There are two components of shear strength:

1.   Friction (ϕ)
2.   Cohesion (c)

Note: pore water has no shear strength.

The values of c and ϕ are known as the shear strength parameters. In 1773 Coulomb developed an expression for these parameters, which defines a straight-line failure envelope (Figure 17.17):

$$\tau_f = c + \sigma_n \tan \phi$$

where:

c = apparent cohesion
ϕ = angle of shearing resistance
σ_n = normal stress on failure plane

In certain soils both cohesion and friction contribute to shear strength; however, in other soil types only one of the shear strength parameters may be present. The failure envelopes in Figures 17.18–17.20 illustrate a range of soil types

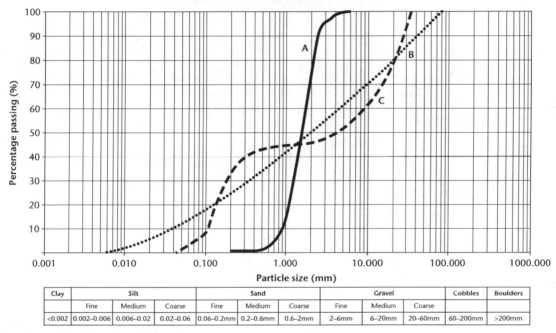

Clay	Silt			Sand			Gravel			Cobbles	Boulders
	Fine	Medium	Coarse	Fine	Medium	Coarse	Fine	Medium	Coarse		
<0.002	0.002–0.006	0.006–0.02	0.02–0.06	0.06–0.2mm	0.2–0.6mm	0.6–2mm	2–6mm	6–20mm	20–60mm	60–200mm	>200mm

Figure 17.15 Particle size distribution

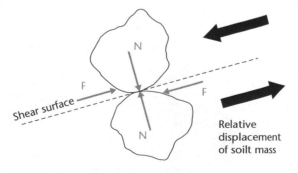

Figure 17.16 Shear strength of soil

17.7 Settlement and consolidation of soil

When a soil is subjected to an applied load, such as from a newly constructed building, a degree of compression will probably happen. There are four main types of settlement that can occur:

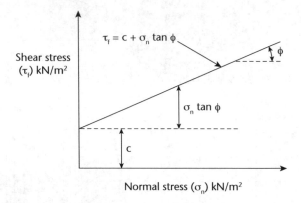

Figure 17.17   Coulomb general failure envelope

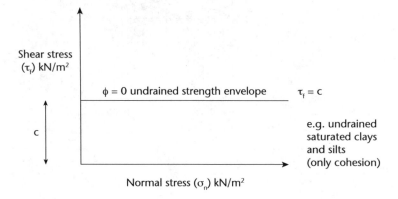

Figure 17.18   Failure envelope for a frictionless soil

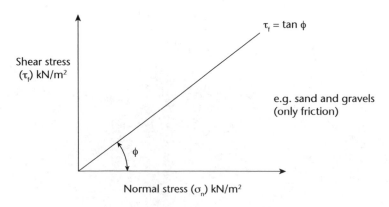

Figure 17.19   Failure envelope for a cohesionless soil

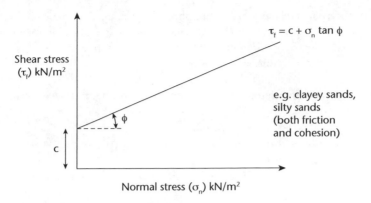

Figure 17.20 Failure envelope for a partially saturated mixture of cohesive and frictional soil

1. *Immediate settlement* occurs instantaneously upon loading and is recoverable (elastic deformation), i.e. it will expand to its original volume if the load is removed. This type of settlement is usually small and occurs during construction.
2. *Consolidation* or *primary settlement* results in a decrease in the volume of voids. This type of settlement is not normally recoverable on the removal of the load. In a saturated soil the decrease in voids also results in the dissipation of excess pore water. The rate at which this type of settlement occurs depends on the permeability of the soil. High-permeability soils (e.g. sands and gravels) settle quickly; thus increase in effective stress occurs over a short time period. Low-permeability soils (e.g. silts and clays) require a longer time to settle or consolidate (years); thus the increase in effective stress is slow.
3. *Plastic deformation* or *secondary consolidation* is due to lateral flow of soil particles and is not recoverable on the removal of the load.
4. *Creep* occurs in organic soil (e.g. peat) due to the collapse of fibrous matter and it is not recoverable on the removal of the load.

Of these four types of compression, the greatest contribution to the settlement of a structure is generally consolidation or primary settlement.

17.8 Site investigation

To establish the suitability of a location for the proposed work, a site investigation is normally undertaken. This typically involves three stages:

1. *Desk study*: a geotechnical engineer will search through relevant maps and documents in order to discover as much about the site and its surroundings as possible. For example, the following information will be studied: geological solid and draft maps, hydro-geological maps, ordnance survey maps (including historical versions), aerial photographs, past activities within the area such as mining and waste tips, public utilities, etc.
2. *Site reconnaissance*: a geotechnical engineer visits the proposed site and undertakes a visual inspection. For example, tree cover will be assessed as this can indicate soil type and saturation levels, e.g. willows trees require very wet ground.
3. *Ground investigation*: the only true way to ascertain the underlying soil conditions of the site. During the investigation accurate information relating to water should be obtained. Suitable soil samples for

both visual classification and laboratory testing should be extracted. These could be in a disturbed or undisturbed state, the latter representing the 'exact' *in situ* state of the soil; for example, in terms of moisture content and degree of compaction. The soil samples are normally extracted from the ground by drilling some form of borehole or excavating a trial pit.

From the information gained from the site investigation, the following, *inter alia*, should be established:

- adequate and economic design to be prepared;
- the best method of construction;
- suitable use (or disposal) of *in-situ* materials;
- the changes that may arise in the ground and environmental conditions as a result of work.

17.9 Critical thinking and concept review

0. List the four types of engineering soil.
1. Define the three types of rocks and cite examples from each type.
2. What is the difference between granular and cohesive soils?
3. Why is soil regarded as a particulate system?
4. List the main tests to determine the fundamental parameters of a soil.
5. What influence does moisture have on a soil?
6. What is the difference between density and unit weight?
7. List three tests to determine the density of a soil.
8. What is the difference between the liquid limit and plastic limit?
9. Name the three type of grading curves for a soil.
10. What are the two components of shear strength?
11. Define Coulomb's equation.
12. Sketch the failure envelope for (1) a dry sand, (2) a pure clay, and (3) a sand/clay mix.
13. What is the difference between immediate and consolidation settlement?
14. Define the three main stages of a site investigation.
15. Why should a site investigation be undertaken?

18

Composite materials

This chapter deals with composite materials, an area of growing importance. The chapter opens with considerations of material selection, illustrating the fact that single natural and man–made materials are not always capable of meeting our design requirements in some situation, and therefore opening the way for composites to be used. Composites made for reasons of strength and thermal insulation will be considered here, and the various types and geometries of composites are set out. After looking at the composites used in construction, the theory of composites made for strength purposes is set out. This is limited to the use of unidirectional, continuous fibres. However, this theory is very useful in many situations.

The theory of composites made for reasons of thermal insulation is then set out, and its application to building wall structures is outlined.

Contents

18.1. Introduction

We have seen in Chapter 1 that man has perhaps 40,000–80,000 different materials at his disposal (Ashby, 1992). This number includes all the pure metals and alloys, synthetic polymers, bituminous materials, glasses, ceramics, fired clay products, etc. as well as naturally occurring materials such as timber, leather, bone, natural stone, etc. It also includes composite materials such as glass-reinforced plastic (GRP), Kevlar, etc. This raises a number of questions:

- In any given design situation, how can we make rational choices when faced with such an overwhelmingly large number of materials?
- What are composites and why are they made?
- How do composites work?

Faced with such a large number of available materials, at first sight it may seem strange that we resort to the manufacture and use of composites, but as we shall see, in meeting the increasingly stringent demands posed by some design situations, none of the huge number of single materials always has the particular combination of properties needed to meet our requirements. We shall therefore pick up the theme of materials selection first broached in Chapter 1, where it was pointed out that the selection of materials was really the selection of the properties that we require.

18.2 Specifying materials

We recall from Chapter 1 that when we specify a material, we are really specifying a required property or combination of properties. We have seen the Ashby diagram shown in Figure 18.1, which shows values of Young's modulus of elasticity (E) plotted against density (ρ) for all classes of single materials (metals, organics – natural and synthetic – and ceramics – natural and man-made). We can see what Ashby discovered, namely that if we plot values of Young's modulus against density, then the various classes of materials group themselves neatly into groups when plotted logarithmically. These groupings are shown as the shaded areas in the diagram, with the contents of each area being marked (engineering ceramics, engineering alloys, etc.).

The data in Figure 18.1 are plotted on a logarithmic scale because the stiffness values span at least four orders of magnitude while the densities span two. Using a linear scale would make the diagram much less clear. Perusal of Figure 18.1 shows that all the shaded areas run approximately from the lower left-hand corner to the upper right-hand corner. The top left and lower right corners are largely clear, with no materials marked. Why is this?

The blank areas indicate that there are no materials with low or very low densities that also have high or very high values of stiffness. Similarly, no dense or very dense materials posses low stiffness values. The latter situation does not concern us because high density (heaviness) and low stiffness is not a desirable or useful combination of properties. However, materials that combine high stiffness (high Young's modulus) with low density (lightness) are very useful in transport systems, and especially in the aircraft and automotive industries, where weight-saving is of paramount importance.

The stability and control of an aircraft in flight depends upon its aerofoil surfaces being presented to the air-stream at the correct angle. A large aircraft can weigh up to 500 tonnes, and the resultant forces on the airframe are very large. If the airframe suffers excessive elastic distortion because of these forces, then the aerofoil surfaces may be presented to the air-stream at the wrong angles, and lift and control may be lost. The airframe needs to be light in weight but it must also possess as high a stiffness, E, as possible. Similarly, with a road vehicle weighing one tonne and travelling at speed, the wheels – and hence the tyres – must have the correct angle to the road surface to maintain adhesion. The attitude of the wheels to the road depends upon the suspension elements, which are attached to the frame or body shell of the vehicle.

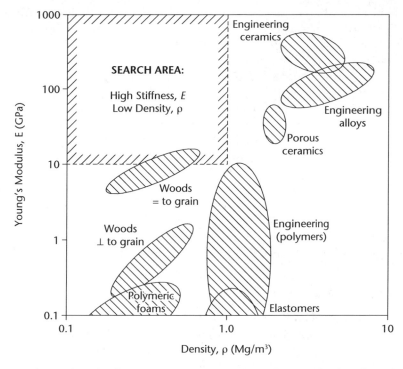

Figure 18.1 Logarithmic plot of stiffness *E*, vs density *ρ*, on logarithmic scales for all single materials (After Ashby, 1992)

If this frame or shell deflects elastically to an excessive extent, then the wheels will assume unsatisfactory angles to the road surface, and adhesion and control may be lost. The frame or body shell of the vehicle needs to be light in weight but must also be as stiff as possible.

In construction, GRP composites have been used in the restoration and refurbishment of listed buildings, and buildings of great architectural merit. Turrets, finials and other decorative features, especially those high on the building and visible only from ground level, have been replaced with GRP items made from architects' drawings. They may be completely fabricated in a workshop, and then lifted into place and securely attached to the structure. One example was the replacement of a finial to a building in Regent Street in London. The item was made and delivered on a Sunday morning. It was lifted into position using a helicopter and installed inside one working day. The saving in cost and time was enormous.

We can see from Figure 18.1 that engineering alloys (metals) and ceramics have high values of stiffness (*E*), but they are also dense (heavy). So despite the fact that we have a huge number of materials to choose from, we sometimes cannot obtain the particular combination of properties that we require. Going back to the examples mentioned above, we often require a material which is both light in weight and very stiff (i.e. has a very high value of Young's modulus *E*). Such materials are usually quite dense. So, if we need a light and stiff material, the solution is to take a light material, such as a polymer, and reinforce it with fibres of a very stiff and strong material, such as glass or carbon fibre. If we make the composite in the correct way, we can obtain a material with a stiffness approaching that of steel, with a fraction of steel's density. So glass or carbon fibre reinforced plastics are examples of composites. *Composites are made because there is no single existing material that will give us the combination of properties that we require.* To restate the definition simply, one material is combined with another in such a way that the resulting composite possesses the required combination of properties.

We recall that in Figure 18.1 we plotted data for single materials. If we now add data for composite materials (as in Figure 18.2), these engineering composites occupy that region of material property space for low-density materials (ρ from 1.0 to 1.2 Mg/m³), and high stiffness (E from 10 up to 200 GPa). No other group of materials occupies this region, offering high stiffness with low density. This illustrates in a graphic way why composites are made; without composites, no other group of materials would offer this combination of properties.

Finally, as with many other of man's inventions, we find that nature developed the idea first, and this is true of composite materials too. The best-known example is timber, in which the cellulose vessel cells act as reinforcement to a lignin matrix. The cellulose has much higher strength and stiffness than the matrix material, and gives timber its characteristic properties, and we shall return to this idea later in this chapter. However, it can be seen from Figure 18.2 that, while timber has a lower density than the engineering composites, it also has lower stiffness.

Another natural example is animal bone. Bones have an outer layer of compact bone; this is dense and is on the outside, i.e. it forms the extreme fibres for the bone, where the applied stresses are highest, and the interior is of cancellous bone – a lightweight but strong honeycombed structure which saves weight. So bones represent a material with a high strength:weight ratio

18.3 Types of composite materials

The example given above of a composite material combining high stiffness and low density is a composite made for reasons of mechanical properties. Many composites are made in order to achieve a particular and

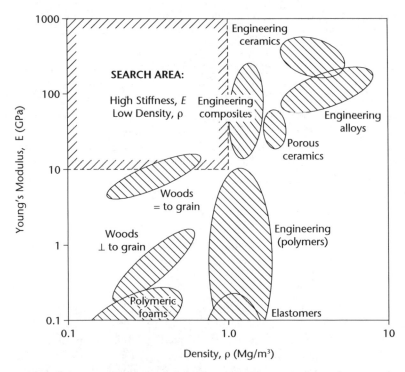

Figure 18.2 Logarithmic plot of stiffness E, vs density ρ, using logarithmic scales for all materials, including engineering composites

(After Ashby, 1992)

282

desirable set of mechanical or structural properties; this is, however, not the only type of composite. We can make composites for optical, thermal and electrical reasons, too. The manufacturers of cameras and optical instruments make composite lenses from two or more types of glass to achieve lenses free from chromatic dispersion (an achromatic doublet), for example. Electrical cable is a composite (sheathing and conductors) made for reasons of insulating electric currents from the structures through which the cables pass or to which the cables are fastened. Similarly, in construction we use composite blocks to obtain a mixture of compressive strength and thermal insulation.

Some purists would argue that metal alloys could be seen as composite materials, but such an argument assumes a familiarity with phase equilibria and the theory of solute hardening, etc. beyond the needs of civil engineers and construction professionals. This book will focus only on the aspects of this subject that are relevant to construction. In practice, this means confining ourselves to composites made for reasons of:

- strength, and
- thermal insulation.

So from the point of view of construction professionals, we can classify composite materials as shown in Table 18.1.

The table gives examples of naturally occurring composites, and timber has, of course, been considered in this text. Timber is an example where fibrous reinforcement is used. Most of the vessel cells run in the longitudinal direction. Some proportion (less than 10 per cent) run in the radial direction. These fibres give the timber its well-known anisotropy; the properties along the grain are very different from those across the grain. This is something that we are all familiar with. We shall examine the reasons and the theory behind this later in this chapter.

In our discussions so far we have looked mainly at composites made for reasons of strength, and examples where the second material is present as fibres aligned mainly in one direction. There are many other arrangements; the second material can be in the form of short fibres randomly arranged, in the form of particulates (spheres, plates, ellipsoids, etc.) in the matrix, as lamellar structures (alternate layers of the materials) or as skeletal or interpenetrating networks. To give a well-known example, plain concrete is itself a composite. Particles of aggregate reinforce a cement matrix. Reinforced concrete is a composite, with steel reinforcement bars aligned in the direction in which the tensile loading is to be carried. Plywood and glulam beams are examples of lamellar composites used in construction. Examples of different composite geometries will be discussed next.

Table 18.1 Broad classification of composite materials

Composite type	Examples
Natural composite materials	Wood
	Compact bone and cancellous bone
	Bamboo
	Muscle and other tissue
Macrocomposites (engineering products)	Reinforced concrete beams
	Galvanised steel
	Helicopter rotor blades
	Wire-wound pressure vessels and gun barrels
	Insulated building blocks.

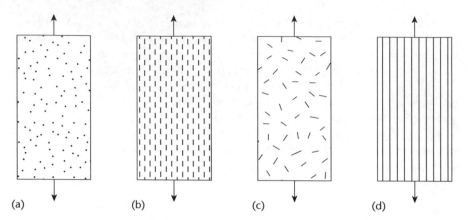

Figure 18.3 Examples of composites. (a) Particulate, random; (b) discontinuous fibres, unidirectional; (c) discontinuous fibres, random; (d) continuous fibres, unidirectional

(After Matthews & Rawlings, 1994)

18.3.1 Possible geometries of composite materials

There are many possible ways of combining the materials in a composite. The reinforcement can be particulate, fibrous or planar. Figure 18.3 illustrates some of the common geometries used.

Concrete is an example of Figure 18.3(a), a random particulate composite, with particles of hard and high stiffness aggregate reinforcing a matrix of hydrated cement. The effectiveness of this reinforcement is illustrated in Figure 18.4, below.

18.3.2 Composites in construction

The use of composites in construction goes back thousands of years. In ancient Egypt, we read in the Bible that the Children of Israel made mud bricks reinforced with straw for their Egyptian masters. This must be one of the first references to man's use of composite materials. To take the example quoted above of a material combining high stiffness with light weight, the aircraft industry and the designers of high-performance Grand Prix racing cars would clearly be interested in such materials. Saving weight is never as critical in building construction as it is in aircraft construction. The materials used in construction are, almost invariably, much lower-value materials than those used in aerospace. However, the need for low–price construction materials throws up requirements that sometimes cannot be met by single materials, and when this happens recourse has to be made to composites.

Steel-reinforced concrete is often quoted as the best-known composite material in use. What is not so often realised is that plain concrete is also a composite. It consists of hydrated cement binding together a mass of aggregates of various sizes, usually ranging from sand to much larger particles 5 mm in diameter up to 20 mm or more. In the construction of massive structures such as dams, large aggregate lumps up to 40 mm in diameter are usually used. In the new Three Gorges Dam in China even larger particle sizes may have been used. The aggregates possess a higher value of Young's modulus than that of cement, and so the resulting concrete when set is stiffer than neat cement would be. Figure 18.4 illustrates this point.

Figure 18.4 is interesting for two reasons. First, it clearly shows that concrete has a higher value of Young's modulus, E, than neat cement (the slope of the initial, linear portion of the graph is steeper than that of neat cement). Second, the concrete appears to show what, in a metal, would be interpreted as ductility, i.e. the graph becomes non–linear and reaches a much higher strain than the aggregate does

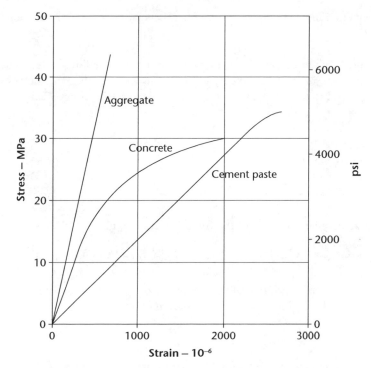

Figure 18.4 Stress–strain relationships for cement paste, aggregate and concrete

(After Neville, 1981)

before it finally fails. This non-linear behaviour is not ductility, but is due to the development of micro-cracks as the aggregate particles begin to de-bond from the cement matrix. The cracks are tiny at first, but as straining proceeds they multiply in number and grow in size until the concrete finally fails. If aggregate or set cement paste are strained singly, they behave elastically up to the point at which catastrophic failure occurs. The composite does not. This gives us a pointer to one of the advantageous properties of composites, i.e. they tend to give ample warning of failure before they actually do so. We shall return to this topic later.

18.4 Theory of composites: 1 – strength properties

This is a large subject with a considerable literature, and a large body of theoretical analysis; no attempt will be made to give a comprehensive treatment here as it would be quite inappropriate. However, the simple theory of fibre reinforcement will be briefly set out and used for some simple examples. In particular, we shall set out the theory for solids *reinforced with uni-directional, continuous fibres*. This theory is useful, as it is an approximate fit for situations where we may have shorter fibres more or less uni-directionally aligned, and it also models timber reasonably well too, as we shall see.

There are a number of requirements that must be satisfied if we are to successfully reinforce one material with another. These are:

* The reinforcement must be stiffer than the matrix material. The reinforcement must have a significantly higher value of Young's modulus E than the matrix.
* The reinforcement must be stronger than the matrix material. The reinforcement needs a significantly higher tensile strength than the matrix material.

- The reinforcement must be properly bonded to the matrix. If there is no bond, we obtain no reinforcement effect.

It is not seen as appropriate to derive the equations for composite properties here; we shall only explore how the simple theory can be used. This theory enables us to calculate the value of Young's modulus for our composite, E_{comp}, and to tailor the resultant stiffness exactly to our requirements, if we need to do so. For this calculation we shall need to know:

- matrix stiffness, E_m,
- fibre or reinforcement stiffness, E_f,
- volume fraction of matrix, V_m, and
- volume fraction of reinforcement, V_f.

In a simple fibre–matrix system, if we know the volume fraction of fibres V_f, the volume fraction of matrix V_m will simply be given by: $1 - V_f$.

The theory enables us to calculate the composite stiffness parallel to the fibre direction, and transverse (i.e. 90°) to the fibre direction.

Parallel to the fibre direction. The stiffness of the composite E_{comp} with a matrix reinforced by continuous, uni-directional fibres, in the direction of the fibres, is given by:

$$E_{comp} = E_f \cdot V_f + E_m \cdot V_m$$

This equation is referred to in some textbooks as 'the rule of mixtures', and in others as the Voigt equation.

Worked example

Calculate the value of Young's modulus E in the fibre direction for a GRP composite where the fibres are uni-directionally aligned, given that the volume fraction of fibres is 40 per cent. The values of Young's modulus for glass fibre and plastic matrix are 80 GPa and 0.2 GPa, respectively.

So, matrix stiffness	= 0.2 GPa
fibre stiffness	= 80 GPa
volume fraction of matrix	= 0.6
volume fraction of fibre	= 0.4

Use the Voigt equation:

$$\begin{aligned} E_{comp} &= E_f \cdot V_f + E_m \cdot V_m \\ &= 80 \times 0.4 + 0.2 \times 0.6 \\ &= 32 + 0.012 \\ &= 32.012 \text{ GPa} \end{aligned}$$

Notice the effects of adding the fibre reinforcement. The plastic matrix initially had a stiffness value of 0.2 GPa. The composite now has a stiffness value of 32.012 GPa, a value 160 times greater. So for a small increase in weight (glass is more dense than plastic), the Young's modulus has increased to a value over 160 times higher. The density of the composite will be about twice that of the unreinforced matrix, but the stiffness gain is very much greater. This represents a stiffness to weight gain of about 80 times.

The end result is that the composite has a stiffness much greater than any other single available material of similar density. This is one of the main reasons that composites are made. We shall now examine the stiffness in the transverse direction to the direction of fibre alignment.

Transverse to the fibre direction. The stiffness of the composite E_{comp} in the transverse direction to the fibres is given by:

$$E_{comp} = \frac{E_m \cdot E_f}{E_f \cdot V_m + E_m \cdot V_f}$$

In some older textbooks this is sometimes called the Reuss equation.

Worked example

Calculate the value of Young's modulus stiffness for the GRP given above for the transverse direction to fibre alignment. Take the same values for stiffness and volume fraction in each case.

So, matrix stiffness	= 0.2 GPa
fibre stiffness	= 80 GPa
volume fraction of matrix	= 0.6
volume fraction of fibre	= 0.4

Using the Reuss equation:

$$E_{comp} = \frac{E_m \cdot E_f}{E_f \cdot V_m + E_m \cdot V_f}$$

$$= \frac{0.2 \times 80}{80 \times 0.6 + 0.2 \times 0.4}$$

$$= \frac{16}{48 + 0.08} = \frac{16}{48.08} = 0.333 \text{ GPa}$$

In this case we are looking at the effect of the reinforcement in the transverse direction. We would not therefore expect to see the kind of dramatic increase in stiffness seen in the fibre direction. In fact, the stiffness of the composite in this direction is 0.333 GPa compared with a value of 0.2 GPa for the unreinforced plastic. This is an increase of 1.66 times, a very modest increase.

18.4.1 Limitations on fibre volume fractions

In the preceding section V_f and V_m were used in the equations to denote the volume fractions of fibre and matrix, respectively. Since we have only fibre and matrix we can write:

$$V_m = 1 - V_f$$

In theory, V_f can have any value between very low values just in excess of zero (if V_f was zero, it would not be a composite) and a high value close to 1 (if V_f was 1, again, it would not be a composite). However, in practice there is an upper limit on V_f. If we assume that fibres are of circular section, and that we are considering continuous, uniformly aligned fibres, then simple geometrical considerations dictate that the maximum value of V_f will around 0.7. If the fibres are discontinuous, the value will be lower, and if the fibres are discontinuous and randomly oriented, then V_f will be lower still, with a maximum not much above 0.3. Figure 18.5 illustrates this point. If four circular section fibres of diameter D are arrayed as shown, they occupy about three-quarters (78.5 per cent) of the space of a square of side $2D$. The ratio will be the same in terms of volume.

$$\text{Area of circles} = \frac{4 \times \pi \cdot D^2}{4}$$

$$\text{Area of squares} = 4 \times D^2$$

$$\text{Ratio:} \frac{\text{Area of circles}}{\text{Area of squares}} = \frac{4 \cdot \pi \cdot D^2}{4 \cdot 4 \cdot D^2} = \frac{\pi}{4} = 0.785$$

Therefore the circles occupy 78.5 per cent of the area of the squares.

The case shown in Figure 18.5 is the extreme case, where the fibres are in contact. Ideally we want the matrix to be bonded to the fibres all around their outer surface, to obtain the maximum stiffening effect. In practical terms, therefore, the fibres would not be in contact, there would be a space between them. In such cases the volume fraction would fall from 78 per cent to nearer 70 per cent.

Closer packing of the fibres can be achieved if the fibres are arranged in a hexagonal pattern, as shown in Figure 18.6. It can be shown by calculation that a maximum fibre volume fraction of $V_f = 0.907$ can be attained with such a hexagonal array. In practice, such an array could only be produced if fibres were laid down individually, to ensure the closest packing. This degree of close-packing is inherently much more difficult to achieve. As before, the fibres would need to touch each other to reach the 90 per cent volume fraction. This would never be achieved in practice; a more realistic value for volume fraction in cases where hexagonal arrays were possible would be nearer to 70 per cent.

With composites made for strength purposes, there is not much point in using a volume fraction much below 0.3, so in practice volume fractions of fibre lie between 0.3 and 0.6–0.65 for the reasons given above.

18.4.2 Random fibre orientation

Besides making composites with the fibres all aligned in the same direction, we can make composites with randomly oriented fibres. The effect of this will be to give a modest increase in Young's modulus (E) in all directions, but the increase will be much less than the 160-fold increase given in the uni-directional case.

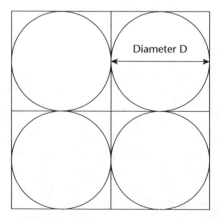

Figure 18.5 Square packing of uni-directional fibres

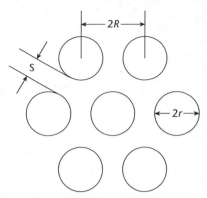

Figure 18.6   Hexagonal packing arrangement of uni-directional fibres in a composite

In such cases shorter fibres are used, rather than the long continuous fibres considered so far. Figure 18.7 shows a two–dimensional schematic of a random fibre oriented structure.

18.4.3 Density of composites

Since composites are frequently made in order to save weight, it is worth looking briefly at the resulting densities of composites, and how these can be calculated. Densities are calculated using a version of the rule of mixtures, thus:

$$\rho_{comp} = \rho_f \cdot V_f + \rho_m \cdot V_m$$

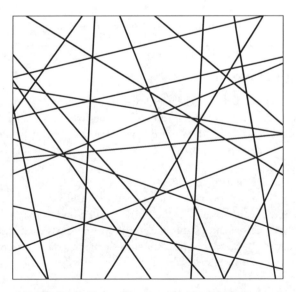

Figure 18.7   Schematic representation of long fibres randomly distributed in two dimensions (random in-plane)

(After Hull, 1981)

where:

ρ_{comp} = density of composite
ρ_f = density of fibre
ρ_m = density of matrix
V_f = volume fraction of fibre
V_m = volume fraction of matrix

We shall examine the case of the GRP composite used in the example in Section 18.4.

Worked example

We have a GRP with a fibre volume fraction of 40 per cent. The respective densities of fibre and matrix are 2,600 kg/m³ and 1,400 kg/m³. Calculate the density of the resulting composite.

So, matrix density   = 1,400 kg/m³
fibre density   = 2,600 kg/m³
volume fraction, matrix = 0.6
volume fraction, fibre   = 0.4

So, using the equation above, overall density of composite:

$$\rho_{comp} = \rho_f \cdot V_f + \rho_m \cdot V_m$$
$$= 2,600 \times 0.4 + 1,400 \times 0.6$$
$$= 1,040 + 840$$
$$= 1,880 \text{ kg/m}^3$$

Therefore, the overall density of the resulting composite will be 1,880 kg/m³.

18.5 How do composites work?

As mentioned above, the reinforcement – usually a fibre – must be properly bonded to the matrix. This is because when the load is applied to the matrix it can transfer this load to the fibre. The fibre, being stronger and stiffer, can easily take the load and it does not deform elastically as much as the unreinforced matrix would do by itself. The load transfer from matrix to fibre occurs across the matrix–fibre interface, hence the need for good bonding.

When the composite is under load, the matrix is transferring the load to the fibre reinforcement via the matrix–fibre interface. The effectiveness of the reinforcement really depends upon the value of shear strength at the interface. Because the stiffness of the fibres is much higher than that of the matrix, it is the fibres that are resisting elastic deformation, and because the fibres are stronger than the matrix it is the fibres that are resisting fracture. Because the shear strength of the interface is so important, it follows that anything that reduces this strength or destroys the fibre–matrix bond will cause the composite to fail. We shall examine failure later.

18.5.1 Stress and strain distribution at fibres

We have already discussed the theory of composites with continuous fibres. But we can also use discontinuous fibres either uniformly aligned or randomly arranged. In both cases we shall obtain enhanced stiffness and strength properties in the resulting composite. It will be worth examining how these shorter fibres work, and in particular the stress and strain distribution around the fibres, especially the fibre ends. The effectiveness of short fibres is a function of their length; longer fibres are more effective than shorter ones, and fibre ends play an important role in the fracture of short fibre composites.

Consider a single fibre of length l embedded in a matrix of lower modulus material, and aligned with the axis of tensile loading. If the matrix is well-bonded to the fibre, the stress applied to the matrix will be transferred to the fibre across the interface. The matrix and fibre will experience different tensile strains because of their differing moduli; in particular, at the ends of the fibre the strain in the fibre will be less than that in the matrix, as shown in Figure 18.8(b). As a result of these strain differences, shear stresses are set up around the fibre end regions in the direction of the fibre axis, and the fibre itself is put into tension.

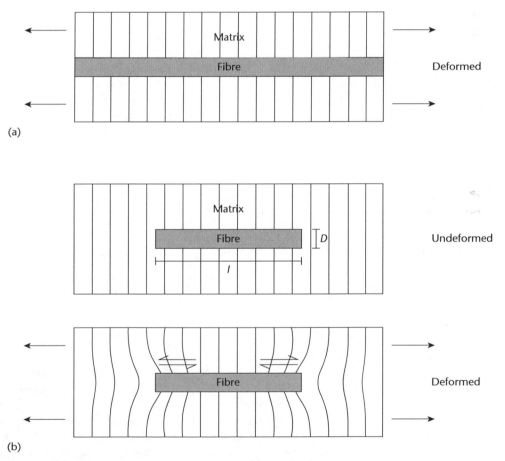

(a)

(b)

Figure 18.8 Effect of deformation on the strain around a fibre in a low modulus matrix: (a) continuous fibre; (b) short fibre

(After Matthews & Rawlings, 1994)

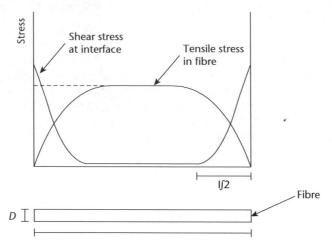

Figure 18.9 Variation of tensile stress in a fibre and shear stress at the fibre–matrix interface
(After Cox, 1952)

The tensile and shear stresses around the fibre and along its length can be represented as shown in Figure 18.9.

Figures 18.8 and 18.9 show that the straining around the ends of the fibres, in both fibre and matrix, is far from uniform. Shear stresses are very high in the end-of-fibre regions, falling from a maximum at the fibre ends to zero in mid-fibre. These stresses will depend upon the overall level of external stress applied to the composite. It will be readily appreciated that if the external applied loading reaches high levels, then these end-of-fibre shear stresses will also become extreme, and can lead to de-bonding of the fibre from the matrix. Such de-bonding will not normally occur in mid-fibre, and de-bonding will often be the precursor to failure. This is because loss of bond strength involves the loss of the stiffening and strengthening effects of the fibrous reinforcement.

If the fibres in a continuous fibre composite break, then the same situation occurs; there is local intense shear stress at the broken fibre ends, leading to de-bonding and strength loss.

18.5.2 Failure and fracture behaviour of composites

The benefits of fibre reinforcement do not end with stiffness enhancement; they can also improve fracture performance. Composite materials acquire a kind of ductility. It is not true ductility, such as that exhibited by, say, the metal copper or a polymer such as nylon, but the composite will strain to a higher value than the failure strain for either the matrix or fibre material. How can this happen?

When a single solid body fails, it usually does so by forming a crack which then propagates right across the body, resulting in its breaking into two halves. The crack may propagate slowly or quickly, but the end result is the same. In the case of a composite the situation is quite different. Let us examine what happens when a composite fails. The mechanism of failure is something that happens inside the composite, and which is not visible externally. The first stage of failure is *de-lamination* or *de-bonding*, i.e. the reinforcement becomes de-bonded from the matrix. Like most other failure mechanisms, de-bonding can nucleate from a tiny area of the fibre–matrix interface. In addition, sometimes individual fibres fracture, and this can also lead to de-bonding. As a de-bonded area grows in size, the matrix can begin to crack; the reinforcement effect is lost due to de-bonding. Figure 18.10 illustrates these effects.

If matrix cracking occurs, the crack will tend to grow in the same way that cracks grow in ordinary solid materials. However, the fibres will tend to hold the two interface surfaces together. Even if the fibres

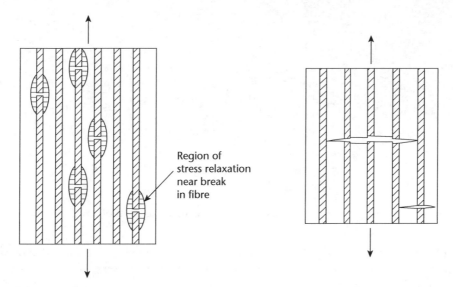

Figure 18.10   De-bonding, matrix and fibre cracking
(After Hull, 1981)

fracture, they will usually fail inside the matrix to one side or another of the matrix crack. The length of broken fibre still embedded inside the matrix will resist 'pull-out' of the length of fibre. Load will have to be applied, or work will have to be done to pull the fractured fibres from the matrix, and this means that overall fracture is delayed until a higher value of strain is reached. So unlike a single, homogeneous solid, where fracture may be rapid, moving quickly through the material without warning, in a composite the progress of fracture will be much slower and more progressive. This is the reason for the behaviour of timber. If timber is over-stressed, we hear cracking noises, but the timber does not fly into two pieces. The cracking sounds are made by individual vessel cells or groups of cells failing, but the timber is still held together by the vessel cells which have not yet failed.

If we consider another class of composite important in construction, namely laminated materials such as plywood, the failure mechanism here is also de-lamination. This is often caused by the effects of damp and moisture. If the wrong grade of plywood has been selected, the glue or adhesive used to bond the sheets of wood together may be soluble in water to some degree, and so dampness, over time, will cause de-lamination to occur. If it is known that plywood will see water or dampness in service then the weather and boil-proof grade should be selected. This means that the adhesive used to bond the sheets together is water-proof, and so the plywood will happily survive such conditions. The same goes for all timber products such as chipboard, block-board, fibreboard, etc. All these products are really composites with fibres reinforcing a matrix of adhesive.

18.6 Theory of composites: 2 – thermal properties

Textbooks on composites tend to concentrate exclusively on those made for reasons of strength and stiffness, and to completely ignore those made for thermal reasons, such as for providing good thermal insulation. Since this is an important topic for those constructing and operating buildings, it will be simply dealt with here. We shall look at how we can calculate the thermal resistance of multi-layered structures, as this is a matter of great importance at the present time, with the need to conserve energy.

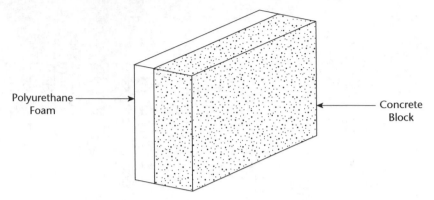

Polyurethane Foam

Concrete Block

Figure 18.11 Composite block made for increasing the thermal insulation of a wall

As the Building Regulations have been progressively updated to improve the thermal performance of buildings, the need has arisen for improved insulation. In the early 1990s composite blocks were available to builders in order to achieve the improved standards of insulation without increasing the thicknesses of exterior walls. Figure 18.11 illustrates the type of blocks marketed in response to the Building Regulation changes.

We need to know the *thermal conductivity (k)* values for the materials in the structure. Thermal conductivity has already been discussed in Chapter 6, and values are readily available in textbooks and tables of property data. We also need to know the thickness (*l*) of the layer of material.

The thermal resistivity of a material is the reciprocal of the conductivity, i.e.:

$$\text{Thermal Resistivity} = \frac{1}{k} \text{ where } k = \text{thermal conductivity}$$

The *thermal resistance* of a layer of material can be calculated using the following expression (Carslaw & Jaeger, 1959):

$$\text{Thermal resistance} = \frac{1}{k} \times t \ (t = \text{thickness, } m)$$

Thermal resistances are similar to electrical resistances. In a simple circuit containing a number of resistances connected up in series, the total resistance to current flow can be found by simply adding up the resistances. Similarly, in a building wall we may have a number of different layers of material, such as brick, block, air cavity, insulation layer, etc. The thermal resistance of the whole wall structure can be calculated simply by adding the values of the various materials together.

The calculation of thermal resistance values is useful, because we can use them to estimate the overall *thermal transmittance* or *U-value* for a building outer envelope. In these days of increasingly expensive energy, it is vitally important to be able to design outer wall structures to retain as much heat as possible.

18.6.1 How do these insulating composites work?

These composites made for reasons of thermal insulation work in an analogous fashion to a simple d.c. electric circuit. If we insert more resistance in series with the existing resistance, we shall reduce the current flow according to Ohm's law. Similarly, if we insert a second layer of material with higher thermal resistance into our wall, we shall reduce the rate at which heat flows through it. The heat flow is analogous to the flow of current in the electric circuit.

Worked example

Concrete blocks 100 mm thick are to be replaced with concrete–polyurethane composite blocks, also 100 mm thick, to increase the thermal resistance of a building wall. The composite will consist of concrete 60 mm thick and a polyurethane foam layer 40 mm thick. Calculate:

1. the thermal resistance of the concrete blocks, and
2. the thermal resistance of the composite blocks.

Thermal conductivity values are as follows:

k for concrete $= 1.4$ W/m K
k for polyurethane foam $= 0.08$ W/m K

1. Resistance of concrete blocks

$$= \frac{1}{k} \times \text{thickness (m)}$$

$$= \frac{1}{1.4} \times 0.1 \text{ m}$$

$$= 0.071 \text{ m}^2\text{K/W}$$

2. Resistance of composites. First find resistance of 60 mm of concrete

$$= \frac{1}{1.4} \times 0.06 \text{ m}$$

$$= 0.043 \text{ m}^2\text{K/W}$$

Next, resistance of polyurethane foam

$$= \frac{1}{0.08} \times 0.04 \text{ m}$$

$$= 0.50 \text{ m}^2\text{K/W}$$

Total thermal resistance of composite

$$= 0.043 + 0.50$$

$$= 0.543 \text{ m}^2\text{K/W}$$

From this we can see that by adding the layer of polyurethane foam we have increased the thermal resistance of the 100 mm layer from 0.071 to 0.543 m²K/W. We have effectively increased the thermal resistance by over 7.6 times, without increasing the wall thickness.

18.7 Critical thinking and concept review

1. What are the *three* requirements that must be met if we are to successfully reinforce one material with another?

2. Give the most important reason for the manufacture of composite materials.
3. Write down the rule of mixtures or Voigt equation, defining all terms.
4. Write down the Reuss equation, defining all terms.
5. Calculate the stiffness in the fibre direction of a composite made using continuous, uni-directional fibres, if the matrix stiffness is 3.2 GN/m^2, the fibre stiffness is 320 GN/m^2, and the volume fraction of fibres is 50 per cent.
6. For the composite given in Question 5, calculate the stiffness in the direction transverse to the fibres.
7. Describe the advantages conferred by the reinforcement on the fracture behaviour of the composite
8. Explain why there is an effective upper limit on volume fraction for continuous fibres in composites of around 60–65 per cent.
9. Calculate the density of the composite given in Question 5, given the following density data:
   • Density of fibres = 1,750 kg/m^3
   • Density of matrix = 1,400 kg/m^3
10. Calculate the thermal resistance of concrete blockwork 100 mm thick if the thermal conductivity of the concrete is 1.4 W/m K.
11. What would be the effect of adding 100 mm of polyurethane foam to the concrete in Question 10, if the thermal conductivity of the foam is 0.08 W/m K?
12. Give examples of three types of composite materials that are commonly used in building construction.

18.8. References and further reading

ASHBY, M.F. (1992), *Materials Selection in Mechanical Design*, Pergamon Press, Oxford.
CARSLAW, H.S. & JAEGER, J.C. (1959), *Conduction of Heat in Solids*, Oxford University Press, Oxford.
GIBSON, L.J. & ASHBY, M.F. (1988), *Cellular Solids*, Pergamon Press, Oxford.
HULL, D. (1981), *An Introduction to Composite Materials,* Cambridge University Press, Cambridge.
MATTHEWS, F.L. & RAWLINGS, R.D. (1994). *Composite Materials: Engineering and Science*, Chapman & Hall, London.
NEVILLE, A.M. (1981), *Properties of Concrete,* 3rd edition, Longman, Harlow.
SZOKOLAY, S.V. (1980), *Environmental Science Handbook,* The Construction Press, Lancaster.

Part III
In-service aspects of materials: durability and failure

19

Failure 1: effects of stress and applied loading

This chapter, together with the two following chapters, deal with the various ways in which materials can fail in service. There are many possible causes of failure, some due to the various loading regimes that are applied to materials in service. In other words, situations where stress is the major factor – this will be the subject of this first chapter on failure. Other component factors that can cause or contribute to failure include environmental factors such as moisture, corrosive agents, chemical attack, ambient temperature and certain components of solar radiation; Chapter 20 will deal with these environmental factors. Another, very extreme, situation that can occur with buildings is the incidence of fire; Chapter 21 will look at the response of building materials to fire.

This chapter considers failures where stress is the main component. The material property known as fracture toughness is also briefly discussed, including ways of measuring fracture toughness by impact testing. The fracture behaviour of various types of materials and the fatigue, impact and creep regimes are reviewed.

Contents

19.1 Introduction

The term failure implies some failure of a material or component that forms part of a structure or system. The structure could be a building or a bridge or it could be part of a transport system, be it on land, sea or in the air. Failure is obviously a highly undesirable event, especially if it occurs suddenly, without warning. Lives can be put at risk and there will always be economic consequences. Not least, there will be interruption to, or loss of service in some way.

As indicated, sudden or catastrophic failure is the most feared eventuality. The failure can be due to a single cause, e.g. high stresses, or it can have more than one cause, e.g. stress *and* corrosion. In this chapter we shall consider failures where stress is the main or important component. However, because of system wear and tear, engineers know that failure in many cases is inevitable. In these cases, wear is ideally monitored and planned routine maintenance can be organised so that vulnerable items and materials are routinely replaced before their condition becomes a cause for concern. In fact, engineers sometimes design certain systems to fail in a 'safe' way, i.e. the system is designed to show that failure is occurring without the failure causing real problems. Such a failure just highlights the need for maintenance. There are innumerable reasons for failure and these can include poor design, incorrect use of materials, use of flawed materials, system misuse, etc. It is therefore the responsibility of the engineer to ensure that the correct materials are used in good designs. The engineer should think about service life issues, and clearly anticipate the possible final failure and take such preventative action as and when appropriate. It will be useful to briefly consider the nature of the various loading regimes likely to occur in buildings.

19.2 Nature of stress

Stress is the intensity of loading inside a material or component brought about by the nature of the loads applied to it, and was discussed in some depth in Chapter 4. We shall not give a strictly mathematical derivation here. The loading can either be static, i.e. unchanging with time, or dynamic, i.e. load that varies with time. Construction and civil engineers talk of 'dead loads' and 'live loads', and these two situations correspond more or less with the static and dynamic regimes mentioned above. For example, the Empire State Building in Manhattan weighs about 305,000 tonnes, and so we would say that its foundations carry a dead load of 305,000 tonnes. However, the building is also subject to live loads due to the weight of its contents, which can vary with time. The building is also subject to wind loading, the value of which can reach thousands of tonnes of force on a very tall building. Again, these live loads are not constant, but vary with time.

It is characteristic of buildings that their structural cores are subject to compressive stresses, and mainly static ones at that. In designing these structures, engineers can safely use the data obtained from laboratory strength tests carried out at quasi-static rates of strain (10^{-3} sec^{-1}). However, this is not invariably the case. For example, we may have a cable-stay or suspension bridge structure carrying heavy vehicular traffic. Each time a heavy lorry passes over the bridge, certain parts of the structure will be put under tension. The duration of the tensile force will be short, but it will be repeated with the passage of each vehicle. This is a dynamic loading situation of the type that can cause *fatigue* problems, and so we shall need to understand the fatigue behaviour of our materials.

Furthermore, if there is an accident, the structure may suffer impact from a moving vehicle, or even, more rarely, a blast wave from an explosion. In such impact loading situations the load may be intense, but of brief duration. Such short duration loads are referred to as impulsive loads. For such dynamic loading situations, knowledge of material behaviour at quasi-static rates of strain will be inadequate to predict structural behaviour. Here we shall need access to high strain-rate data for out material response.

Finally, it is worth pointing out that even with monotonic (unvarying) compressive stress regimes, we may observe time-dependent phenomena such as *creep*. Creep involves a slowly developing, small, permanent strain in material under the action of a constant load. Not all materials exhibit creep at room temperature – for example, steel does not suffer creep at ambient temperature, but the metal lead does. Non-metallic materials such as concrete, glass, timber, brick masonry, polymers, etc. can all exhibit creep, and creep behaviour will therefore be considered here.

19.3 Fracture

Fracture is the most obviously apparent form of failure; quite simply, it involves the separation of a whole body into two or more pieces when it is subjected to an applied load at temperatures well below the melting point of the material concerned. The applied loading can be static or dynamic. Dynamic loading is the term used to describe situations where the applied load varies with time, and it can be used in situations where the loading is suddenly applied – for example, by an impact. Such short duration loading is also called *impulsive loading*.

There are many different types of fracture; however, the most common types are brittle and ductile fracture. These are differentiated by the amount of plastic deformation absorbed by the breaking material, and they will be examined in the next sections of this text. We have looked in previous chapters at the strength properties of materials, and in particular at the stress–strain characteristics of materials. We know that when materials are put under load, especially tensile loading, they initially deform elastically, with a linear stress–strain graph. Eventually, they cease to deform elastically, and then they become plastic, and any deformation is then permanent. In the case of ductile materials such as metals, this plastic deformation regime can be extensive, with the stress–strain record increasing to a maximum value (the ultimate tensile strength). Thereafter, the stress–strain line exhibits a decreasing slope until final fracture occurs. However, whenever the load vs extension (or stress vs strain) record shows a decrease, this usually indicates that some kind of failure or cracking is occurring within the deforming material.

In Chapter 3, Section 3.5 we looked at the process of fracture at the atomic level. We saw that the bonds between atoms could be broken one by one, leading to the production of a micro-crack. It was pointed out that all the rows of atoms would transfer their load to the last bond at the root or tip of the crack, and the result of this would be that the actual stress at the crack tip will be many times higher than the nominal stress on the component. We shall take up this idea again when we come to the section on fatigue.

19.3.1 Brittle fracture

Brittle materials exhibit little or no plastic deformation before they fracture. In other words, for brittle materials the yield strength is almost the same as the ultimate tensile strength. This is shown in the stress–strain plot for a brittle metal in Figure 19.1. The shaded area ABCD represents the work done per unit volume of material in straining it to fracture. The work done per unit volume is the same as the energy put into the material in taking it to fracture. The area under the elastic portion of this curve, taking the yield

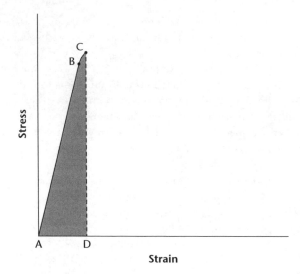

Figure 19.1   Stress vs strain plot for a brittle material

point B as the upper limit, is the elastic strain energy per unit volume stored in the metal under load. This can be evaluated as follows:

Stored elastic strain energy per unit volume = ½ $\sigma \cdot \varepsilon$

where

σ = yield stress
ε = elastic strain at yield.

If the volume of metal being strained is large, if we multiply the strain energy per unit volume by the total volume of strained metal, then the total stored energy can be very large. This stored energy, when released at fracture, can help to drive the fracture process. The creation of fracture surfaces will require energy to drive the process, and the stored strain energy can help to drive the fracture.

The other point to make is that brittle fractures occur very quickly; the fracture can proceed at velocities of up to 1–2 km per second. This means that a fracture can run through a very large structure such as a ship's hull in milliseconds. The case of the Second World War Liberty ships is mentioned later. Brittle fractures involve the process of *cleavage*, which means that the crack goes through individual grains, cleaving them into two halves. The cleavage planes are flat and smooth, the crystal splits into two along the close-packed planes, so that the fracture surface is made up of reflective surfaces. Because of all the tiny flat surfaces, such brittle fractures often twinkle when examined under a light.

19.3.2 Ductile fracture

In contrast with brittle materials, ductile materials exhibit gross plastic deformation before fracture. Figure 19.2 shows a tensile stress–strain curve for a ductile material; a curve for a brittle material is superimposed onto the ductile curve. It is immediately apparent that the area under the ductile curve is very large in comparison with the brittle curve (AB'C'D' compared with ABCD). The extra area under the ductile curve represents the work done per unit volume (or energy expended per unit volume) in plastic

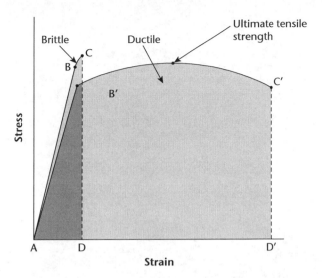

Figure 19.2 Typical stress vs strain plots for brittle (ABCD) and ductile (AB'C'D') materials loaded in tension to failure

deformation in taking the metal to failure. This energy can be evaluated using numerical integration. The extra energy required for ductile fracture means that ductile materials are much tougher than brittle ones. There is a specific material property known as *fracture toughness* and this will be mentioned again later.

Fracture in ductile materials occurs more slowly than in brittle ones. Since micro-cracks tend to form at grain boundaries, ductile fracture involves pulling grains apart. In other words, the fracture surface goes around grains, through the grain boundary zones where there is a high density of micro-cracks.

Ductile fracture surfaces are quite different in appearance from brittle fracture surfaces. A ductile fracture in steel will appear a dull grey in colour, with a rough surface texture, whereas a brittle fracture surface will often twinkle in the light. This is due to the fact that in ductile fracture, the individual metals grains are pulled apart, whereas in brittle failure the individual grains will be split apart. The individual grain facets will be smooth and reflective, and this is what causes them to twinkle in the light. Ductile and brittle fracture surfaces appear distinctly different at both microscopic and macroscopic scales. At the macroscopic scale, ductile fractures show more pronounced necking regions, with a rougher surface texture. Brittle fractures show little or no necking, with relatively smoother fracture surfaces. This is illustrated in Figure 19.3.

At the microscopic level, ductile fracture surfaces appear to be dimpled, whereas brittle fracture surfaces are smoother and faceted, as shown in Figure 19.4.

19.3.3. Ductile–brittle transition

In terms of failure, ductile fracture is much preferable to brittle fracture as it is not sudden and catastrophic. We can observe the commencement of failure, and close down the machine or system well before final failure occurs. With brittle fracture, the failure is sudden, complete and often catastrophic, and occurs without prior warning.

We need to be aware that with some materials, and under certain ambient and loading conditions, we can obtain a ductile-to-brittle transition. When this happens a material which is usually ductile suddenly becomes brittle and can fail without warning. This situation can pose a major hazard. We need to ask: what

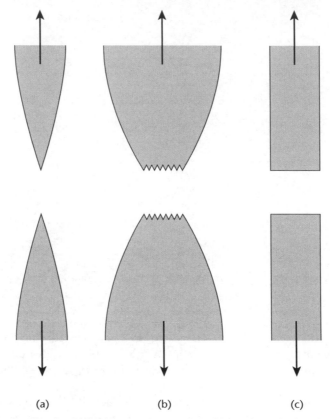

Figure 19.3   Fracture profiles for (a) highly ductile material, (b) moderately ductile material and (c) a brittle material

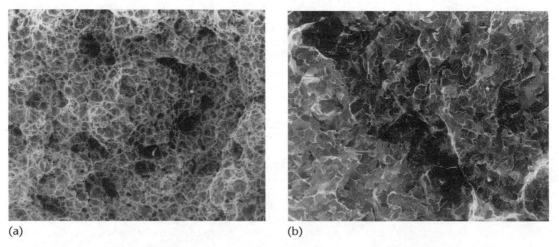

(a)

(b)

Figure 19.4   Scanning electron micrographs of (a) ductile fracture surface and (b) brittle fracture surfaces

are the conditions that can lead a material to change its mode of failure from ductile to brittle? In fact there are two conditions that can bring about such a transition:

1   Low ambient temperature (typically below 0 °C).
2   Dynamic loading (impact, with strain rates above 10^3 sec^{-1}).

Both sets of circumstances can arise in building and civil engineering situations. We have already seen that steel is by far the most important and widely used metal in construction and civil engineering, for the reason that it possesses high strength, ductility and toughness. At normal ambient temperatures this is true. However, in situations where the ambient temperatures fall to 0 °C and below, steel can suddenly become brittle. It was this fact that led to the famous problem with the Liberty ships in the Second World War. Some steels become brittle in freezing water, while others retain their ductility until much lower temperatures are reached (see Section 19.4).

Regarding the effects of impact or dynamic loading, many materials show differences from their behaviour at normal, quasi-static rates of strain. In general, when most materials are deformed at high rates of strain (10^3 sec^{-1} or above) they tend to exhibit higher strengths and reduced strains to failure (i.e. reduced ductility). However, some steels suffer a ductile-to-brittle transition when subject to intense dynamic loading. This is much more serious than merely showing reduced strain-to-failure. This is the reason that the lunar modules used in the Apollo Moon Programme were not made of steel, but of aluminium alloys. If a steel module had been struck by a meteorite moving at high velocity (typically 20 km sec^{-1}), the object would possibly have punched a hole in the module with fatal consequences for the astronauts inside. Aluminium suffers no ductile–brittle transition no matter how intense the loading, and so an impact would merely have dented the module.

Of course, on Earth, meteorite strikes are very rare events, but buildings can suffer impact loading for various reasons; they can be accidentally struck by vehicles of various kinds, by aircraft and also deliberately by projectiles and blast waves as a result of terrorist activities. Because steel is such an important material, we need to be aware of the possible effects on it of low temperatures and impact loading situations.

In fact, materials scientists and engineers have found that in addition to properties of strength and ductility, materials also possess an intrinsic property called *fracture toughness*. This measures the propensity of a material to fracture rather than to deform when subject to load, and it is a property distinct from those of strength and ductility.

19.3.4 Fracture of timber

Timber is an interesting material. It is really a natural composite, with cellulose fibres reinforcing a lignin matrix. The strength and stiffness of timber is provided by the cellulose fibres, most of which lie in the longitudinal direction, i.e. they are aligned parallel to the tree trunk. Timber is used structurally to make beams to support floors, for example. Traditionally it was tested by the process of *stress grading*; this involved using a three-point bend test at intervals along the length of timber being checked. This simulates the stress pattern to which the timber is subject in service, and a three-point bend test can be used to test timber to destruction. Figure 19.5 illustrates the load vs displacement graph obtained in a typical test on a small timber beam.

We can see from Figure 19.6 that initially the timber behaves in an elastic manner – the graph is linear, rising to a maximum. At this point in the test, cracking was heard and the load showed a sudden fall. However, as loading was continued, the load graph began rising again, and this continued until more cracking was heard. The piece of timber now had a permanent bend at the point of application of the load, but remained unbroken. As loading continued, further cracking and sudden load drops were observed. The initial area under the graph was effectively doubled as the load approached zero.

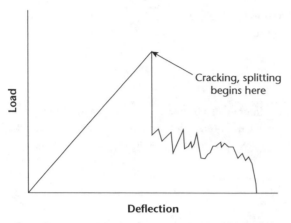

Figure 19.5 Load vs displacement graph obtained in the testing of a small timber beam

What was the explanation for this behaviour? We have seen that timber is a composite, and when it is overloaded some of the fibres will begin to crack and to break. However, since the timber consists of many fibres in a matrix the cracking and breakage of a few fibres will not bring about fracture of the entire piece of wood. Some fibres may be broken and the matrix may be cracked, but the piece of timber will remain in one piece, held together by the remaining fibres. So the falls in load seen in Figure 19.5 will correspond to some of the fibres breaking, but the remaining fibres will take up the load.

19.4 Fracture toughness testing

Experience with many materials over many decades has taught engineers that, besides strength, there is another important material property known as *fracture toughness*. Some materials always fail in a ductile manner, even when tested at very high strain rates – for example, pure aluminium. For this reason the Apollo lunar modules were made from aluminium alloys. Some materials are ductile under normal conditions, but can exhibit brittleness when subjected to low temperatures or at very high rates of strain – steel can behave like this. Some materials are always brittle.

We have seen that the presence of cracks, sharp changes in section, etc. cause local high stresses which can lead to failure in circumstances when failure might not be expected. Very ductile, tough materials do not become embrittled, even when placed under normally adverse conditions. Such materials are said to have high fracture toughness. This is actually a fundamental property of a material as it relates to the energy required to make a crack propagate in the material. For example, copper has high fracture toughness as it takes about 10^6 J m^{-2} to make a crack propagate. Glass, on the other hand, has very low fracture toughness, and requires only 10 J m^{-2} to propagate a crack. We know from everyday experience that copper is tough and glass is not, and the fracture toughness is the property that accounts for this.

The problems of fracture toughness and its measurement were highlighted dramatically during the Second World War. The need for a large merchant fleet of ships to convey war materiel, food, machine tools, etc., from the United States to its allies in the UK and Soviet Russia prompted the Americans to use faster shipbuilding techniques. Using conventional methods it took 9–10 months to build a ship. However, prefabricating hull sections using welding methods, and then welding the sections together, reduced the time to build a complete ship to one month. Using these new methods the United States built several thousand Liberty ships in a fraction of the time it would have taken using traditional methods.

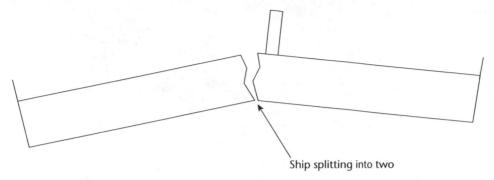

Ship splitting into two

Figure 19.6   SS Schenectady: the crack began at the sharp corner of a hatchway, and ran right around the hull at high speed

However, a welded ship's hull is effectively one piece of material, whereas a riveted hull is not. If one plate in a riveted hull suffers brittle cracking, the crack will run from one edge of the plate to the other, and then stop. If the same plate in a welded hull suffers brittle cracking, the crack will run across the plate and then continue through the entire structure as it is effectively a single piece of material. The ship's hull will thus break into two, and this is what happened to some Liberty ships, especially those in the Arctic convoys en route to Russia. Figure 19.6 shows the ship *Schenectady*, which was split into two; the crack began at the sharp corner of a hatchway and ran right around the hull at high speed. The exact mechanisms of low temperature brittle fracture were not fully understood at the time, but the problem was solved by using the Charpy test as a quality control test for selecting the steels for ships' hulls. Batches of steel that failed in a brittle manner when tested at 0 °C were rejected for use.

Of course, steel plates (and sometimes rivets) had failed before the Second World War, but when only one plate fails the ship's integrity is not threatened. Any water seeping in through the cracks can be pumped out by the bilge pumps. The problem only became apparent once riveting was abandoned in favour of welding.

19.4.1 Impact testing techniques

The usual standardised tests for measuring fracture toughness are the *Izod* and *Charpy* tests; they are also sometimes called notch toughness tests. Historically, the Izod test has been most favoured in the UK, while the Charpy test is widely used in Europe and North America. Both types of test machine are often called pendulum machines because they both have a swinging arm arrangement as shown in Figure 19.7.

In these tests we used notched specimens. The specimen is supported as a horizontal beam in the Charpy test, or is held in the vertical position in the Izod test. The specimens are clamped in position, and during the test they are impacted by the heavy, swinging pendulum. The pendulum is always held at height (h) above the specimen before each test. To carry out the test, the pendulum is released and it swings down and strikes the notched specimen. The specimen is then forced to bend and/or fracture under the impact; the strain rate will be of the order of 10^3 per second. The specimen is notched to provide a stress raiser and site for failure. If the specimen does not break, the pendulum is brought to a halt and we can see that the specimen has absorbed the kinetic energy of the pendulum in bending, but not failing. If, on the other hand, the specimen fractures in a brittle way, the pendulum will swing past the specimen site and gain some height on the opposite side (h'). From the difference in the starting height h and the finishing height

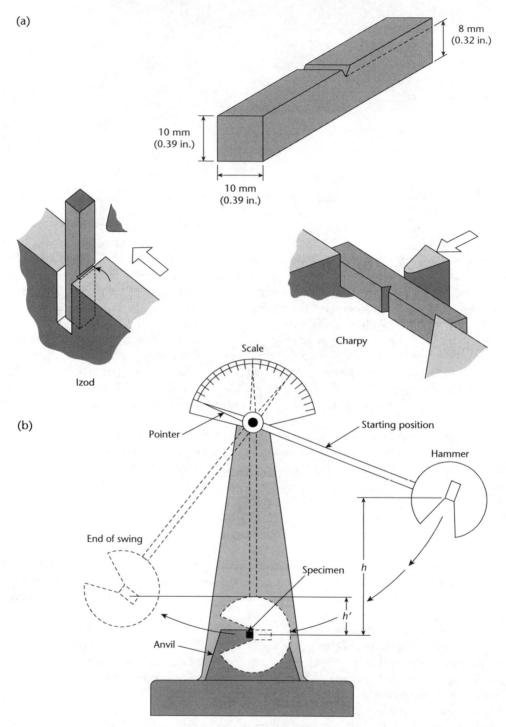

Figure 19.7 The Charpy/Izod impact testing arrangement: (a) impact specimen; (b) impact testing machine

(After Callister, 1994)

h' we can calculate how much energy the specimen has absorbed in breaking. If the difference in the heights is small, little energy has been absorbed and the specimen material is obviously brittle. If there is a greater difference in the heights, then more energy has been absorbed and the material is tougher. If the pendulum is brought to a standstill without failure, then the material has absorbed all the kinetic energy and is obviously very tough.

This test can be very useful when it is carried out over a range of temperatures, as it can uncover whether the material being tested is subject to a ductile–brittle transition. We have already mentioned the Second World War Liberty ships. These pendulum impact tests were routinely employed to test samples of steel plate for use in the all-welded hulls of the Liberty ships. The specimens can be cooled to various temperatures immediately prior to testing, so that the influence of temperature can be measured. If necessary, steels showing ductile–brittle transition at too high a temperature can be rejected for service in cold weather or cold water conditions.

19.4.2 Significance of transition–temperature curve

The chief engineering use of the Charpy test is in selecting materials which are resistant to brittle fracture, and this is done using the transition temperature curves. The design philosophy is to select a material which has sufficient notch toughness when subjected to severe service conditions. Figure 19.8 shows a ductile–brittle transition brought about by low ambient temperatures.

It should be borne in mind that impact rates of strain (10^3 sec^{-1}) can also bring about a ductile to brittle transition in some materials, including steels. A plot of fracture energy versus strain rate will show a similar profile when certain susceptible materials are tested. Fortunately, we have at our disposal a vast range of engineering steels, a sufficient number of which have higher resistance to rate-induced brittleness.

19.5 Fatigue failure

Fatigue is brought about by a repeated cycling of an applied load, and this load has to cause tensile stresses in part of the loading cycle. Fatigue occurs commonly in metals; most people are familiar

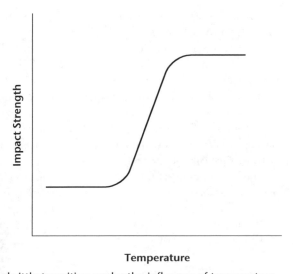

Figure 19.8   A ductile to brittle transition under the influence of temperature

with the term 'metal fatigue', and the phenomenon has been known about for over 100 years. However, metal fatigue has only been properly understood since the Second World War. The incidence of fatigue was feared because it occurred without warning, and at stresses apparently well below the yield strength of the metal concerned. We now know that the process of fatigue occurs in three stages:

1   crack initiation
2   crack growth
3   final, sudden failure.

Of course, the crack initiation and growth stages are nearly always not visible to the naked eye, and the final, catastrophic failure is the first indication that anything is amiss. It is sudden failure that made fatigue such a fearful phenomenon, and it is only since the Second World War that we have gained an understanding of the causes and mechanisms of fatigue failure. Research into fatigue was stimulated by a number of high-profile failures, including a number of bridge failures and, perhaps best known of all, the failures of the de Havilland Comet I jet airliners in the early 1950s.

19.5.1 Initiation, growth and failure

Studies of fatigue carried out after the Second World War indicated that fatigue cracks had first to be *initiated*. The growth stage was understood, but the mechanisms of initiation awaited the development of scanning electron microscopy, for example. Work on the fatigue failure of thick-walled pressure vessels showed that cracks could be initiated at small inclusions present in the metal, no matter how tiny. These inclusions can be tiny fragments of refractory materials from the steelmaking process; and they have different stiffness values than the steel matrix.

If the inclusion is uncovered during machining, and is subject to high pressure (tensile stress), the first time the vessel is pressurised the particle can become de-bonded from the steel matrix. It thus forms a small micro-crack in the steel surface. As we saw in Chapter 3, Section 3.5, the stress at the crack tip can be many times the nominal stress, and well in excess of the steel's yield strength and tensile strength. Each time the component is put under stress, the bonds at the root of the crack will be broken, and the crack will grow in length.

Let us assume that ten atomic bonds are broken during each stress cycle. If we assume an atomic bond length of, say, 0.25 nm (0.25×10^{-9} m), each stress cycle will cause the crack to grow by a distance of 0.25×10^{-8} m. One thousand such cycles will cause the crack to enlarge by 0.25×10^{-5} m, and one million cycles will cause a growth of 0.25×10^{-2} m. A crack 2.5 mm long will be visible to the naked eye. However, in most situations these cracks will be located in components and positions not open to view. We can easily see from these simple calculations why the growth stage is so dangerous.

Classically, the crack grows until a significant part of the load-bearing section has failed. The stress in the remaining material is now raised to dangerously high levels, and it is during this growth stage that the classic beach marks are formed. The final fast fracture zone has a different appearance, and once final failure has occurred it is possible to see the location of the origin of the fatigue crack. Figure 19.9 shows a schematic diagram of a fatigue fracture surface showing the point of initiation, beach marks and final fast fracture zone.

Figure 19.10 shows an electron micrograph of a fracture surface in an aluminium specimen, and the striations indicative of fatigue failure can be clearly seen.

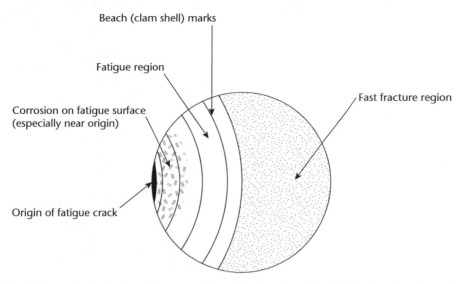

Figure 19.9 Schematic of a fatigue fracture surface showing the point of initiation, beach marks and final fast fracture zone

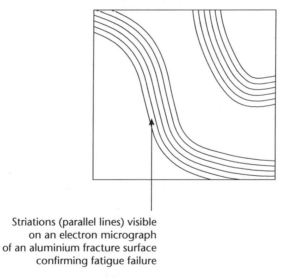

Striations (parallel lines) visible
on an electron micrograph
of an aluminium fracture surface
confirming fatigue failure

Figure 19.10 Striations on an aluminium fracture surface

19.5.2. Types of fatigue behaviour

Metals show two types of fatigue behaviour, and these are illustrated in the fatigue (S–N) curves shown in Figures 19.11 and 19.12. An S–N curve is a plot of stress (S) against stress cycles to failure (N). The curve tells us the approximate duration (lifetime in terms of number of stress cycles) of a component for given mean stress values. Simply put, the higher the mean stress, the fewer stress cycles the component will endure before failure.

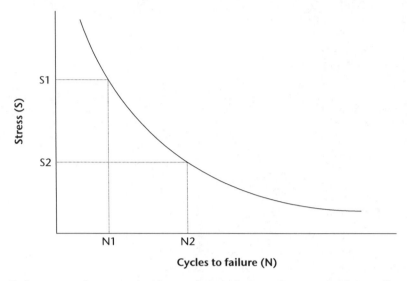

Figure 19.11   Fatigue curve for a metal without a fatigue limit, such as an aluminium alloy

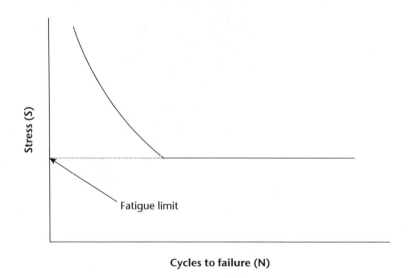

Figure 19.12   Fatigue curve for a metal that displays a fatigue limit, such as steel

Figure 19.11 shows fatigue failure occurring at higher number of cycles to failure with reducing stress level, with no minimum stress level. Non-ferrous metals tend to show fatigue behaviour like this, and it is interesting that the de Havilland Comet was built from aluminium alloys, i.e. materials without a fatigue or endurance limit.

Figure 19.12 is the S–N curve for a material showing a fatigue limit. In this case we can see that below a certain value of mean stress the component will sustain an infinite number of stress reversals without failure. This type of S–N curve is obtained with many ferrous materials and some titanium alloys. Materials

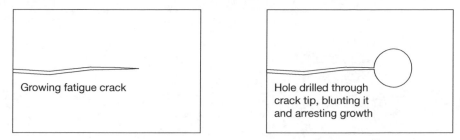

Figure 19.13   Crack tip blunting by drilling a hole through the cracked plate

with a fatigue limit would be ideal candidates for use in making reactor vessels for chemical plant, because they would give the best guarantee of a very long service life, without potential failure problems. The designer would design for operating stress levels below the fatigue limit, and this would ensure that the plant could be operated indefinitely without the risk of failure from fatigue.

Finally, it is worth pointing out that in certain situations, when fatigue cracks are discovered in sheet and plate materials, they can sometimes be arrested by crack tip 'blunting'. It is the sharpness of the crack tip that acts as a great stress intensifier. If we drill a hole through the crack tip as shown in Figure 19.13, the hole increases the radius of the crack tip by orders of magnitude to the point where the stress intensity is reduced so the crack no longer grows with each stress cycle.

19.6 Creep failure

Creep is the term given to plastic deformation that occurs in materials subject to stress over very long time periods. It can occur at elevated temperatures in some metals, but it can also occur at ambient temperatures in other metals. Materials such as concrete, polymers and timber can also suffer creep at ambient temperatures, and these will be discussed later.

Creep is generally an undesirable phenomenon and is often the limiting factor in the lifetime of components; it is usually a concern for engineers and metallurgists when evaluating the life of components that operate under high stresses and/or elevated temperatures. Creep is not necessarily a failure mode, but is often a damage mechanism. However, in some cases it can be the cause of failure, and such an example with concrete is discussed at the end of Chapter 20.

Examples of creep failure for non-civil engineering applications include turbine blades in jet engines, boilers, nuclear power plant elements, incandescent light bulb filaments, etc.

19.6.1 Creep behaviour

In a creep test, a constant load is applied to a tension specimen maintained at a constant temperature. Strain is then measured over a period of time. The slope of the strain vs time curve is the strain rate of the test during the steady state phase of the test. Put another way, the slope is the creep rate of the material.

Figure 19.14 is a schematic representation of the typical creep behaviour of a metal. Primary creep, stage I, is a period of decreasing creep rate ($\Delta\varepsilon/\Delta t$) and is transient creep; during this stage deformation takes place and the resistance to creep increases until stage II. Secondary creep, stage II, is a period of roughly constant creep rate, and is referred to as steady state creep. The strain rate eventually reaches a minimum and becomes near-constant. Tertiary creep, stage III, occurs when there is a reduction in cross-sectional area due to necking or effective reduction in area due to internal void formation. In tertiary creep, the strain rate increases exponentially with strain. Rather than failing suddenly with a fracture, the material

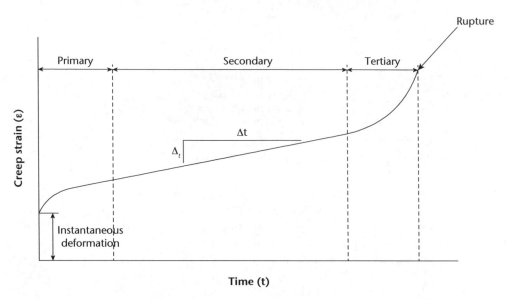

Figure 19.14 Typical creep curve of strain vs time at constant stress and constant, elevated temperature, for a metal showing the three stages of creep

permanently strains over a longer period of time until it finally fails. Creep does not happen upon sudden loading but the accumulation of creep strain over longer times causes failure of the material. Creep deformation is therefore a 'time-dependent' property of the material.

Although it is widely assumed that metals creep at elevated temperatures, a few do creep at room temperature – for example lead: $0.5 \cdot T_m$ where T_m is the melting temperature in degrees absolute (K).

19.6.2 Creep of concrete

Creep in concrete is defined as deformation under sustained load, and it takes place at normal ambient temperatures. Long-term pressure or stress on concrete can make it suffer a permanent deformation (strain). Concrete is usually used for structural, load-bearing applications in building construction and civil engineering and so it is usually placed under compressive stress. A load-bearing column will suffer a permanent reduction in height over time.

Creep of concrete has both good and undesirable aspects. On the one hand it is a good thing as it gives the concrete the ability to relieve stresses that may otherwise lead to cracking. On the other hand, creep is often responsible for excessive deflections under service loads, and this can lead to instability of arch or shell structures, creep buckling of long columns and loss of pre-stresses.

Creep is a diverse and rather complex process; to elaborate further would be beyond the scope of this text; however, sources for further reading are included at the end of the chapter.

19.6.3 Creep in polymers

Many polymers experience time-dependent deformation when subjected to constant stress levels – this is known as *viscoelastic creep*. This is due to the fact that applied forces cause bonds in polymer chains to rotate about an axis in order for the chain to unfold, and this takes time. If a load is slowly applied to a polymeric body, the chains in the polymer have time to unfold and stretch. Under these conditions, polymers undergo large deformations and the elastic component of the polymer's behaviour (i.e. the spring-like response)

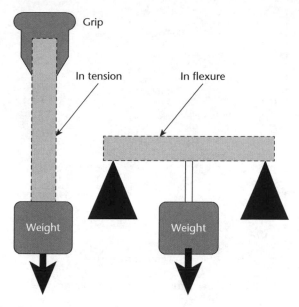

Figure 19.15   Experimental creep test set-up for testing at constant stress and temperature

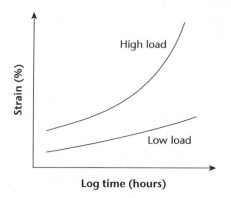

Figure 19.16   Typical creep curves for polymers

plays only a minor part in the deformation. However, if the load is applied very rapidly, the chains do not have sufficient time to react. Instead of rotating, the bonds within a chain only have time to stretch and bend. Large deformations will not occur and, consequently, the polymer appears more elastic and much less flexible. This dependence of deformation behaviour on the duration of the applied load, or strain rate (i.e. how fast deformations take place) is referred to as *viscoelasticity*.

Generally, test pieces are placed in a constant temperature environment and a load is applied to them (Figure 19.15). Accurate measurements of elongation or deflection (depending on the mode of loading) with sensitive devices/gauges are made over a period of time. Typical creep strain versus the logarithm of time elapsed curves are shown in Figure 19.16 (curves at constant load or stress).

The temperature at which a polymeric body is loaded is very important with respect to its mechanical behaviour. Low temperatures imply low internal energy within the molecules. Polymer chains which are less energetic are also more reluctant to move under the action of a force, and this makes it more difficult

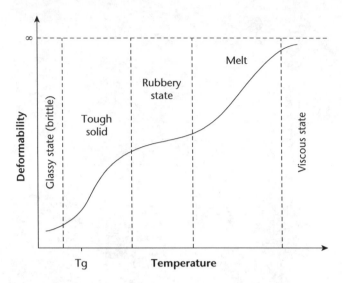

Figure 19.17 Effect of temperature on the deformability of a polymer showing the five different regions of viscoelastic behaviour

for them to unfold and their ability to undergo large deformations is restricted. In this state, polymers are more likely to resist the applied load and have a higher modulus of elasticity (stiffness). At higher temperatures, the energy level of chains facilitates movement, so unfolding is easier. In contrast to lower temperatures, a given amount of deformation requires a lower force and, conversely, a force of a given magnitude produces a larger deformation. Figure 19.17 shows the variation of the deformability of polymers over a wide range of temperatures.

With increasing temperature and above the glass transition temperature, T_g, solid polymers become softer and progress through the rubbery state to finally become a viscous melt capable of flow. The term 'rubbery' refers to the ability of the polymer to deform sluggishly, but the deformations recover when the load is removed. The term 'glassy' refers to the hardness, stiffness and brittleness of the polymer at low temperatures. The glass transition phenomenon is a very important topic in polymer science; its understanding is therefore crucial and is dealt with in more detail in Chapter 14.

19.7 Failure avoidance through good design

We have reviewed the principles of failure due to stress in this chapter, and we can draw certain conclusions from what we have learned. These conclusions can be used to promote good design, and by this we mean design that avoids catastrophic and premature failures. Good design involves using the correct geometries in the structures and components that we design, and it also means selecting and employing the correct materials.

19.7.1 Good design

Good design means avoiding features that raise local stresses, such as very sharp section changes, sharp angles, etc. It is much better to use properly radiused corners, gradual section changes and so on. Screw threads can also cause fatigue problems; it has been frequently observed that fatigue cracks are initiated at the root of the first thread when a screw-threaded component is subject to cyclical tensile stresses. Surface condition can also play a part in initiating fatigue. A smooth, polished surface is much better than a rough, machined surface because it contains no machine marks or 'stress raisers'.

In a bridge structure it is always preferable to use multi-strand cables rather than bars to carry intermittent or variable tensile loads. If a crack is initiated in a solid bar it may propagate and the bar may fail, whereas in a multi-strand cable failure of one strand will reduce the strength of the cable by a negligible amount. Chain-link suspension bridges are not constructed any longer for this reason, as the failure of one link would result in failure of the whole structure.

19.7.2 Good material selection

If the component or structure is to be exposed to corrosive conditions, then it may be necessary to specify single-phase material. In aluminium alloys, for example, an alloy in the solution and precipitation-hardened condition may appear to be attractive because of its high strength. However, precipitation-hardened materials are two-phase materials, and will be susceptible to corrosion (the precipitate phase will have a different electrode potential from the matrix). If the component is under stress while in service, it will suffer *stress corrosion*, i.e. it will corrode at a faster rate than it would do unstressed. The same considerations apply to components likely to be subject to fatigue – *corrosion fatigue* is the name given to failure under these conditions.

These lessons regarding fatigue will not often apply to buildings where constant compressive stresses are the norm, but they may well apply to, e.g. suspension bridge structures, where cyclical tensile loading conditions are encountered.

19.8 Critical thinking and concept review

1. Briefly explain the difference between ductile and brittle fracture.
2. Describe the difference between inter-granular and trans-granular fracture in metals.
3. What is the difference between 'dead loads' and 'live loads' in buildings, and what are their causes?
4. What is the difference between monotonic stress and dynamic stress?
5. In general, what effect does dynamic loading have on the mechanical response of materials in terms of: (a) strength and (b) strain to failure?
6. Explain how the Charpy and Izod tests measure the energy of fracture.
7. What is meant by an impulsive load? Give examples of impulsive loading situations.
8. Some early suspension bridges were built with chain-link catenaries. Nowadays we use multi-strand steel rope with up to 25,000 individual wires, as in the Humber Bridge. Why are rope catenaries better than chain ones?
9. Why was metal fatigue regarded as such a fearful phenomenon in the early to mid twentieth century?
10. Crack tip blunting can sometimes be used to arrest the growth of a fatigue crack.
    Explain how it works.
11. What is the difference between endurance limit and fatigue limit when describing fatigue failure of materials?
12. In what type of structures are fatigue problems likely to be encountered?
13. Explain why creep effects should be taken into account when building steel-reinforced concrete structures?
14. What is meant by creep failure?
15. Why does the metal lead creep at room temperature, whereas steel does not?

19.9 References and further reading

ASHBY, M.F. & JONES, D.R.H. (1980), *Engineering Materials: An Introduction to their Properties and Applications*, Pergamon Press, Oxford.

CALLISTER, W.D. (1994), *Materials Science and Engineering*, 3rd edition, John Wiley & Sons Inc., New York.

NEVILLE, A.M. (1981), *Properties of Concrete*, Longman Scientific & Technical, Harlow.

SMALLMAN, R.E. and BISHOP, R.J. (1995), *Metals and Materials*, Butterworth- Heinemann, Oxford.

VAN VLACK, L.H. (1982), *Materials for Engineering: Concepts and Applications*, Addison–Wesley, Reading, MA.

20

Failure 2: environmental degradation of materials

This chapter covers the corrosion of materials and their failure by degradation in service. It opens with the corrosion of metals, and an explanation of various ways in which metals can be protected against corrosion. The corrosion of ceramics is covered next, followed by the chemical attack of concrete, a very important subject in view of the huge amount of concrete used in construction. The mechanisms of degradation in polymers, timber and clay brickwork masonry are also dealt with.

Contents

20.1 Introduction

To some degree or another the environment in which a material is serving will have an impact upon it. There will be an interaction of some kind, and the result of the interaction will often be to impair the performance and usefulness of the material. The material's mechanical properties such as strength and ductility will be reduced, other physical properties may be adversely affected, or its appearance will suffer. If these effects are ignored the results can be serious, sometimes even catastrophic.

The response of the three main classes of material (metals, ceramics and polymers) differ markedly. Metals suffer actual material loss, either by a dissolution process known as *corrosion*, or by the formation of a surface coating of scale or film in the process of *oxidation*. Many ceramic materials are stable oxides that are resistant to corrosion processes, although glass can suffer corrosion by certain acids, and by highly alkaline water running off freshly set cement. Other ceramic and inorganic materials such as concrete and clay brick are porous, and can suffer from *chemical attack* by salts dissolved in water which is absorbed by their pores. Alternatively, absorbed water can be responsible for *freeze–thaw* damage/erosion when they are exposed to the weather for prolonged periods in service. Polymers are susceptible to the action of the sun's ultraviolet radiation, because they are made of carbon chain molecules. Carbon molecules are naturally susceptible to solar radiation, because life is carbon-based and the photosynthesis that occurs in green leaves is a photochemical process. Natural hydrocarbon materials such as timber are prone to insect and fungal attack. Fungal attack is a naturally evolved process for recycling dead timber. In this chapter we shall examine these various mechanisms of degradation.

20.2 Corrosion of metals

The subject of corrosion is large, and we shall not explore all of it here; we shall restrict ourselves to considering that part of it that is of interest and relevance to construction professionals. The most important metal used in construction is steel, which is susceptible to oxidation at high temperatures, and wet corrosion at room temperature. Copper, lead and aluminium are all more resistant to corrosion, and so we shall mainly consider here what is known as *wet corrosion*. Wet corrosion occurs when the metal is simultaneously in contact with both air (oxygen) and water, hence the name. While remembering that corrosion is a complex process, it will be instructive to consider a simple picture of what happens when rusting occurs when a drop of water is left on a bare steel surface. Figure 20.1 shows a drop of weak salt solution in water on a bare steel surface.

Oxygen can access the region around the edge of the droplet, but not that at the centre. The central region becomes anodic to the outside region, which is cathodic. Corrosive attack commences in the

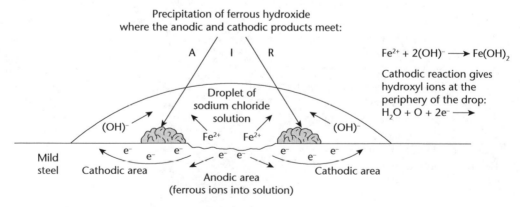

Figure 20.1 The mechanism of rusting of mild steel

anodic zone near the centre of the liquid, with iron going into solution as iron ions (Fe^{2+}); the lost electrons move away through the metallic iron from the anode to the cathodic zone. In the cathode zone, water and oxygen combine with the migrating electrons to form hydroxyl ions (OH^-). The two reactions are interdependent, and may be summarised thus:

Anodic reaction: $Fe \rightarrow Fe^{2+} + 2e^-$

Cathodic reaction: $H_2O + O$ (from air) $+ 2e^-$ (in steel, from anode) $\rightarrow 2OH^-$

Because we have effectively a current flow from anode to cathode, the process is said to be *electrochemical*. Note that for each iron atom that dissolves, liberating two electrons, two hydroxyl ions are formed. Although this water drop model is very simple, it points to certain principles found in wet corrosion processes. The principle that any region of metal with relatively freer access to oxygen becomes cathodic is a general one. It is called the *differential aeration effect*, and it is encountered many times in any study of wet corrosion in a wide variety of situations. Under these circumstances, the anodic zone is where corrosion is concentrated, where the metal loss occurs, and where the characteristic 'pitting' is observed after the process has started. It clearly shows that both water and air are necessary for corrosion to occur. Because the water is weakly saline (contains dissolved sodium chloride, NaCl) ferrous chloride will be found in the anode area, and sodium hydroxide will be found in the cathode area. Where these two species meet, ferrous hydroxide will be precipitated. The ferrous hydroxide in turn will turn into the familiar hydrated oxide commonly known as rust. This process is illustrated in Figure 20.1.

$FeCl_2 + 2NaOH \rightarrow Fe(OH)_2\downarrow + 2NaCl$

In Table 20.1, gold and the metals immediately below it are classed as noble metals because they are very difficult to corrode. Indeed, gold can only be corroded with great difficulty, and requires a mixture of boiling acids to do this. The metals at the foot of the table, magnesium, sodium and potassium are known as base metals because they are corroded with ease. This is shown by the explosive reaction of sodium when

Table. 20.1 Some normal electrode potentials: the e.m.f. series

Metal	Electrode reaction	Standard electrode potential (volts)
Gold	$Au^{3+} + 3e^- \rightarrow Au$	+ 1.420
Platinum	$Pt^{2+} + 2e^- \rightarrow Pt$	+ 1.200
Silver	$Ag^+ + e^- \rightarrow Ag$	+ 0.800
Copper	$Cu^{2+} + 2e^- \rightarrow Cu$	+ 0.340
Hydrogen	$2H^+ + 2e^- \rightarrow H$	0.000
Lead	$Pb^{2+} + 2e^- \rightarrow Pb$	−0.126
Tin	$Sn^{2+} + 2e^- \rightarrow Sn$	−0.136
Nickel	$Ni^{2+} + 2e^- \rightarrow Ni$	−0.250
Cadmium	$Cd^{2+} + 2e^- \rightarrow Cd$	−0.403
Iron	$Fe^{2+} + 2e^- \rightarrow Fe$	−0.440
Chromium	$Cr^{3+} + 3e^- \rightarrow Cr$	−0.744
Zinc	$Zn^{2+} + 2e^- \rightarrow Zn$	−0.763
Aluminium	$Al^{3+} + 3e^- \rightarrow Al$	−1.662
Magnesium	$Mg^{2+} + 2e^- \rightarrow Mg$	−2.363
Sodium	$Na^+ + e^- \rightarrow Na$	−2.714
Potassium	$K^+ + e^- \rightarrow K$	−2.924

contacted by water. These are the extremes, and the metals in the middle of the table corrode with intermediate levels of difficulty.

Why is this? To answer this question, we have to look at the thermodynamics of oxidation of these various metals. At room temperature, metallic gold is in a lower-energy state than its oxide. Therefore we find metallic gold in nature. If we consider another familiar metal such as iron, at room temperature, iron oxide is in a lower-energy state than metallic iron. Therefore we find iron oxide in nature.

Another way of looking at the series is to say that the metals in the top half of the table are cathodic to those in the bottom half. Another way of putting it is to say that those in the lower half are anodic to those in the upper half.

Because steel is such a widely used metal in construction and engineering, it must be protected against corrosion. There are various ways of doing this; some are quick and relatively cheap, others are more expensive, but they usually involve giving the steel a coating to protect it from corrosion. These coatings can be of plastic, paint or metal. Metallic coatings used to protect steel include cadmium, tin, zinc and chromium. Cadmium and chromium are applied by electroplating; the steel, with cleaned surfaces, is made an electrode in a plating bath containing salts of cadmium or chromium. A layer of cadmium or chromium is thereby deposited on the surface of the steel. In the case of zinc, the cleaned steel item is dipped into a bath of molten zinc (zinc melts at 400 °C), and as the zinc wets the steel surface a layer of the zinc freezes onto the steel, and the steel item is withdrawn from the bath. This process is known as *galvanising* and it is effective as well as being the cheapest type of corrosion protection applied to steel components.

The other metal applied by hot dipping is tin. Tinned steel strip is used in the food canning industry. However, the mechanism of protection is different between zinc and tin. Tin is above iron in the e.m.f. series shown in Table 20.1, whereas zinc is below iron. Tin is an example of cathodic protection and zinc is anodic protection. We shall examine these mechanisms next.

20.2.1 Anodic protection

Anodic protection involves coating the steel with a layer of metal which is anodic to steel, such as zinc. Zinc is anodic because it falls below iron in the table of electrode potentials. This means that in a corrosion situation, the zinc will corrode in preference to the steel, and this remains true even if a breach is made in the zinc layer. Therefore, application of the layer of zinc is not critical, and will still be effective even if the covering is not 100 per cent perfect. Furthermore, protection will still be effective even if some of the zinc coating is lost due to corrosion.

20.2.2 Cathodic protection

Cathodic protection involves coating the steel with a layer of metal that is cathodic to steel. Tin is cathodic because it lies above iron in the e.m.f. table. In this case the protection mechanism is different from zinc coating. Tin is a more noble metal than iron (steel), and so is much less likely to corrode. However, if there is a hole or breach in the tin layer then the underlying iron becomes liable to corrode. In practice, the situation is worse than this, because the corrosion rate is sensitive to the relative areas of anode and cathode. If the area of anode is very small compared to the area of cathode, then corrosion proceeds at a very rapid rate. For this reason it is very important that the layer of tin contains *no* holes, however small they may be.

20.2.3 Stainless steel

The secret of making stainless steel was discovered by accident in Sheffield, not long before the First World War. The addition of not less than 12–13 per cent of the metal chromium makes steel stainless, i.e. proof against wet corrosion under most circumstances. One of the commonest grades of stainless steel is the so-called 18/8; this contains 18 per cent chromium and 8 per cent nickel. Because more than 25 per cent of the alloy is of chromium and nickel, both costing thousands of pounds per tonne, stainless steel is at least ten times more expensive than ordinary carbon steel.

While wet corrosion is responsible for financial losses in the region of hundreds of millions of pounds per year, the cost of stainless steel precludes its use from all but a fraction of potential applications where corrosion resistance would be desirable. If we look at the electropotential series above, we can see that chromium actually comes below iron, i.e. it is more anodic than iron. This being so, how can chromium confer the well-known corrosion resistance?

20.3 Corrosion of ceramics and glass

Many ceramic materials are oxides of metals and non-metals and are chemically and physically highly stable and resistant to corrosive influences. Nevertheless, there are examples of susceptible materials, one of which is glass. Glass is an important material used in construction, based on silica (SiO_2). We have already discussed the structure and properties of glass, but here we will briefly mention how it can suffer from corrosion.

The two main agents of corrosion are hydrofluoric acid (HF) and cementitious water, i.e. water running off recently placed cement or concrete. HF is unique in its corrosive properties. High-silicon cast irons, stoneware and glass are generally resistant to corrosion by most acids, but all of them are readily attacked by fluoric acid. HF may not be stored in glass bottles, unlike other acids.

In building, freshly set cement and concrete is highly alkaline because of all the calcium hydroxide produced during the hydration processes involved in the setting and hardening processes. The $Ca(OH)_2$ is highly alkaline (pH of 12–13). One hazard to new glass windows is that in wet weather rainwater may run off fresh mortar or concrete and onto panes of glass in the building's windows. This highly alkaline solution can corrode window glass if it remains in contact for a prolonged period.

20.4 Degradation of Portland cement concrete

Concrete is often portrayed as a versatile material, which it is, and also as a very durable material. This latter attribute may or may not be true, depending upon how well the concrete has been placed, compacted and cured. We must remember that concrete is a material which is *made on site* (unless it is pre-cast concrete). It may be delivered in ready-mixed form, but it is not made until it has been placed, compacted and cured, and these operations are carried out on site.

It is well-known that concrete has excellent compressive strength, but has little or no tensile strength. The key to understanding the strength properties and long-term behaviour (including durability) of concrete is, first, to remember that it will always be *porous* to some degree. In fact, it is impossible to produce concrete without some porosity, and the chemicals that attack concrete enter it, usually dissolved in water, via this porosity. Second, we must also briefly consider the simple chemistry of Portland cement so that we can form a picture of how the various forms of attack occur. Armed with an understanding of these factors, we shall then be in a position to both protect our concrete and to make it more durable.

20.4.1 Chemistry of Portland cement

The use of Portland cement in the manufacture of concrete has been discussed in Chapter 12, but it will be useful to look in more detail at its chemistry here.

Let us refresh our knowledge of what happens when concrete sets. Concrete is made from Portland cement, aggregates and water, together with any additives that may be necessary to give us the properties we require. The active ingredient is the Portland cement, which reacts with the water in a hydration reaction, and acts as the binder to hold the concrete together. The cement particles are ground to a small particle size, typically around 20 μm in diameter. Mixed with water, this cement forms a slurry which wets the aggregate surfaces and binds the whole together. To understand this further, we shall briefly look at the chemistry of Portland cement.

Portland cement is made from *two* main raw materials – limestone ($CaCO_3$) and clay (SiO_2). These are crushed and ground, and water is added to form a slurry, and they are blended in the correct proportions. Dried slurry cake is then fed into the cement kiln, being progressively heated to temperatures in excess of 1,400 °C. At this temperature, the material is completely anhydrous (water-free), and the limestone has become quicklime (CaO).

$$CaCO_3 \quad \rightarrow \quad CaO \quad\quad\quad + CO_2\uparrow$$
$$\text{Limestone} \quad\quad \text{Quicklime} \quad\quad \text{Carbon dioxide gas}$$

The quicklime reacts with the silica clay (SiO_2) to produce the *four* compounds that make up ordinary Portland cement and its variations. The details of these four compounds are set out in Table 20.2.

By varying the relative proportions of these four compounds, cements with varying properties can be made. Alite, belite and aluminate are all white, whereas ferrite is grey. It is the presence of ferrite that gives ordinary Portland cement its characteristic grey colour. White cement is made by omitting the ferrite.

20.4.2 Hardening (setting) of Portland cement

When Portland cement is used in the making of concrete or mortar, it is mixed with aggregate materials and water. The water and cement powder form a slurry which ideally completely wets the surfaces of all the aggregate particles, both large and small, and which binds the whole together as it sets. It is the Portland cement which is the active ingredient. We need to examine this setting reaction in more detail if we are to understand the behaviour and properties of concrete, including its long-term durability.

When Portland cement is mixed with water, the four anhydrous compounds listed above all begin to react with the water in a *hydration* reaction. This reaction occurs slowly, and involves the evolution of heat. The alite and aluminate compounds react most quickly, and evolve most of the heat, while the belite and ferrite hydrate more slowly, evolving much less heat. Hydration means that the compounds, which are crystalline, combine with a certain amount of water to form new hydrated crystals exothermically.

What does all this mean in physical terms? How does the mixture go from being a slurry with no strength to a hardened solid? This is not explained in many of the textbooks. The hydration reaction sees hydrate crystals growing outwards from the surfaces of the cement particles. Some of the hydrate crystals are needle-like, others are plate-like. A picture of such growing crystals taken using a scanning electron microscope is shown in Figure 20.2.

Figure 20.3 shows the setting and hardening of Portland cement schematically.

Figure 20.4 shows a sequence of three scanning electron micrographs which record three stages during hydration, approximately corresponding to the schematic depicted in Figure 20.3. The first picture

Table 20.2 Names and formulae of the principal compounds in ordinary Portland cement

Full name of compound	Short name	Full formula	Short formula
Tricalcium silicate	Alite	$3CaO.SiO_2$	C_3S
Dicalcium silicate	Belite	$2CaO.SiO_2$	C_2S
Tricalcium aluminate	Aluminate	$3CaO.Al_2O_3$	C_3A
Tetracalcium aluminoferrite	Ferrite C_4AF	$4CaO.Al_2O_3.Fe_2O_3$	C_4AF

Figure 20.2 Scanning electron micrograph of plate-like crystals of monosulphate ($3CaO,Al_2O_3$. $CaSO_4.12H_2O$) and needle-like crystals of calcium silicate hydrate and ettringite ($3CaO$. $Al_2O_3.3CaSO_4.32H_2O$). The area shown is about 30 micrometres wide (30×10^{-6} m)

(After E. Breval *et al.*)

(a) shows unreacted grains, while the second picture (b) reveals the hydrate crystals beginning to grow after about five minutes from mixing with water. The final picture (c) shows the extensive needle-like structures that typify the final microstructure. The mutual interlocking of these provides the mechanical strength of the final concrete.

It is important to recognise that these needle-like crystals grow from their ends, and do not become fatter. Therefore, small spaces will always exist between the hydrate crystals. Modern Portland cement is ground to a finer particle size nowadays than before the Second World War. Nowadays, the particle size will typically be 10–20 microns in diameter (1 μm = 10^{-6} m). Therefore, we must recognise that concrete will *always be porous* to some degree. In the final section of this chapter we shall briefly look at the ways by which concrete can be made less porous.

20.4.3 Capillarity

Before we proceed further, it will be useful to examine the reasons why water will be drawn into a porous material. The mechanism is called *capillarity*. Figure 20.5 illustrates capillarity effects with fine-bore glass tubes in water and mercury. The finer the tube bore, the greater will be the height of the column of water drawn into the tube.

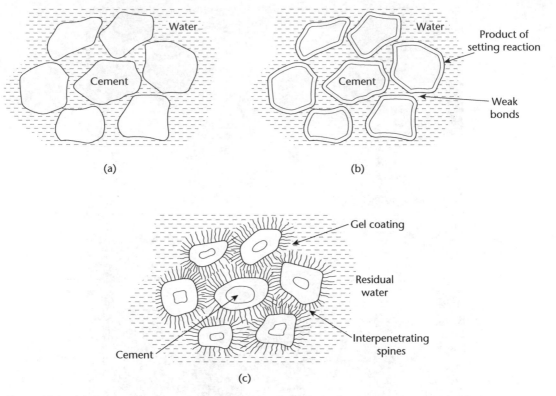

Figure 20.3 (a) Cement grains are mixed with water, (b) water begins to react outside of cement grains, (c) spiny hydrate crystals grow and begin to interlock, setting and hardening as further interlocking takes place.

(*Source:* Nelkon & Parker, 1979)

Figure 20.4 A sequence of micrographs recording the hydration of cement particles. These correspond to the schematic shown in Figure 20.3

(After E. Breval *et al.*)

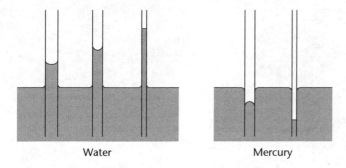

Figure 20.5   Showing capillarity in fine tubes of various diameters
(After Nelkon & Parker, 1979)

20.4.4 Role of porosity in the durability of concrete

One of the principal agents causing the deterioration and failure of concrete is chemical attack. How do the chemicals attack the concrete? Quite simply, chemicals enter the concrete dissolved in water, and the water enters the concrete via the porosity which exists in all concrete. However, it is not only the amount of porosity that is important. For attack to occur, the porosity must be:

- connected to the outer surface of the concrete;
- highly interconnected.

Only if the pores are highly interconnected will the water that soaks into the surface be able to penetrate more deeply into the mass of concrete. Figure 20.6 illustrates this. The two diagrams show two situations of similar percentages of porosity.

There are several types of attack, as follows:

- freeze–thaw damage
- leaching
- carbonation
- chloride attack
- sulphate attack
- alkali–silica attack

It cannot be stressed too strongly that it is the porosity in the concrete which renders it vulnerable to all these forms of deterioration. These will now be considered in turn.

20.4.5 Freeze–thaw damage

This is a physical form of damage, and it is caused by a combination of two facts:

- concrete is porous, and
- water undergoes a 9 per cent increase in volume when it freezes.

However, water has a maximum density (minimum volume at 4 °C). As it cools towards freezing it contracts to a minimum at 4 °C, before expanding slightly as it cools to 0 °C. At 0 °C it freezes and undergoes the 9 per cent volume expansion. This is shown in Figure 20.7, and is called the anomalous expansion of water.

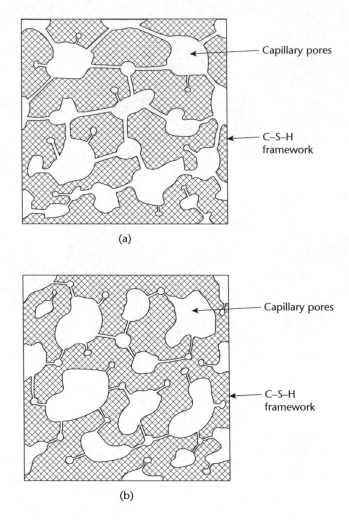

Figure 20.6 Schematic representation of materials of similar porosity, but (a) shows high permeability – capillary pores interconnected by large passages – and (b) shows low permeability – capillary pores segmented and only partly connected

(After Neville & Brooks, 1987)

It is the large expansion brought about by freezing that causes disruptive cracking of porous materials, including concrete.

20.4.6 Chemical attack in concrete

As outlined above, the main forms of chemical attack in concrete are leaching, carbonation, chloride attack and sulphate attack, and these will now be dealt with in turn.

Leaching

We have seen that in the setting reactions, the lime (CaO) in the compounds is converted to calcium hydroxide $(Ca(OH)_2)$. This substance is soluble in water to a small extent. So if water soaks into the

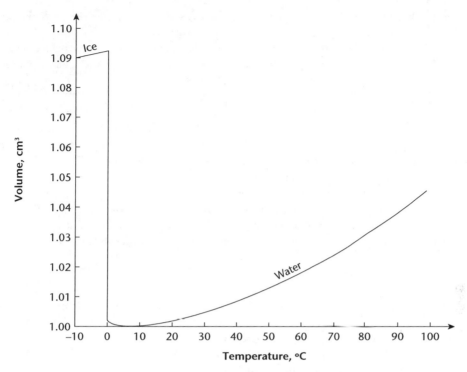

Figure 20.7 The anomalous expansion of water, variation of volume with temperature
(*Source:* Nelkon & Parker, 1979)

concrete via the porosity, then a small quantity of the calcium hydroxide can be dissolved. When the concrete dries out, the water plus the dissolved hydroxide will be drawn to the surface of the concrete, and the water will evaporate off. The next time it rains and the concrete gets wet, the hydroxide deposit will be washed away, and water will again soak into the concrete, dissolving yet more hydroxide, and so the process of dissolution (leaching) will continue. It can be appreciated that the long-term effect of this repeated process will be to increase the porosity of the concrete, thus weakening it.

Carbonation

Carbonation is only a problem for steel-reinforced concrete, not plain concrete. The atmosphere contains a certain, small amount of carbon dioxide (CO_2), and this carbon dioxide can cause a chemical reaction to occur in the surface layers of moist, porous concrete. The CO_2 reacts with the calcium hydroxide ($Ca(OH)_2$) in the cement and forms calcium carbonate – hence the term carbonation. This reaction is shown below:

$$Ca(OH)_2 \quad + \quad CO_2 \quad \rightarrow \quad CaCO_3 \quad + \quad H_2O$$
Calcium hydroxide $\quad$ Carbon dioxide $\quad$ Calcium carbonate $\quad$ Water

Calcium carbonate is limestone or chalk, so why is this a problem? There is no problem if we have plain concrete, but it is a major problem if we have steel-reinforced concrete. The reason for this is that calcium hydroxide is *very alkaline* (pH = approx. 13), and calcium carbonate is only weakly alkaline (pH = 8–9). The high alkalinity of fresh, wet cement containing freshly formed calcium hydroxide passivates the surface of all steel reinforcement, and protects it against corrosion. This is true even if rusty rebars are used.

However, if carbonation occurs with conversion of calcium hydroxide to calcium carbonate, this protection is lost, and the rebars become susceptible to corrosion. Since carbonation starts at the surface of the concrete, and slowly works its way inwards, we have problems if the rebars are too close to the concrete surface, i.e. if they have inadequate cover. The problem becomes even more acute, of course, if the concrete is very porous.

Chloride attack

Like carbonation, this is only a problem for steel-reinforced concrete. The problem is caused by dissolved chlorides that are present in some groundwaters, in seawater and in de-icing salts put down on road surfaces, to name three common sources. Chloride ions (Cl^-) are acidic, and they will enter porous concrete; if they do, they will neutralise and destroy the passivating effect of the highly alkaline calcium hydroxide surrounding any steel reinforcing bars. The presence of chloride ions will also increase the electrical conductivity of the moisture, and so will accelerate corrosion of any reinforcing steel. So the presence of chloride ions leads to corrosion of reinforcing steel, an effect that proceeds even faster if the steel is pre-stressed. In fact, the rate of corrosion will depend partly on the state of stress of any reinforcement.

Sulphate attack

Sulphate attack is a very serious matter, whether or not the concrete is steel-reinforced. Sulphates can occur in many groundwaters and soils, but the concentration of these sulphates can vary widely. A good site investigation should provide the answers to the following two questions:

1   Are sulphates present in the soils and/or groundwater?
2   If so, what is the concentration of these sulphates?

What exactly is sulphate attack, and why is it so serious? The reason is that most sulphate solutions react with the calcium hydroxide and calcium aluminate (C_3A) of hydrated cement to form calcium sulphate and calcium sulpho-aluminate compounds. These compounds are very similar to the naturally occurring mineral ettringite, and their formation involves a *volume expansion*. In other words, the ettringite compounds occupy a larger volume than the material that went to make them. Their formation within the hardened cement therefore has a disrupting effect. By expanding they put stress on the surrounding materials, causing cracking and fracture. Ultimately, sulphate attack will completely break up a mass of concrete that it attacks, whether or not the concrete contains reinforcement. So it is perhaps the most serious form of chemical attack.

Magnesium sulphate is the most vigorous type of sulphate to cause attack, and obviously the permeability of the concrete and the presence of cracks will also affect the severity of any attack. Ordinary Portland cement (OPC) is very susceptible to sulphate attack. If concrete needs to be placed in ground containing sulphates, then sulphate resisting Portland cement can be used. This is a form of Portland cement containing less than 5 per cent aluminate. Since it is the aluminates that react with the sulphates, by keeping the cement largely free of aluminate we can obviate the problem. If the aluminates are not present, no attack can occur. Examples of what can happen if concrete is not protected against sulphates include heaving of concrete floors in houses where no impermeable (water-proof) membrane was placed on top of the sand blinding before placing the concrete floor. Undersailing of brickwork below the damp-proof course has also occurred because of expansion of the concrete.

Alkali–silica reaction

This is sometimes called 'concrete cancer' and is caused by a reaction between alkalis in the cement paste, and certain forms of active silica in the aggregate. The reactive forms of silica are:

opal – amorphous structure
chalcedony – cryptocrystalline fibrous structure
tridymite – crystalline structure

Figure 20.8 shows a particle of opal, a reactive form of silica, which has been transformed into a gel by reacting with the alkaline constituents of the surrounding cement paste.

PROTECTION OF CONCRETE AGAINST SULPHATE ATTACK

If concrete is to be placed in ground where sulphates are present, then there are a number of measures that can be taken to protect it from chemical attack. The measures taken will depend upon the concentration of sulphates, since this will determine the potential severity of any attack. The porosity of the concrete will also determine the severity of attack. The protective measures are set out in BS 8110 Part 1:1985, and are also discussed by Neville and Brooks (1987).

There are a number of factors that should be borne in mind. Magnesium sulphate is the most damaging of all sulphates, and the soil survey should give an indication of their presence and concentration. Improved performance from the concrete can be obtained by increasing the cement content of the mix by up to about 10 per cent. This makes for a stronger concrete; the extra strength helps resist the disrupting effect of any sulphates. For low concentrations of sulphate, this may be sufficient.

Since it is the tricalcium aluminate (C_3A) constituent that is attacked by sulphates, the vulnerability to attack can be reduced by the use of cement low in C_3A, i.e. by the use of sulphate-resisting Portland cement. Improved resistance can also be by the addition of pulverised fuel ash (Portland–pozzolan type cement) and ground–graded blast furnace slag (Portland blast furnace cement).

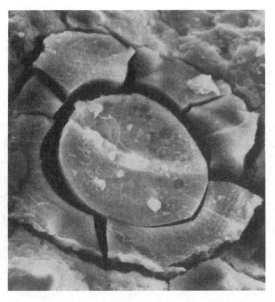

Figure 20.8 Stones (aggregates) are added to cement in the production of concrete, but certain minority minerals can be detrimental. This photo-micrograph shows a particle of opal (silica) which has been transformed into a gel by reaction with the alkaline components in the surrounding cement. The gel expands due to the absorption of water, and this causes cracking of the surrounding concrete

(After E. Breval *et al.*)

In situations where the highest concentrations of sulphates occur, in addition to the measures described above, the concrete may be protected by the application of non-permeable membranes. These are applied to the surface of the placed and cured concrete after the formwork is struck and before back-filling with soil.

20.6.7 Design of modern high-strength concrete

It was first appreciated about 40 years ago that concrete's low strength was due to the presence of a large population of voids and cracks in the average sample of concrete made with 'normal' mixes. The application of fracture mechanics concepts showed that if concrete could be made with a much smaller void and crack size it could in principle be much stronger. It was at this time that detailed studies were first made of the hydration reactions involved in the setting of concrete, using electron microscopy. A research and development programme was therefore initiated, with the result that much stronger concrete became available.

The two photographs in Figure 20.9 (first published in the early 1980s) illustrate the successful achievement of this reduced porosity, high-strength concrete. The materials scientists who developed it called it micro-defect free (MDF) concrete.

Figure 20.9(a) shows a cement spring which was made by winding the freshly made cement around a cardboard former prior to setting. In Figure 20.9(b) this spring has been placed in tension and shows an extension of about 30 per cent.

Normal structural concrete with compressive strengths in the range 30–50 N/mm² used to be the norm. Indeed, much of the concrete used today still has strength values in this range. However, by the end of the 1980s structural concrete with strengths of 120 N/mm² were first used in the construction of tall buildings in the United States. Indeed, one building in Chicago contains concrete with a strength of 160 N/mm².

High-strength concrete was developed for strength reasons. However, the same factors that cause low strength – the cracks and porosity – also give rise to poor durability. The high-strength concrete with reduced pore size and reduced pore population should also exhibit improved durability. Since the first use of this high-strength concrete in a structural application was less than 20 years ago, the evidence of improved durability is still accumulating.

Figure 20.9 Concrete and cement perform well in compression, but in situations involving tension or shear the results are poor, or at best mediocre. The traditional way of compensating for this inadequacy has been to reinforce the material with a second material possessing good strength in tension, usually steel bars. The latest method is to use fine aggregates, reduced water, super-plasticizers and the addition of micro-silica (very fine particles of silica) in an effort to fill up potential cracks and pores. The pictures show how successful this method has been

(After D. Birchall, 1982)

The mix design for high-strength concretes involves the following measures:

- Use of small-diameter aggregates (maximum particle size 10 mm);
- use of less water (water/cement ratio 0.26–0.3:1 maximum);
- addition of super-plasticisers to compensate for use of less water;
- addition of fine pulverised fuel ash;
- addition of microsilica or silica fume (fine silica particles of aerosol dimensions).

Microsilica is removed from the fumes produced by, for example, electric arc steel furnaces. The microsilica is mixed with water and supplied in slurry form for addition to concrete mixes. The silica fills the porosity in the concrete, reducing both the percentage porosity and the average pore size. Figure 20.10 illustrates the difference between the porosity in normal (average pore size *c*.1 mm) and high-strength concrete (average pore size 1 μm).

20.5 Degradation of polymers

Polymers are water-proof and do not corrode as metals do. Furthermore, they are not porous, and so are not susceptible to chemical attack as concrete is. However, they are subject to degradation from ultraviolet (UV) radiation. Why is this? Life on Earth is carbon-based, and carbon compounds are susceptible to the effects of incoming solar radiation, in particular the short-wave UV part of the spectrum. The photochemical processes involved in photosynthesis necessary for life are driven by this UV radiation. So if man uses carbon-based materials they, too, will be susceptible

The reason that UV is so damaging and visible light is not is due to the fact that UV rays contain much more energy than visible light rays. The short wavelengths are particularly energetic and penetrating. Energy is like matter, and just as matter is divided into individual atoms, so energy is divided into individual packets called *quanta*; this was discovered by Max Planck in 1900. In the same way, light energy is received as a series of small energy pulses called *photons*. These photons have

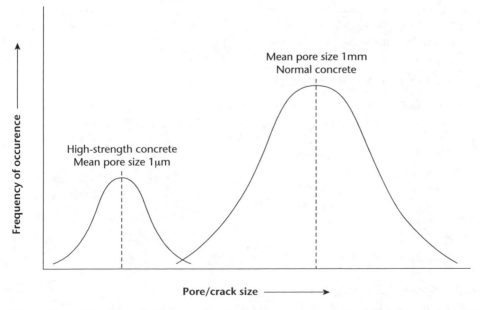

Figure 20.10 The difference in porosity distributions in normal and high-strength concrete

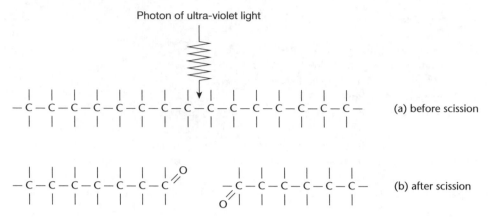

Figure 20.11   Photo-oxidation of a long-chain polymer

(After Dinwodie, 1996)

different wavelengths depending upon their colour. They also have different energies, and photons of short-wave UV are particularly energetic. In fact, photons of UV have sufficient energy to break C–H bonds and even C–C bonds. The cutting of the bonds is called *scission*. The exact physics and chemistry of this process is complex, but basically, if an incoming photon strikes the polymer molecule, it has enough energy to raise an electron in the molecule into an excited state. If the energy state of the excited electron is high enough, it can transfer energy into a bond dissociation reaction (bond breakage). The whole process of photon absorption and bond breakage is called *photodecomposition*. Similarly, carbonyl groups (C=O) are also efficient absorbers of UV. If the free radicals produced subsequently react with oxygen, the degradation process is known as *photo-oxidation*; this process is illustrated schematically in Figure 20.11.

The effects of photo-oxidation include progressive embrittlement of the polymer, making it increasingly susceptible to impact and frost damage. Plastics can be protected against photo-oxidation by adding UV absorbers to the mix before polymerisation. The two commonly used are carbon black (black) and titanium dioxide (TiO_2) (brilliant white). Without UV absorbers, the life of materials such as uPVC rainwater goods and window frames would be limited to a few months.

20.6 The degradation and decay of timber

With the passage of time and the influence of the environment, all materials exhibit some loss of properties and performance, and timber is no exception. Some species offer more resistance to degradation (often hardwoods) than others. The agencies of degradation can be grouped under five headings:

1 biological
2 chemical
3 photochemical
4 thermal
5 mechanical.

20.6.1 Biological degradation

This may be divided into two types: fungal attack and insect attack.

Fungal attack

Most forms of decay and sap-stain in timber are caused by fungi that feed either on the cell wall tissue or cell contents of woody plants. Fungi are simple plants that cannot synthesise chlorophyll and so they must obtain nutrients by taking organic material from other sources. They belong to a class of organisms called *saprophytes*. They live on dead tissue and they are really nature's way of recycling dead wood.

Saprophytes need oxygen and a supply of food and water, and a minimum moisture content of 20 per cent is necessary for their germination and growth in timber. The optimum temperature for growth varies for different species of fungus, but is usually in the range from 20–30 °C. Little growth takes place below 5 °C and fungi are usually killed by prolonged heating to 40 °C. Certain timbers, particularly the heartwoods of some hardwoods, are resistant to fungal attack because they contain minor constituents which are toxic to fungi. The life cycle of a fungus is illustrated in Figure 20.12.

Insect attack

Within the UK insect attack is limited to a small number of species of insect and tends to be less serious than fungal attack. Outside of the temperate climate regions, and in the tropics, *termites* or *white ants* cause more damage than all other forms of insect attack put together. The main damage in the UK comes from beetles which lay their eggs on timber, and then during the larval stage bore through the timber, eating the organic material – mainly the sapwood – causing loss of mechanical strength. Strength is lost because the cellulose vessel cell walls are cut through, and this can be very serious if the wood becomes badly infested. The life cycle of a typical wood-boring beetle is illustrated in Figure 20.13.

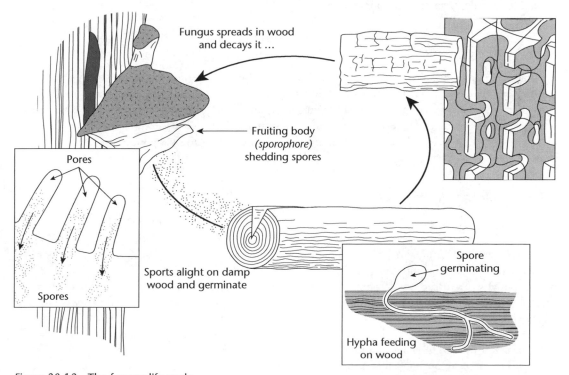

Figure 20.12 The fungus life cycle

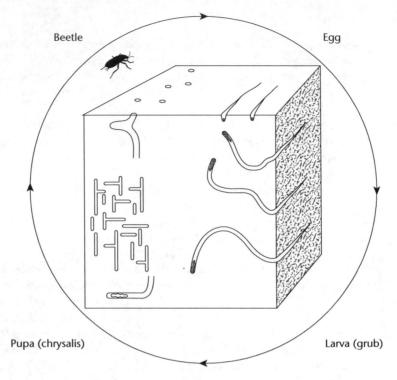

Beetle

Egg

Pupa (chrysalis)

Larva (grub)

Figure 20.13   Life cycle of wood-boring beetle

20.6.2 Chemical attack

All timber is porous because of the fact that it is permeated with a set of vessel cells. However, in general the higher the density and the impermeability of the wood, the greater is its resistance to chemical degradation. The resistance of impermeable timbers to acids is greater than that of most common metals; iron begins to corrode at pH levels below pH 5, whereas attack on wood commences below pH 2, and even at lower values proceeds at a slow rate. In alkaline conditions, wood has good resistance at values up to pH 11.

20.6.3 Photochemical degradation

We are all familiar with the fact that exposure to sunlight causes the coloration of the heartwood of most timbers to lighten. Oak, for example, turns a silvery grey colour, while a few such as Rhodesian teak and Douglas fir will darken. Indoors such changes take a long time, whereas outdoors the changes are more rapid, and effects are noticeable after just a few months. These colour changes are but an initial stage in the whole process whereby light, wind and rain result in a complex degradation mechanism in which the lignin degrades and the cellulose vessels shorten. The breakdown products are washed away by rain, imparting a silvery grey colour to the timber surface; which consists of a thin layer of loosely matted fibres between a large number of small slits. This degradation can be appreciably slowed down by careful application of various finishes to the wood. Obviously, degraded surface layers need to be removed before applying any surface coatings, otherwise they will not adhere to the surface.

20.6.4 Thermal degradation

Prolonged exposure to elevated temperatures (other than in a fire) results in a reduction in strength and a very marked reduction in toughness or impact resistance of timber. There is some uncertainty about what minimum temperature will degrade timber, but it would appear that temperatures as low as 60 °C can induce degradation over many years of exposure. The rate of degradation will rise markedly with increase in temperature and time of exposure; hardwoods appear to be more susceptible to thermal degradation than softwoods. Thermally degraded timber develops a brown colour and breaks easily with a brittle failure.

20.6.5 Mechanical degradation

The most common type of mechanical degradation that occurs is that which happens in timber subjected to continuous loading for long periods of time. For example, after the lapse of 50 years, the load which a piece of timber can withstand is less than half the stress it can carry at the outset of loading. Similarly, there is a reduction in stiffness (Young's modulus of elasticity, E) with time that shows itself as an increase in deflection in timber with time under constant stress. This is a form of creep in timber.

A second and less common form of mechanical degradation is the induction of compression failure in the cell walls of timber. Timber that is over-stressed in compression in the longitudinal direction forms kink bands and compression creases. Such defects can reduce the tensile strength of the wood by 10–15 per cent, but the loss of toughness under impact conditions can be as high as 50 per cent. In the days when the Royal Navy used wooden ships, many battle casualties and injuries were inflicted by shards of timber flying from the point of impact of enemy shot on the ship's hull. Mistreatment of wooden ladders and scaffold boards on building sites has led to injury and loss of life on numerous occasions.

20.6.6 Prevention of degradation

To combat insect and fungal attack, there are various chemical treatments available; these involve impregnating the timber with substances that are toxic to insect larvae and to fungus spores. While these treatments can be effective, it should be borne in mind that substances that are toxic to these forms of life may also be toxic to humans. There have been many documented cases of people renovating old properties and having the timbers treated with chemical preservatives, who have subsequently developed serious neurological problems after living in their properties. One preservative, Lindane, was banned by the United States Food and Drug Administration (FDA) because it was deemed to be toxic to humans. The use of chemical preservatives must therefore be viewed with caution. The best form of prevention is good design; if the timber does not become damp, it cannot decay. Fungus spores require a minimum moisture content of about 20 per cent to germinate.

20.7 Failure of clay brickwork masonry

Clay brickwork has been employed for building in England for centuries, and has proved to be a remarkably durable material with many centuries-old brick buildings surviving to the present day (Clifton-Taylor, 1987; Brunskill, 1990). Bricks are similar to concrete in that they are porous, and it is this porosity that renders them liable to degradation and failure because of moisture ingress by capillarity. Again, as with concrete, it is the amount of porosity and its connectedness that determines their durability.

Clay consists of silicate platelets, as illustrated in Figure 20.14. This shows crystals of kaolinite ($Al_2Si_2O_5(OH)_4$). These platelets easily stack on top of each other like a pack of cards. Water molecules, H_2O, can be adsorbed onto the surfaces or absorbed between adjacent crystals. This moisture plasticises the

Figure 20.14 Clay crystals with plate-like structure at very high magnification (×33,000)

(After W.H. East, *J. Amer. Ceramic Soc.*)

mass of platelets so that they easily slide over each other, and so a mass of wet clay becomes very plastic, i.e. it is easily shaped and moulded.

After clay bricks have been moulded, they are dried and then fired. Drying is carried out to remove excess water; if the moulded bricks were fired immediately, this water would turn into steam and cause bloating, resulting in mis-shaped bricks. The sequence of diagrams in Figure 20.15(a–c) shows what happens in the drying and firing stages. The clay mass shrinks as the water between the particles is removed (a–b). Further drying removes the pore water and introduces porosity (b–c). The final diagram shows the fired brick structure, and it can be seen that the mass of the brick is permeated by a network of air passages between the platelets. It is this porosity that renders the brick liable to deterioration. Deterioration can most commonly be by freeze–thaw action, but other reactions can also occur due to the ingress of moisture.

The durability of bricks depends upon the porosity of the brick concerned. Bricks are made by a number of processes, including moulding, extrusion and wire-cutting, and compaction in moulds.

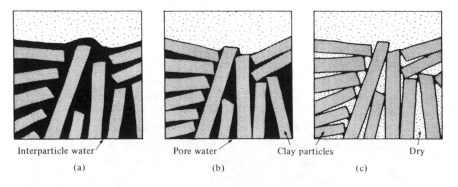

Figure 20.15 Schematic of the removal of water during the drying and firing processes

(After F.H. Norton, 1952)

Unpublished work by Sturges (2012) shows that the porosity of bricks and their ability to absorb water varies very widely. Water absorption tests were carried out on ten samples of bricks of four different types:

1   moulded facing bricks with frogs,
2   extruded and wire-cut facing bricks,
3   paviours, and
4   engineering class B bricks.

All the bricks were initially dried in a kiln set at 105 °C for 24 hours, and then weighed. In each case the bricks were then placed in a water tray with just the stretcher face in contact with the water. They were then re-weighed at regular intervals over the following 24 hours to measure the water absorption. The results showed very clear trends, as illustrated in Figure 20.16. It can be seen that the class B engineering bricks absorbed very little water, about 0.5 per cent by weight after 24 hours. The moulded and frogged facing bricks, by contrast, had absorbed about 22 per cent water by weight 24 hours later. Furthermore, the initial rate of absorption in the facing bricks was very high indeed, with almost 10 per cent absorption in the first 30 minutes, on average. In fact, one or two bricks exceeded 10 per cent inside 30 minutes.

These differences in porosity ensure that there will be extreme differences in the durability of these bricks in service. First, it is essential that the facing bricks are only used in vertical walling. If such bricks are used to cap a wall with their stretcher faces uppermost, the surfaces will probably be lost during the first winter's frost. Indeed, such high porosity means that these bricks should not be used for vertical walling in locations subject to driving rain. Periods of driving rain followed by periods of frost will cause progressive damage to even vertical walls.

In such circumstances, special quality bricks can be used. These are more expensive, made with reduced porosity and with controls on soluble salt contents, etc. and are much less susceptible to weathering and frost damage. Ultimately, of course, engineering bricks may be employed, and these will obviate any danger of frost damage.

For horizontal surfaces, bricks are often used for aesthetic reasons and for ease of laying, and paviours are designed for this horizontal application. The data in Figure 20.16 show that they absorb only 2–3 per cent water at most, and this fact protects them against frost damage.

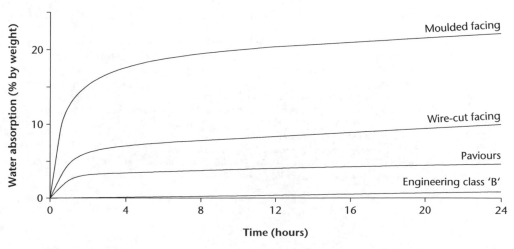

Figure 20.16   Water absorption vs time for moulded facing bricks, extruded and wire-cut bricks, paviours and engineering class B bricks

Finally, engineering bricks possess much higher strengths than other types, and so for situations where strength and durability are necessary, engineering bricks should always be used.

20.8 Critical thinking and concept review

1. What is meant by *wet corrosion*?
2. In wet corrosion, which region of the corroding metal becomes anodic and which region becomes cathodic?
3. The standard electrode potentials for gold and potassium are +1.42 and −2.924 volts, respectively. By how many volts is the potential for gold above that for potassium?
4. What is meant by a 'noble' metal and what by a 'base' metal'?
5. What gives stainless steel its corrosion resistance?
6. Name *two* fluids that will corrode normal plate glass.
7. Which property of concrete makes it liable to chemical attack and freeze–thaw damage?
8. What is meant by *capillarity* and why does it occur?
9. Steel reinforcing bars may be rusty when placed in fresh concrete. Why is this not a cause for concern?
10. Name the two types of chemical attack which cause problems with steel- reinforced concrete.
11. Name the *four* compounds found in Portland cement.
12. What causes sulphate attack in concrete?
13. Why is sulphate attack such a serious matter?
14. What is the main agent of degradation in polymeric materials exposed to the weather?
15. Name the two biological causes of decay in timber.
16. Describe the process of photo–oxidation of polymeric materials.
17. How can plastics be made more resistant to photo–oxidation?
18. Why is wet clay so plastic?
19. Explain why fired clay is porous.
20. In what ways are the degradation processes of concrete and clay bricks similar?

20.9 References and further reading

ASHBY, M.F. and JONES, D.R.H. (1986), *Engineering Materials 2*, Pergamon Press, Oxford.

BRUNSKILL, R.W. (1990), *Brick Building in Britain*, Victor Gollancz Ltd, London.

BYE, G.C. (1983), *Portland Cement: Composition, Production and Properties*, The Institute of Ceramics, Pergamon Press, Oxford.

CLIFTON-TAYLOR, A. (1987), *The Pattern of English Building*, 4th edition, Faber and Faber, London.

COTTERILL, R. (1989), *The Cambridge Guide to the Material World*, Cambridge University Press, Cambridge.

EAST, W.H. (1950), 'Fundamental Study of Clay: X, Water Films in Monodisperse Kaolinite Fractions', *Journal of the American Ceramic Society*, 33 (7) 211–218.

ILLSTON, J.M. (ed.). (1994), *Construction Materials*, E. & F.N. Spon, London.

JACKSON, N. and DHIR, R. (eds.) (1988), *Civil Engineering Materials*, MacMillan, Basingstoke.

NELKON, M. & PARKER, P. (1979), *Advanced Level Physics*, Heinemann, London.

NEVILLE, A.M. (1991), *Properties of Concrete*, Longman, Harlow.

NEVILLE, A.M. and BROOKS, J.J. (1987), *Concrete Technology*, Longman, Harlow.

NORTON, F.H. (1952), Elements of Ceramics, Addison-Wesley, Reading, Massachusetts.

STURGES, J.L. (2012), Unpublished work, Leeds Metropolitan University.

21

Failure 3: effects of fire on building materials

This chapter examines in some detail the effects of fire on building materials. Of the principal hazards to buildings that exist, fire presents the greatest threat to life and to buildings in the UK. It is an extreme cause of failure, and it occurs sufficiently often to justify a chapter to itself. The principles of burning or combustion are briefly reviewed to give an idea of the conditions faced by building materials in a fire. The thermal properties of materials (covered in Chapter 6) are briefly reviewed, and their application to fire situations is set out. Then the various classes of building materials and their response to fire are considered. The concept of fire endurance is put forward, and a simple calculation method that can be used to estimate the fire endurance of timber and concrete is explained with an example.

Contents

21.1 Introduction

A very wide range of materials is used in building construction, including:

- metals – primarily steel, but also non-ferrous metals like aluminium alloys, copper and copper alloys and lead;
- fired clay products – primarily bricks and tiles;
- cement and concrete;
- timber – both as solid timber and as timber products like plywood, chipboard, fibreboard, glulam, engineered wood, etc.
- natural stone;
- gypsum plaster and gypsum products;
- glass – both flat glass as windows and also glass fibre-reinforced composites, as well as glass wool insulating materials;
- polymers and polymeric materials;
- bituminous materials.

This is a wider range of materials than that used in any other industry, and includes both naturally occurring materials like timber, stone and aggregates, as well as manufactured or man-made materials such as polymers, glass and metals. As we have seen, some of these materials are porous and permeable to water, whereas others are non-porous and impervious to water.

In a fire they will be subjected to heating. The heating may be mild, moderate or very intense, depending on the location of the materials with respect to the fire. Perusal of the above list shows that some of the materials represent fuel, and will burn in a fire (e.g. timber and plastics). Some of them will melt, others will soften and weaken and lose their strength, and others will be prone to cracking and shattering because of thermal shock. Just a few will do none of these things, but will retain most of their strength.

To understand what happens in a fire, we need to know a little about burning or the process of combustion, including the temperatures that might be encountered in a typical building fire; this will be briefly considered next.

21.2 Burning, the process of combustion

What is fire? The answer is that fire is *burning gas*. We see flames because of the millions of tiny (carbon) particles inside them which glow because they are hot. In fact, only gaseous fuel will burn. We are all familiar with solid fuels such as coal, coke and timber, and gaseous fuel such as propane, butane and especially North Sea gas. If only gas will burn, how do solid and liquid fuels burn?

The answer is that liquids *evaporate*, i.e. the liquid becomes a vapour, and so it can intimately mix with the oxygen in the air and burn. Evaporation is aided by heating. Solids undergo *pyrolysis*, a process of chemical breakdown (caused by heating), and one result of pyrolysis is that flammable gases are liberated and given off by the solid. These gases mix with the oxygen in the air and burn.

The process of burning is given a special name by chemists – it is called *combustion*. Combustion involves the reaction of fuel (the material that burns) with oxygen to form reaction products, usually carbon dioxide and water vapour, and *heat*. Because heat is given off by combustion it is called an *exothermic* reaction. The burning or combustion of North Sea gas, with which we are all familiar, may be represented by the chemical equation:

$$CH_4 + 2O_2 = CO_2 + 2H_2O + x.kcals$$

$$\text{METHANE} \quad \text{OXYGEN} \quad \text{CARBON DIOXIDE} \quad \text{WATER VAPOUR} \quad \text{HEAT}$$

In this equation, the complete combustion of one molecule of methane with two molecules of oxygen (reactants) gives one molecule of carbon dioxide, two molecules of water vapour and a certain amount of heat (reaction products).

Since we require fuel, oxygen and heat to sustain combustion, once burning has started, i.e. *ignition* has occurred, the burning will continue without the need to add any further heat. In effect, it supplies its own heat. The other class of chemical reactions is called *endothermic*; here, heat has to be supplied to sustain the reaction. If the heat is removed, the reaction stops.

Buildings can be partly made of combustible materials, and they will also usually contain combustible materials. We must remember that human beings are made from hydrocarbons, and therefore represent fuel, along with their possessions such as clothing, furniture, books, foodstuffs, etc. Since oxygen is necessary for combustion to occur and the Earth's atmosphere contains 21 per cent oxygen by volume, fire is an ever-present hazard.

21.2.1 Temperatures in building fires

What temperature levels are obtained in building fires? We need to remember that fact the fire is contained by the building, and this containment has an influence on the fire. If we burn two identical piles of fuel, one inside a building and the other in the open air, we shall observe two different fires. The fire inside the building will burn faster and reach higher temperatures than the same fire out of doors. This is caused by an effect called *thermal feedback*, which we shall examine later. However, feedback means that fires inside buildings can attain temperatures of 600, 700 or even 800 °C. In very major fires like the one at the Bunsfield fuel depot in December 2005, temperatures can reach 1,000 °C and above. This gives us an indication of the temperatures to which building materials may be exposed in a fire.

21.3 Importance of the thermal properties of materials

We have seen that heat will affect materials in different ways (melting, softening, burning, cracking, etc.), and we now have a feel for the temperatures to which they might be subjected. We have also examined the thermal properties of materials in Chapter 6.

In terms of the response of various materials to fire, we have a potentially complex picture here. However, we must remember that it is the heat that causes the effects – burning, melting, softening, cracking, etc. outlined above. To bring about these effects, *the heat has to soak into the various materials*. Some of the materials, particularly metals, are highly conductive to heat, whereas other materials, such as timber, are highly effective insulators. *Therefore it follows that how materials respond to heat and fire will depend on how easily heat can soak into (permeate) them.* It is very important that we appreciate this fact if we are to make sense of how materials behave in fire. Table 6.1 in Chapter 6 lists the thermal properties of a range of the materials commonly used in building construction. Because of this, some materials that might be expected to perform poorly in fire, such as timber, actually perform well, while others like steel – which is very strong and has a high melting temperature (1,450 °C) – perform much less well.

The other point is that if we understand the thermal behaviour of materials we can use them more intelligently. In fact, we can use certain materials to protect other materials against the effects of fire. The three quantities of thermal conductivity, density and specific heat capacity determine the response of materials to heat, and the important derived property that governs how easily heat soaks or diffuses into a material is its *thermal diffusivity*. We have already looked at this property in Chapter 6, on the thermal properties of materials, but it is such an important concept with regard to fire performance that we shall look at it again in the context of building fires.

21.3.1 Thermal diffusivity

The property of a material which determines how easily and quickly heat soaks into it is its *thermal diffusivity*. This property depends in part on the thermal conductivity (k) of the material, as well as on its density (ρ) and specific heat capacity (C_p). The thermal diffusivity is calculated as follows:

$$\text{Thermal diffusivity} = \frac{k}{\rho \times C_p}$$

Values of conductivity, density and specific heat capacity for a range of construction materials are given in Table 6.1. It can be seen that the conductivity is the dividend in the above expression. In fact, the thermal conductivity is the main factor that determines diffusivity. So, in general terms, those materials which are good conductors of heat will be those that heat through most quickly and easily. Figure 6.4 in Chapter 6 illustrates the influence of thermal diffusivity.

21.3.2 Thermal inertia

Thermal inertia is the property that determines how quickly a material surface heats up when impacted by radiant or convected heat. It is a very important material property when good passive design is being considered. As with diffusivity, it depends on the same three properties of thermal conductivity (k), density (ρ) and specific heat capacity (C_p). Thermal inertia is calculated thus:

$$\text{Thermal Inertia} = k \times \rho \times C_p$$

Values of thermal inertia for a range of building materials are given in Table 21.1. These data show that materials having a high thermal diffusivity also have high values of thermal inertia. This is logical; if a material rapidly diffuses heat away from the surface and into the interior, then raising the surface temperature will be difficult and a slow process. If the heat remains in the surface layer, then heating the surface will be quick and easy, but through-heating will be difficult and slow.

Table 21.1 Thermal diffusivity and thermal inertia data for some common building materials

Material	Thermal diffusivity $m^2\ s^{-1}$	Thermal inertia
Aluminium	73.050×10^{-6}	443.520×10^6
Brick	0.385×10^{-6}	1.664×10^6
Copper	111.980×10^{-6}	$1{,}323.649 \times 10^6$
Plasterboard	0.198×10^{-6}	0.127×10
Plaster finish	0.385×10^{-6}	0.650×10
Cellular insulation	1.000×10^{-6}	0.0009×10
Fibrous insulation	0.167×10^{-6}	0.015×10
Steel	15.959×10^{-6}	225.770×10
Concrete	0.663×10^{-6}	2.957×10
Timber	0.074×10^{-6}	0.163×10
Glass	0.574×10^{-6}	1.742×10
Rubber (polyisoprene)	0.066×10^{-6}	0.341×10
Polyethylene (high density)	0.255×10^{-6}	1.059×10^6

21.3.3 Thermal expansion effects

All materials tend to expand when they are heated. The amount of expansion differs from material to material, and is governed by the coefficient of thermal expansion for the particular material. These expansion effects can cause problems in a fire, including masonry walls bulging and being pushed over, buckling of steel columns and beams, etc., and these effects will be discussed later in this chapter.

Table 21.2 gives the coefficients of thermal expansion for a range of common building materials, as well as the amounts of unrestrained expansion that would occur in each material for a 50 °C rise in temperature. We can see from this data that materials have widely varying coefficients of thermal expansion; clay bricks and natural stone expand by only a small amount when heated, whereas polymers expand a great deal. However, we cannot consider expansion in isolation. Materials that heat through will expand uniformly, whereas those that are only heated in the surface layers will only expand at the surface, and they will tend to bend, or the surface layer may spall away from the rest.

These expansive effects can have serious consequences in a fire. For example, with steel-framed structures the expansion of beams can push columns outwards and cause them to buckle, steel beams resting on or abutting masonry walls can cause bulging of the walls, or even push them over. We shall revisit this later.

21.4 Fire performance of building materials

We shall consider the effects of fire in more detail on building materials in three groupings:

1. metals
2. non-metals – ceramic-type materials
3. non-metals – hydrocarbons and organic materials

Table 21.2 Thermal expansion data for some common building materials

Material	Approximate coefficient of linear expansion per °C ($\times 10^{-6}$)	Unrestrained movement for 50 °C change (mm/mm)
Clay bricks and tiles	5–6	0.25–0.30
Limestone	6–9	0.30–0.45
Glass	7–8	0.35–0.40
Marble	8	0.40
Slates	8	0.40
Granite	8–10	0.40–0.50
Asbestos cement	9–12	0.45–0.60
Concrete	9–13	0.45–0.65
Mild steel	11	0.55
Bricks (sand–lime)	13–15	0.65–0.75
Stainless steel (austenitic)	17	0.85
Copper	17	0.85
GRP	20	1.00
Aluminium	24	1.20
Lead	29	1.45
Zinc (pure)	31	1.55
PVC (rigid)	50	2.50
PVC (plasticised)	70	3.50
Polycarbonate	70	3.50

Table 21.3 Melting temperatures of some important metals and alloys used in construction

Metal	Melting temperature °C
Steel	1,450
Lead	323
Copper	1,083
Aluminium alloys	550
Zinc	400
Brass (an alloy)	1,027

21.4.1 Metals

The metals commonly used in construction include steel, copper, lead, aluminium and zinc. Metals and alloys have well-defined melting temperatures (pure metals) or temperature ranges (alloys); these are listed in Table 21.3.

Steel

Structural steel is an extremely important material for providing strength in modern buildings as it possesses a unique combination of high strength and stiffness, as well as good ductility, and all for a very low price. Because it can be shaped into an almost infinite variety of shapes and sections, including plate, sheet, structural sections such as I-beams, channels, angles and rod, wire and tubes, and is easy and quick to erect, it has become a dominant material.

However, it is adversely affected by the high temperatures produced by fire. The reason is that the crystal defects known as dislocations, required for plastic deformation to occur, multiply exponentially with increase in temperature. Blacksmiths have known for centuries that iron can easily be forged if it is first heated to be red hot in their fires. Dislocation multiplication dramatically reduces the yield strength of the steel, and it is this fact that causes steel structures to suffer *plastic collapse* in building fires.

Structural steel is non-combustible and does not contribute to a fire. It has a high melting temperature (Table 21.3) and does not melt in fire. However, because of the dislocation multiplication effect mentioned above, it suffers a marked reduction in yield strength (σ_y) and stiffness (E) when subjected to elevated temperatures. Yield strength is very important, and any reduction in its value leads to a corresponding reduction in load-carrying ability. The combined effect of a reduction in these two properties has a major effect on structural performance. Figure 21.1 shows the influence of temperature on yield strength and elastic modulus E.

Steel is a metal alloy and so it has a high value of thermal diffusivity (Table 21.1) compared with almost all other construction materials (only copper and aluminium have higher values). Therefore, in a fire any heat impinging on the surface of exposed steelwork will quickly soak into the steel, so it will soon be heated through. Therefore, all heat-related properties will quickly be affected.

Steel has a moderate value of thermal expansion coefficient, and it is this fact, together with the reduction in stiffness and yield strength, that lead to the plastic buckling effects often seen in unprotected steel structures after fires. Expansive effects in structural steelwork can cause problems with masonry walls; these effects are illustrated in Figures 21.2 and 21.3.

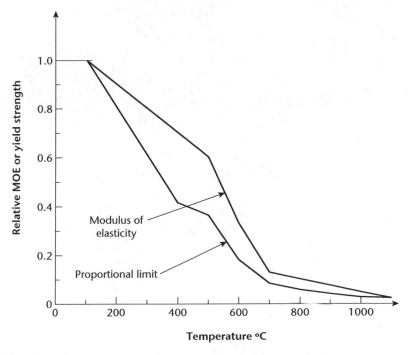

Figure 21.1 The effect of temperature on the strength and stiffness of steel

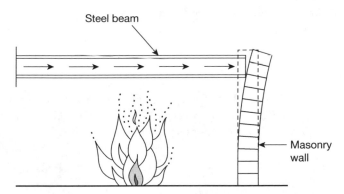

Figure 21.2 How expansion of steel beams can result in bulging or pushing over of a brick wall
(After Gosselin, 1987)

Aluminium

Aluminium is principally used for glazing systems, cladding panels and in the form of foil for damp/ moisture proofing. As a metal it has a high value of thermal conductivity, and aluminium will rapidly heat through if it is exposed to heat. Unfortunately, aluminium alloys also have low melting points, typically around 550–600 °C (Table 21.3). This means that in most fires aluminium products will melt, and therefore aluminium cannot be used for structural purposes in building.

Aluminium-based glazing systems will usually be destroyed in a bad fire, and so we cannot rely on aluminium-based materials to offer any extended fire endurance.

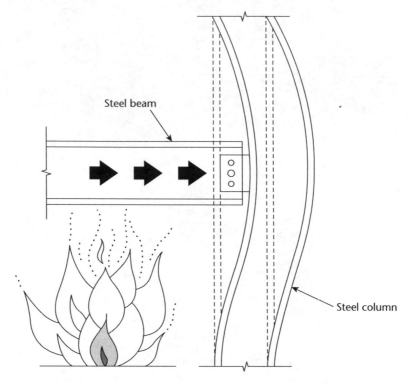

Figure 21.3 How the expansion of a steel beam can 'assist' the buckling of a steel column
(After Gosselin, 1987)

Copper

Copper has excellent corrosion resistance, so it is used in sheet form for roofing. It also finds wide application in plumbing systems, and also in electrical systems because of its excellent electrical conductivity. It has a fairly high melting point (1,083 °C); this means it will melt in severe fires, but not necessarily in less serious cases.

Lead

Lead is used in sheet form for roofing purposes, either as a complete covering or for flashings. It is easily worked and jointed by soldering, and has good corrosion resistance. However, it has a very low melting temperature (323 °C) and so it usually melts in a fire. The use of lead in plumbing systems was common before the Second World War, but its use was discontinued because of its toxic nature and its tendency to dissolve and pollute water supplies (plumbo-solvency).

Zinc

The main use of zinc is for galvanising steel to protect it from wet corrosion; zinc is rarely used by itself. When it is used, it is in the form of zinc sheet, for the waterproofing of roof structures. Zinc has a very low melting temperature (400 °C), and it will easily melt in a fire.

21.4.2 Non-metals, ceramic-type materials

This group includes materials such as concrete, clay bricks, glass and plaster.

Concrete

Concrete has a low value of thermal conductivity; in part, this is due to all the porosity that normal concrete contains. Therefore it will take a long time to heat through if heat is applied to it. Concrete is composed of cement, sand, aggregates and water, and when exposed to heat it will lose water and also strength.

Figure 21.4 shows the strength at elevated temperature compared as a percentage of the normal temperature strength for both siliceous and calcareous aggregates. The ultimate strength F_c and stiffness E are plotted against temperature.

Lightweight concrete can be made using lightweight aggregates or as aerated concrete. Either way, such concrete has a greater fire endurance than normal weight concrete, and it therefore provides a useful material for the protection of structural steel. The low thermal diffusivity of concrete is illustrated in Figure 21.5.

Low diffusivity means that heat does not diffuse quickly into a material when it is heated. Such materials will therefore sustain *steep temperature gradients*. Materials with high thermal diffusivity values therefore exhibit shallow temperature gradients because the heat diffuses rapidly into them.

In common with other materials, concrete will expand when heated. Reinforced concrete will increase in load-bearing capacity when the ends are thermally restrained by the structure. Where the thermal restraint is low, as at the exterior parts of the building, thermal expansion may cause cracks to open and supporting members to suffer excessive distortion.

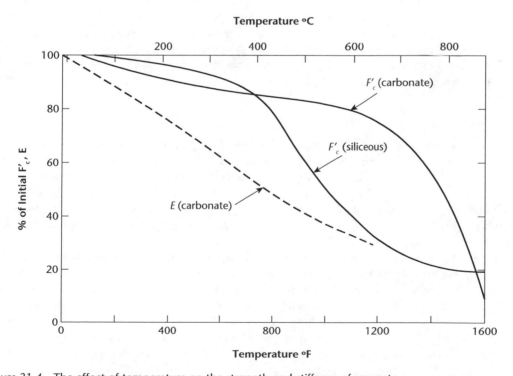

Figure 21.4 The effect of temperature on the strength and stiffness of concrete

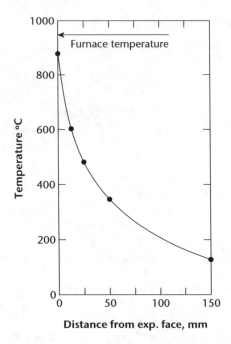

Figure 21.5 The temperature gradient in a 150 mm slab of concrete after two hours of standard fire exposure

(National Fire Protection Association, 1986)

The fire resistance of concrete is reduced by its tendency to spall. Spalling is the disintegration or cracking and falling away of part of the concrete surface, thus exposing the steel reinforcement or the interior of the concrete to the fire. Heat causes the concrete to dehydrate, and spalling is caused by the explosive build-up of steam from the expanding water that cannot escape or otherwise move to the unexposed surface.

Clay bricks

Clay bricks possess low values of thermal conductivity and modest thermal expansion coefficients. They are not combustible, and furthermore, they are made by firing in a kiln at very high temperatures (900–1,000 °C) – even higher in the case of engineering bricks. This is the kind of temperature encountered in a bad fire so, as a class, clay bricks survive well in building fires. Little research has been carried out into the strengths of bricks at elevated temperatures, but such experimental data as do exist indicate that bricks retain a large proportion of their strength (perhaps 75 per cent) at these high temperatures.

There has been very little published on the effects of fire on clay bricks. Nevertheless, they are so important in construction that we must attempt an outline of their fire performance. We have seen that bricks possess low values of thermal conductivity; this means that heat will not soak through them very quickly. Therefore, if a mass of flame were to heat the face of a wall in front of a fire, the other side of the wall would remain cool for quite a long time. We have also seen that bricks have a modest coefficient of thermal expansion. The result of these two properties will mean that a brick so heated will expand on the side facing the flame, while the other side remains cool and unexpanded; this causes the brick to distort, as shown in Figure 21.6.

Heat

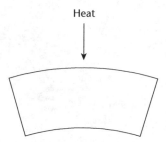

Figure 21.6　The 'bowing' of a brick in response to being heated on one face

In practice, of course, bricks are not used singly, but are built into walls, and so realistically we must consider the effects of fire on wall structures. Walls are composites, consisting of the bricks and the mortar joints. Because of the low thermal diffusivity and the expansion on the heated side of the bricks, unrestrained walls tend to bulge towards the direction of heating by the fire, as shown in Figure 21.7.

Plaster

Gypsum products, including plaster and gypsum board (often called plasterboard, sheet rock or drywall) perform very well in fire. Hydrated (set) plaster contains a substantial amount of chemically combined water, and during a fire this water becomes unbound and slowly evaporates off so the gypsum is calcined and slowly disintegrates into a fine powder. Since it is only water vapour that is given off, the gypsum does not contribute to the fire load or emit any toxic fumes. It protects the material lying underneath it and so makes a very positive contribution to the fire safety of the building.

Gypsum board consists of a layer of plaster sandwiched between sheets of paper. Products of various thicknesses – between 6 mm and 25 mm – are available. In the United States, drywall sheets 50 mm thick were used to encase the steel columns in the World Trade Center in lower Manhattan. While the regular gypsum board is most common, improved fire resistance can be obtained by using glass fibre reinforced gypsum. The fibres improve the fire performance by holding the calcined gypsum in place for a longer period of time during a fire.

Heat

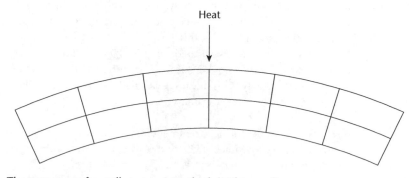

Figure 21.7　The response of a wall structure to the heat from a fire

Glass

Three forms of glass are used in building construction:

1. flat glass for glazing
2. glass wool insulation
3. fibre–glass reinforced products

Ordinary window glass has little resistance to breakage in a fire; glass has low fracture toughness, and thermal shock will usually cause it to shatter. Thick plate glass or tempered glass used in curtain walls can withstand high temperatures for substantially longer times before cracking and breaking. Wired glass is often specified to provide greater integrity after cracking in a fire. However, when considering the fire resistance of glazing the entire glazing system, including the framing, etc. must be taken into the analysis.

Glass wool insulation is often made by binding the glass fibres together with a resin binder, and while the glass fibres will not burn, the resin binder is often flammable. If so, the insulation will propagate flames when they are in an open, unconfined space.

Fibre-glass reinforced products such as prefabricated bathroom units and translucent window panels have the advantages of economy and aesthetic appeal. The glass fibres are used to reinforce a thermosetting resin, with fibres and resin present on an approximate 50/50 basis. The glass fibres will not burn or support combustion, but the resin matrix is often highly flammable, and so these units can be very combustible. Furthermore, when they burn they will emit substantial volumes of smoke, and smoke is a major component of the hazard posed by a fire.

21.4.3 Non-metals, hydrocarbons and organic materials

Generally speaking, as a group, hydrocarbons and organic materials (naturally occurring hydrocarbons) represent fuel and so will burn in a fire. However, in the case of timber in thick sections, this is not as disastrous as it sounds. Massive sections of timber will survive and perform surprisingly well in fire. Other hydrocarbons such as polymers perform poorly in fire, and because of the various chemical species that they contain, they can give rise to the production of toxic smoke. Because of this, their use should be specified with care.

Timber

Timber is a hydrocarbon material and is therefore a fuel. However, it has very low values of thermal conductivity and so heat will diffuse into it only very slowly. Many textbooks say that the char layer which forms on timber when it burns 'protects' it; this is erroneous. The thing that makes thick sections of timber surprisingly good in fire is this low value of thermal conductivity. To burn, timber must pyrolyse. To pyrolyse it has to be heated to 160–200°C. Because the heat takes a very long time to soak into the wood, it takes a long time to pyrolyse, and the end result of this is that the timber will burn only very slowly. This means that thick sections of timber survive very well in a fire, and retain their load-bearing capability far longer than exposed structural steel sections.

On the other hand, timber does add to the fuel load in a building, and thin timber sections such as floor boards, panelling, etc. will burn through fairly quickly. Char rates will vary from timber species to species, but an average value of $0.7–0.8$ mm min^{-1} is often used as an approximation.

Polymers

Polymers as a class have very poor fire performance. Because they are hydrocarbons they represent fuel and therefore most will burn in a fire. Some burn much more readily than others, but this is generally true. Polymers have low strengths, stiffness values (E) and densities, and they quickly soften and lose what little strength they have at quite low temperatures (c.200 °C). As a group they tend to have high coefficients of thermal expansion, and so those heat-affected plastic fittings which do not burn will suffer massive distortion in a fire.

These comments remain broadly true even when plastics are reinforced with glass or carbon fibres, as they increasingly are nowadays. Their performance can be improved by the addition of flame retardants during manufacture. The aim here is to make them *endothermic*.

Bituminous materials

Bitumen consists of a complex mix of hydrocarbons of varying molecular weights. Because it is a hydrocarbon it represents fuel, and will burn in a fire. Bitumens are used for their moisture- and water-proofing properties, often in the form of a thin layer. Traditionally they were used in roofing felt, and because they are combustible they will contribute to a fire. When heated they will soften and 'melt', and this behaviour can contribute to the spreading of the fire.

Composite materials

Composite materials are increasingly finding applications in fire-sensitive, load-bearing structures, in both the off-shore industry and also in the on-shore chemical industry, because of their lightness and resistance to corrosion. A key factor in prolonging their fire integrity is the decomposition of the resin matrix, which is a highly endothermic process (i.e. it absorbs a great deal of heat). Resins can be of various types – polyester, vinyl ester, epoxy and phenolic – and the performance of these will vary. This topic is the subject of current research.

21.5 Fire endurance

An important aspect of fire-safe design in buildings is the principle of *compartmentation*. This is the subdivision of the volume of the building into separate compartments. This is done as part of the passive fire-safe design. The objective is to prevent the spread of fire by containing it inside the room or enclosure in which it started for a pre-determined amount of time. This allows time for the building's occupants to make good their escape from the building, and time for the fire service to make their way to the building and to commence fighting the fire.

21.5.1 Fire endurance calculations

For design purposes we need to be able to estimate the fire endurance of structural members. We shall need to ensure that the structural integrity of a building is not compromised before the occupants have had time to escape. Designers use various formulae and equations for this purpose, some of which are simple, while others are more complex. The following specimen calculation gives us a feel for the methods involved.

Fire endurance: specimen calculation

A wooden column of square cross-section measuring 450 mm by 450 mm is engulfed in flame. Assume that the timber of which it is made chars at a rate of 0.6 mm min^{-1}. If the column was designed with a safety factor of 2 to 1, calculate its approximate fire endurance.

>   Original cross-sectional area = 450 mm $\times$ 450 mm
>   Cross-sectional area = 202,500 mm^2

The column has a 2:1 safety factor, therefore the minimum cross-sectional area = 101,250 mm^2

The timber chars at a rate of 0.6 mm min^{-1}. So in one hour it will char by

>   0.6 mm min^{-1} $\times$ 60 min hr^{-1}
>   = 36mm

What will be the effective, i.e. unburned cross-sectional area, of timber after one hour?

>   Effective cross-sectional area = (450 mm – 2{36mm}) $\times$ (450 mm – 2{36mm})
>   = (450 mm – 72 mm) $\times$ (450 mm – 72 mm)
>   = 387 mm $\times$ 378 mm
>   = 142,884 mm^2

The cross-sectional area is still greater than the minimum (101,250 mm^2) – the column is okay.

How would this look after 1.5 hours? How far in would the char layer be after 1.5 hours?

>   Char layer thickness would be 1.5 $\times$ 36 mm = 54 mm
>   Effective cross-sectional area = (450 mm – 2{54 mm}) $\times$ (450 mm – 2{54 mm})
>   = (450 mm – 108 mm) $\times$ (450 mm – 108 mm)
>   = 342 mm $\times$ 342 mm
>   = 116,964 mm^2

The column cross-sectional area still exceeds the minimum, so the column is still okay.

Picture after two hours. Char layer would have moved in by 2 $\times$ 36 mm, i.e. 72 mm.

>   Effective cross-sectional area = (450 mm – 2{72mm}) $\times$ (450 mm – 2{72 mm})
>   = (450 mm – 144 mm) $\times$ (450 mm – 144 mm)
>   = 306 mm $\times$ 306 mm
>   = 93,636 mm^2

This is less than the minimum cross-sectional area, so the column would fail; the fire endurance is less than two hours. Therefore, in this case the fire endurance would be taken as 1.5 hours.

In practice, a slightly more complex equation would be used to calculate the fire endurance, but this example illustrates the principle. Note that a similar calculation method could be employed for estimations of the fire endurance of concrete, providing that the rate of dehydration, in millimetres per minute, was known.

21.5.2 Case study: World Trade Center, heat release on 11 September 2001

As an example of how destructive fire can be, it is instructive to examine what happened to the World Trade Center buildings on 11 September 2001. These buildings were designed in the late 1960s to withstand the impact of a large aircraft, which at that time would have been a Boeing 707, weighing 160 tonnes. In fact, they were each struck by a 200-tonne Boeing 767, and in both cases the impact did not cause the subsequent collapse. The towers succumbed to the devastating fires that followed the impacts caused by the burning of the aviation fuel released (both aircraft were full of fuel for a transcontinental flight to the Pacific side of the United States). We know the main details of these aircraft, and so it is possible to calculate both the kinetic energy dissipated in each impact, and also the thermal energy released by the burning of the fuel loads of the aircraft.

Each aircraft weighed around 180 tonnes, and were each carrying around 10,000 kg of fuel for a transcontinental flight to Los Angeles. It is thought that the first aircraft was travelling at around 430 mph, while the second one was travelling about 100 mph faster. For the purposes of estimating the respective energies we will assume that they were each travelling at 500 mph.

Total energy dissipated on impact = kinetic energy of aircraft + heat due to burning of fuel
$= \frac{1}{2}m{\cdot}v^2 + M{\cdot}\text{calorific value}$

The relevant data are as follows:

- mass of Boeing 767 (m) = 180,000 kg
- Mass of fuel (approx.) (M) = 10,000 kg
- Speed of aircraft (v) = 223 m sec^{-1} (500 mph)
- Calorific value of fuel (C.val.) = 43.55 MJ kg^{-1}

Kinetic energy dissipated on impact ☺

Kinetic energy $= \frac{1}{2}m.v^2$
$= 0.5 \times 180{,}000 \text{ kg} \times 223^2$
$= 0.5 \times 180{,}000 \times 49{,}729$
$= 0.5 \times 180 \times 10^3 \times 49.7 \times 10^3$
$= 4{,}473 \times 10^6 \text{ J}$
$= 4.473 \text{ GJ}$

Heat energy released

Heat $\Delta H = \text{mass} \times \text{calorific value}$
$= 10{,}000 \text{ kg} \times 43.55 \text{ MJ kg}^{-1}$
$= 435{,}500 \text{ MJ}$
$= 435 \text{ GJ}$

We can immediately see from these calculations that the heat energy released in the burning of the fuel loads is about 100 times greater than the kinetic energy dissipated during the impact. This result is in line with many fire situations involving transport systems (Johnson, 1984, 1986). In view of the fact that the initial impact destroyed the drywall fire protection around the central core steel columns, and thereby exposed the steel to the fire, it becomes easy to understand why the buildings suffered plastic collapse.

21.6 Critical thinking and concept review

1. Define *thermal diffusivity* and show how it is calculated.
2. Define *thermal inertia* and show how it is calculated.
3. Explain the difference between thermal diffusivity and thermal inertia.
4. Why are the thermal properties of materials so important in determining their fire behaviour?
5. Compare and contrast the respective fire performances of:
   - steel
   - concrete
   - timber
   - clay bricks

   In your discussion, refer to the thermal diffusivity and thermal inertia values of these materials.
6. Using the data given below, show by calculation which of the following construction materials has the lowest value of thermal diffusivity.

Material	Thermal conductivity (k) (W/m K)	Density (ρ) (kg/m³)	Specific heat capacity (C_p) (J/kg K)
Steel	60.00	7,800	482
Concrete	1.40	2,400	880
Timber	0.11	545	2,720
Clay Brick	0.80	2,000	800

7. 'For reasons of fire safety, the internal walls of rooms and enclosures inside buildings should be covered with a material having a high value of thermal inertia.' Is this statement true or false? Give the reasoning for your answer.
8. 'Steep temperature gradients will be produced when heating materials with low thermal diffusivity.' Is this statement true or false?
9. A steel beam spans a workshop and is 12m in length. If there is a fire in the workshop, and the temperature of the beam is raised by 550 °C, calculate the linear expansion likely to occur in the beam.
10. A concrete beam 600 mm deep and 450 mm wide is engulfed in flame on both sides and on its soffit. It is designed with a 2.5 to 1 safety factor. If the concrete dehydrates at a rate of 0.6 mm per minute, estimate the fire endurance of the beam.

21.7 References

DESCH, H.E. and DINWOODIE, J.M. (1996), *Timber: Structure, Properties, Conversion and Use*, MacMillan, London.

FITZGERALD, R.W. (2004), *Building Fire Performance Analysis*, Wiley, Chichester.

GOSSELIN, G.C. (1987), *Fire Compartmentation and Fire Resistance*, IRC, National Research Council of Canada, Ottawa.

JOHNSON, W. (1984), 'Vehicular Impact and Fire Hazards', *Proceedings of the International Conference on Structural Impact and Crashworthiness*, pp. 75–114, Elsevier, London.

JOHNSON, W. (1986), 'The Circumstances of Fire and Damage after Vehicular Impact', *Proceedings of the International Symposium on Intense Dynamic Loading and its Effects*, pp. 36–48, Science Press, Beijing.

NATIONAL FIRE PROTECTION ASSOCIATION (1986), *Fire Protection Handbook*, National Fire Protection Association, Quincy, MA.

SHIELDS, T.J. and SILCOCK, G.W.H. (1987), *Buildings and Fire*, Longman, Harlow.

Part IV

Conclusion: sustainability of materials

22

Environmental impact of materials

This chapter examines the nature and extent of the impact that the extraction, production and use of building materials has on the Earth's natural environment. It examines the steps involved in producing the materials, and the total life cycle impact from cradle to grave. Most importantly, various ways of quantifying the impact are set out in this chapter, as well as suggestions for mitigating the various adverse effects described.

Contents

22.1. Introduction

We have already alluded to the fact that construction uses a wider range of materials than any other industry, as well as the fact that it consumes more material than all other industries put together. Since 1900 the Earth's population has more than trebled, from around 1.8 billion to around 7.0 billion people (as of 2012). All of these people need somewhere to live, and since construction exists to provide buildings, the industry is having a huge influence on life on Earth. In Chapter 1 we saw that the unprecedented growth in demand for materials by the construction sector of industry since the Second World War is having a major impact. In this chapter we shall examine the reasons for this impact and also the ways in which it is being felt. If we are seriously intending to mitigate the extent of the impact, we need to be able to measure it, and so we shall also examine ways in which the impact can be quantified or measured.

22.2 Materials

In terms of demand for materials, the construction sector's growth has eclipsed that of all other industries (see Figure 1.1 in Chapter 1). The figures speak for themselves, with annual production of aggregates currently over ten billion tonnes, steel production around 1.45 billion tonnes, cement production over 2.4 billion tonnes, etc. Furthermore, the annual production of all of these materials is rising quite sharply. However, none of these three materials require elements that are scarce. Aggregates contain a lot of silica (SiO_2), steel is mainly iron, cement requires calcium carbonate ($CaCO_3$) and silica. The elements silicon, iron, oxygen and carbon are all plentiful in the Earth's crust, and are not in imminent danger of being exhausted. So why are the environmentalists becoming concerned?

To understand this, we need to view materials in their historical context, and to examine the factors that make possible their extraction and production. Our present society has become utterly dependent on materials, in ever larger quantities. They are so important that we designate the various ages of mankind by the materials that were used, i.e. Stone Age, Bronze Age, Iron Age, etc. In the Stone Age, weapons and tools were made from stone – flint and quartz – and bone, simple structures were constructed from stone, mud bricks and timber, and these could be weather-proofed using natural plant fibres and animal skins. Note that all these are natural materials requiring no chemical processing, and note also that they are all *renewable* materials.

Going back 7,000 or 8,000 years, the first metals that mankind used were those that occur in native form, i.e. copper, gold and silver. These can all be work-hardened, but none of them are suitable for making tools. The effect of adding about 10 per cent tin to copper to produce bronze was probably discovered around 3000 BC. Bronze is a very hard and strong material well suited to the making of tools. Bronze then gave way to iron around 1000 BC following the discovery of iron-making in about 1450 BC. The Iron Age was a long one, lasting until the advent of the Industrial Revolution in AD 1750–1800. The Industrial Revolution could not have happened without the availability of the inexpensive, strong, stiff, tough material that we know as steel. In the later nineteenth century, processes to mass-produce steel were developed and the Age of Steel lasted until the 1950s. Light (low-density) metals including aluminium, magnesium and titanium were also developed, but they are much more expensive, and much less widely used than steel.

The Age of Steel gave way to the Polymer Age during the 1950s, and this lasted until the end of the twentieth century. Notable advances in the use of ceramic materials also took place during the second half of the twentieth century. We are now entering the nano-technology or molecular age. Back in 5000 BC we had only a handful of different materials; today we have tens of thousands of different materials if we count each alloy and polymer type separately, together with all the ceramics and other materials. However, in our journey from 5000 BC to AD 2000 we have gone from natural, renewable materials to *non-renewable* materials that require a great deal of processing (and increasingly large inputs of energy); the cross-over from renewable to non-renewable materials probably occurred at the beginning of the Industrial Revolution in England in around AD 1750. This transition represented a move from a sustainable to an ultimately non-sustainable technology. During the same time interval, the Earth's population has increased by a factor of nearly ten, and the root of many of our current problems lies in these two factors. This is illustrated in

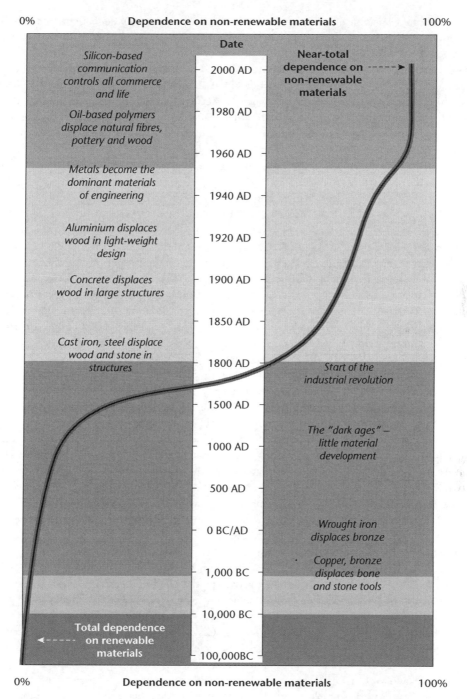

Figure 22.1 There has been an increasing dependence on non-renewable materials over time
(After Ashby, 2009)

Figure 22.1. We can see graphically how our society has become increasingly dependent on non-renewable materials. Such dependence is unimportant while these materials are plentiful, but it becomes an emerging problem when the materials become scarce.

Classical economics failed for many years to understand the reasons for the spectacular growth in the world's economy after the Second World War ended in 1945. Adam Smith, in 1776, published his famous volume entitled *Inquiry into the Nature and Causes of the Wealth of Nations*, in which he saw the division of labour as the main engine of economic growth rather than land or money (capital). This was a work of pre-eminent importance, and it laid the foundations for our modern study of economics. Smith's analysis was undoubtedly correct at the time of writing, in the 1770s. At this time the Industrial Revolution was gathering speed in Britain, and as a result another factor not considered by Smith was coming into play. This factor was the use of energy.

Energy consumption is a matter of great concern in 2014 because so much of our energy is derived from fossil fuels, such as oil, coal and natural gas. When these fuels are burned to produce their energy, carbon dioxide (CO_2) is released as a by-product into the Earth's atmosphere. Carbon dioxide is a greenhouse gas, and since the advent of the Industrial Revolution in around 1750, the CO_2 content of the Earth's atmosphere has risen from a pretty stable 280 parts per million (ppm) to around 380 ppm today. Furthermore, this figure is rising steadily, and the level of 380 ppm is the highest value for the past 420,000 years. There is now general agreement that this is an anthropogenic (man-made) effect, and this is a serious matter because, CO_2 being a greenhouse gas, the increased level of CO_2 is giving rise to global warming. Today, in 2014, mankind's activities cause the emission of around 7,000 billion tonnes of CO_2 per year. This is not the place to discuss in detail the likely effects of global warming – they have been dealt with at length elsewhere – but the production of materials for construction and the heating of buildings make a very large contribution to global warming, and this is of concern to us all.

There is currently concern about how much energy buildings consume for heating and lighting purposes. Over the years since the first oil crisis in 1973, the UK Building Regulations have been modified to reduce energy consumption by reducing outside wall U-values, increasing air-tightness and improving performance in respect of water consumption, handling of waste, etc. This is important, but the ramifications of construction activities for energy consumption are much wider than this. The amount of energy consumed in extracting and transporting raw materials, producing finished materials and transporting these to site is enormous, and increasing rapidly as the construction industry uses more and more material.

Another factor causing concern is the matter of waste. Western civilisation is characterised by the massive amount of waste it produces. This question was recognised over 20 years ago in the construction industry with the publication of the report 'Materials for Construction and Building in the UK' by the Institute of Metals (1987). This report was written by a working party of the Materials Forum and the Institution of Civil Engineers. One of the conclusions reached by this report was that enough material to build around 13,000 houses per year was being wasted by the UK industry. These losses occurred mostly on site due to damage, theft, surplus due to inaccurate ordering, etc.

At the other end of the building cycle, a huge amount of demolition waste is generated each year in the UK. Traditionally, buildings have been demolished wholesale; this results in a great pile of mixed waste of concrete rubble, bricks, plaster, timber glass, metal, etc. Such a mixture has no real economic value, and much has been put into landfill. Increasing population levels means that pressure on land is increasing, and tying up valuable land for the disposal of waste is not a tenable strategy in the long term. In the 1987 report, wastage was seen as a materials problem, although the report stressed the need for reducing the amount of energy needed to manufacture construction materials. However, the problem is more serious, and each pile of waste material should be viewed as a pile of waste energy because in discarding a material we are also discarding the energy consumed in its manufacture. With regard to long-term sustainability, energy and water threaten to be bigger problems than materials, and we shall examine these questions later; but first we shall examine how much material is consumed in construction.

Table 22.1 World production of principal materials, and the approximate proportion going to the construction industry

Material	Annual world production (tonnes)	% of world production used in construction
Cement	2,400,000,000	95–100
Aggregates	12,000,000,000	95–100
Steel	1,450,000,000	Up to 50
Timber	1,000,000,000	*c.*65
Polymers	150,000,000	*c.*20–25
Total	17,000,000,000	

22.2.1 Construction materials and their importance

Construction uses a wider range of materials than any other industry. Materials used include cement and aggregates to make concrete, metals – primarily steel, but with significant amounts of copper, copper alloys and aluminium alloys – timber, fired clay products, glass, gypsum products, polymers, bituminous materials, etc. The global consumption of materials is shown in Table 22.1.

Table 22.1 does not include materials such as fired clay products and glass, both of which are consumed in large amounts by the construction industry.

Construction materials do not usually occur naturally in a useable form, except for natural stone and timber. Even timber needs a certain amount of processing, including seasoning and conversion, before it can be used by the industry. In the case of other timber products such as plywood, chipboard, etc., the amount of processing can be considerable. The conversion of metal ores into metals and raw hydrocarbons into plastics are highly energy-intensive processes. Material consumption therefore has large energy ramifications. Their impact does not stop there, however. To win the ores, and to extract the raw hydrocarbons, etc., means displacing huge volumes of material, much of which is ultimately discarded. The discard will include topsoil and various overburdens removed to gain access to the ores and minerals, as well as various types of dross and waste from the manufacturing process. Furthermore, such waste can be gaseous, liquid or solids, both bulk solids and dusts and fines.

The third class of impacts that construction material production will have is the factories and infrastructure created solely for their production. Such facilities will have a large footprint, i.e. they will take up a fair amount of increasingly precious land area, which will not then be available for other uses. When we look for ways to measure environmental impact, these three factors – energy, wastes and land area footprint – prove useful and powerful indices to use, and we shall look into these in more depth later.

22.3 Energy

Once he made the transition from hunter–gatherer, man has made increasing use of energy, first in providing himself with food, and then in providing all the other requirements for living, including clothing and buildings to live and work in. Wind and water power had been harnessed for the grinding of grain. However, the Industrial Revolution saw man exploiting energy on an increasingly large scale to power industrial, mechanised processes. This involved using steam power on an ever increasing scale. Improved transportation systems were developed – first, a network of canals that enabled supplies of raw materials, fuel and also manufactured goods to be distributed. No sooner was the canal network in place than the rail network began to be built. In its developed state at the end of the nineteenth century, rail provided a comprehensive transport network serving all the urban areas in Great Britain. It is true that a road network also existed, but most road traffic was horse-drawn at the close of the nineteenth century. The internal combustion engine supplanted the horse for road traffic between the wars, but the real explosion in the number of vehicles on the roads occurred after the Second World War.

All these industrial and transport activities saw an accelerating growth in per capita consumption of energy both in the UK and the wider world. Transport by land, sea and air have all grown enormously in the years since the Second World War, and the building and expansion of all the associated infrastructure and facilities for land, sea and air transport has had a major impact too. We have seen the approximate figures giving the annual production of various classes of materials, and it is pertinent to ask which materials consume the most energy in their production. According to Allwood and Cullen (2012), the five materials of steel, cement, plastics, paper and aluminium are responsible for 55 per cent of industrial emissions of CO_2. All of these materials are used in construction.

22.4 Stages in material production and use

Some materials like natural stone, sand and aggregates are found in nature in a form that requires little processing. These are inorganic materials. Some organic materials such as timber grow naturally, and trees can be felled. The timber then requires seasoning and conversion, processes that involve no chemical processing. These are low energy intensity materials; that is to say that comparatively little energy is required to bring them into a useable form. Because minimal, low-energy processing is all that is necessary, the production of these materials has a correspondingly low environmental impact.

Other materials are found on or under the Earth's surface, chemically combined with other elements and compounds from which they must be separated before they can be produced in a useable form. The digging out or extraction of the raw materials can have a large impact, as can their processing; large amounts of energy are used, and this gives rise to correspondingly large impacts. Many metals and polymers are highly energy intensive for these reasons.

However, the extent of the impact does not always cease with the extraction and production of the material. Considerable transportation can be involved, both of raw materials and finished goods. For example, iron ore may be extracted in Australia and transported to Japan or China, and finished steel is exported to countries far distant before being incorporated into a building. The transportation has a major impact, in that burning of fuel or expenditure of energy will be involved. So if we are to fully estimate the impact that the material has, we need to examine its complete trajectory from raw material right through to final use, including demolition and disposal. A trajectory for a material such as steel might look as shown in Figure 22.2 below:

Extraction → Processing → Shaping → Use → Maintenance → Demolition → Disposal/Recycling

Figure 22.2 Stages in the life cycle of steel as a construction material

Each step in this trajectory involves the expenditure of energy, the displacement of material, or the emission of gas, dust or waste liquids, and so has a potential impact upon the environment. If we wish to be exhaustive, each step in the above trajectory can be further subdivided into more detail. For example, consider the extraction of iron ore. This may involve the stripping of an overburden, i.e. there may be soil lying on top of the ore which must first be removed to allow access to the ore deposit. This will consume energy. Then there will be the digging out of the ore itself. This may be followed by a crushing process, and then the ore material will be loaded onto vehicles to transport it to the steel plant.

What will be the environmental impacts from these extraction operations? Removal of the overburden can involve disruption to plant life growing in it. The excavating machines will consume fuel and emit CO_2, and will probably produce some dust. The machines that dig out the ore will also emit CO_2 and produce dust, as will the transport vehicles onto which the ore or mineral is loaded. Transport to the factory that processes the ore will involve CO_2 emissions and consumption of fuel.

Once the ore is at the factory, then another series of processes with environmental impacts begins. As we shall see, during manufacturing processes not all of the input material emerges as finished product. Some waste residues are formed as by-products, and these are either discarded or, more hopefully, used as feedstock in another manufacturing process. The material production and processing will involve the consumption of water, a valuable and finite resource. This represents a major impact. Finally, the fact that we have built the factory on the Earth's surface is a major impact in itself. Nothing can be done with the ground on which the factory stands; we call it the *footprint* of the factory. If we now examine the production of steel from iron ore to finished steel in more detail, it will enable us to pinpoint the impacts that such operations have, and also help us devise ways of evaluating and quantifying these.

22.5 Measurement of environmental impact

If we are to assess the impact of our activities on the environment, we need ways of measuring it. There is no single index that gives the complete picture; we shall require several indices. As we have seen in the preceding paragraphs, impact can be felt in various ways, and we need to account for all of them.

If we consider the production of a construction material such as steel, we can get an idea of the various impacts that manufacturing has on the Earth's systems. To make steel we require iron ore, limestone and coke. These are the initial ingredients that are fed into the iron blast furnace, together with the energy supplied to the furnace via the hot blast, and these represent the input impacts. There are also output impacts, including all the dust and hot gases emitted by the furnace during smelting.

Considering the inputs in more detail, we may need to strip away an overburden of topsoil to get at the iron ore. This may be replaced after removal of the ore, but energy will be used in handling it. Dust and exhaust fumes will result from this. Similarly, the limestone and coal will also have to be mined or extracted, with the expenditure of energy and production of fumes and dust. The coal will be distilled to produce coke, entailing more energy expenditure, emissions of fumes and dust and the production of other by-products from the coke ovens. The ore, coke and limestone will be crushed and may be turned into fairly uniformly sized pellets for charging to the blast furnace. There are advantages in having a blast furnace burden of uniformly sized particles. However, we can see that even before we have turned the ore into iron we have expended a lot of energy, produced a lot of dust and gas emissions and handled and transported a lot of material.

The products tapped from the furnace are the pig iron and the slag by-product. The molten pig iron or cast iron is transferred to a steel furnace, where a stream of pure oxygen will be blown through it to burn out the excess carbon and turn it into steel. Some adjustments to the composition can be made to the steel at this stage. Molten steel will then be transferred to a continuous casting machine, the solid output from which is fed into a sequence of rolling mills, to give the final steel product. This might be steel strip, sheet, plate or engineering sections (I-beams, angles, etc.). The rolling mill will consume large amounts of energy to drive the rolling stands, and intermediate re-heating might also be necessary, depending on the amount of plastic deformation required. At the end of the rolling process, the product may need to be rolled into coils or cut into sections, or otherwise finished. All these steps will consume energy. Mill scale and contaminated water will be produced.

This is a very brief summary of what happens in steel production, but we can get an idea of the impact if we add up all the inputs and outputs from the process.

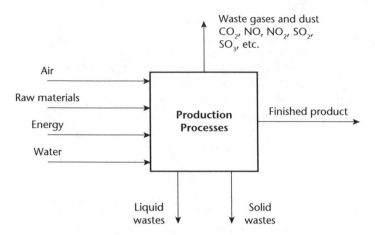

Figure 22.3 Production process considered as a 'black box'

Figure 22.3 looks at the production of a material considered as a 'black box'. Details of the process are not given, only the various inputs and outputs. This can be a very helpful way to look at production processes because we look at the solid, liquid and gaseous wastes given off, as well as the final product. In the nineteenth century, as the Industrial Revolution took off in the UK, manufacturers looked only at finished products and ignored the various wastes being generated. The aim in any industrial process should not only be the achievement of energy efficiency, but the minimisation of solid liquid and gaseous wastes. A truly sustainable process would be one where no waste materials were made, or where such wastes were used up in other processes.

Turning back to steel production, we know that the input is iron ore, limestone and coke, together with a lot of electricity, oil and gas fuels. The outputs from all the various production steps outlined above will be gaseous, liquids and solids. The gaseous products will include carbon dioxide and sulphur dioxide gases, which will be emitted to the atmosphere. There may be quantities of other gases too, oxides of nitrogen, etc. With these gases will be large amounts of heat, and solids in the form of particulates, i.e. dust. These dust particles will range in size from a millimetre down to very fine particles of colloidal dimensions called aerosols. We may need to take steps to collect these from the furnace fumes to prevent them from getting into the atmosphere.

In terms of liquids produced, these may include acidic pickling liquids, oily water, etc. These may not be discharged into the normal drainage because of their toxicity. Solids produced (besides the dust already referred to and the final steel products) will include tonnes of mill-scale (iron oxide) flakes, slag particles, soil and rock from the mining stage, scrap steel off-cuts and trimmings, etc. All in all, therefore, if we do a mass balance we can get a clear picture of what is going on. The mass balance is an accounting procedure whereby we set down the total weights of all the inputs, and also the weights of all the outputs, including emissions of gases, liquids and solids, including pollutants and by-products as well as finished steel. If we do this we may find that we dug out or mined or otherwise disturbed a total of over ten tonnes of material to obtain our one tonne of finished steel.

To improve performance we need to consider the whole raw material gathering and processing route as a system, along the lines shown in Figure 22.4.

The advantage of the 'black box' or systems approach is that a proper material balance or audit has to be undertaken, and so no outputs can be ignored. It is much easier to assess likely environmental effects. Furthermore, the materials input vs output balance enables us to quantify undesirable outputs. We can evaluate the impact of a given volume of production. A quantitative approach is essential if we are to minimise impact and move towards sustainable operation. In the next section we shall look at various indices that can help us put exact values on the impacts of the things we do.

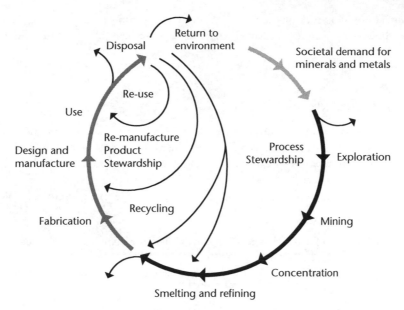

Figure 22.4 Mineral extraction, processing and product manufacturing cycle
(After Davies, 2009)

Another advantage of the systems approach is that it causes us to look at *all* of the outputs, and it is these waste outputs that can cause so many environmental problems. If we are not to modify the Earth's ecosystems to the point where they will no longer support human life, there are certain limits or boundaries that we must not exceed. Lynas (2011) has indicated what these boundaries or limits might be, and he identifies nine in all. These are: climate change (carbon emissions), biodiversity, nitrogen, land use, fresh water, toxic emissions, aerosols, ocean acidification and ozone. The extraction and use of materials has impacts in most of these areas.

However, not all of the materials of construction will have equally large environmental effects; Allwood and Cullen (2012) point out that just five materials are responsible for 55 per cent of all industrial emissions. These materials are steel, cement, plastic, paper and aluminium, and all of them are utilised in building construction.

22.6 Indices of environmental impact

If we are to precisely measure the extent of environmental impact of the manufacture of building materials, we need to take account of all the various types of impact that we have identified so far. In Section 22.4 we identified various impacts, and indices have been developed by engineers and environmentalists for five of these impacts. We shall look at these five first; they evaluate the impact of energy consumption, emission of carbon dioxide, production of waste and spoil, assessing the footprint of the manufacturing plant and process, and water usage.

22.6.1 Embodied energy

This index measures the amount of energy consumed in making a unit quantity of the required material (usually 1 kg or 1 tonne). It is given in the units of joules per kilogram or joules per tonne. It is *not* a measure of energy 'contained' within the material, but its value will depend upon the efficiency of the production plant. For example, during the Second World War it probably took 100 GJ of energy to produce a tonne of steel. In 2011, in a modern plant it will take no more than 22–23 GJ of energy to produce a tonne of steel.

Units: GJ/tonne or MJ/kg.

22.6.2 Embodied carbon dioxide

This index measures the emissions of CO_2 in making a unit quantity of material. It is measured in terms of tonnes of CO_2 emitted per tonne of product. CO_2 is not the most potent greenhouse gas, but it is the most important because of its emission every time we burn a fossil fuel. It is a natural by-product of combustion.

Units: kg of CO_2/kg of material.

Values of both embodied energy and embodied CO_2 for a range of building materials are given in Table 22.2.

Table 22.2 Ecological impacts of building materials: embodied energy and embodied CO_2 values

Material	Embodied energy (MJ/kg)	Embodied CO_2 (tonne CO_2/tonne)
Aggregate (general)	0.10	0.01
Aluminium (virgin)	191.0	14.33
Extruded (e.g. window frame section)	201.00	15.08
Sheet (e.g. for cladding)	199.00	14.93
Aluminium (recycled)	8.10	0.61
Asphalt (paving)	3.40	0.26
Bitumen	44.10	3.31
Brass	62.00	4.65
Cement	7.80	0.59
Clay brick	2.50	0.19
Clay pipe	6.30	0.47
Clay tile	2.50	0.19
Concrete block	0.94	0.07
Concrete (GRC)	7.60	0.57
Concrete paver	1.20	0.09
Concrete – ready mix 17.5 MPa	1.00	0.08
Ready mix 30 MPa	1.30	0.10
Ready mix 40 MPa	1.60	0.12
Concrete roof tiles	0.81	0.06
Copper	70.60	5.30
Glass – float	15.90	1.19
Glass – toughened	26.20	1.97
Laminated	16.30	1.22
Insulation		
Fibreglass	30.30	2.27
Polystyrene	117.00	8.78
Lead	35.10	2.63
Linoleum	116.00	8.70
Paint		
Solvent based	98.10	7.36
Water based	88.50	6.64
Paper – building paper	25.50	1.91
Plaster (gypsum)	4.50	0.34

Material	Embodied energy (MJ/kg)	Embodied CO_2 (tonne CO_2/tonne)
Plasterboard	6.10	0.46
Plastics		
HDPE	103.00	7.73
LDPE	103.00	7.73
Polypropylene	64.00	4.80
Polystyrene	117.00	8.78
Polyurethane	74.00	5.55
PVC	70.00	5.25
Rubber		
Natural latex	67.50	5.06
Synthetic	110.00	8.25
Steel (virgin) general	23.00	1.73
Galvanised	25.00	1.94
Structural	25.00	1.93
Recycled	10.10	0.76
Wire	12.50	0.94
Stone		
Local	0.79	0.06
Imported	6.80	0.51
Timber, softwood		
Air dried – rough sawn	0.30	0.02
Kiln dried – rough sawn	1.60	0.12
Air dried – dressed	1.16	0.09
Kiln dried – dressed	2.50	0.19
Mouldings	3.10	0.23
Hardboard	24.20	1.82
MDF	11.90	0.89
Glulam	4.60	0.35
Plywood	10.40	0.78
Shingles	9.00	0.68
Timber, hardwood		
Air dried – rough swan	0.50	0.04
Kiln dried – rough sawn	2.00	0.15
Vinyl flooring	79.10	5.94
Zinc	51.00	3.83
Galvanising	2.80	0.21

Source: Acorn, A. (1998), *Embodied Energy Coefficients of Building Materials*, Centre for Building Performance Research, NZ.

22.6.3 Water usage

There is only a finite quantity of water in the world – 1.4 billion cubic kilometres – and this amount is fixed and cannot be increased. Furthermore, this is all the water, 97 per cent of which is saline, which is not potable (i.e. it cannot be drunk). So only 42 million cubic kilometres are fresh water, and more than half of this is tied up in ice caps and glaciers. While the amount of water is fixed, the Earth's population is increasing dramatically, and this puts water resources (scarce in many parts of the world) under great and increasing pressure.

Water is essential for the continuance of life, but it is also used in many industrial processes, including the manufacture of materials. So in addition to the world's population growth, the rapid industrialisation of countries such as China, India and Brazil is also putting an increasing strain on our limited resources. This pressure on water supplies is set to become ever more acute as the twenty-first century progresses. Since water is such a vital resource, its consumption is an important index of sustainability. Specific water usage values for a range of building materials are given in Table 22.3, the water consumption index being the consumption of water in litres per kilogram in the manufacture of any material. From this data we can immediately see that the manufacture of certain materials, including non-ferrous metals and plastics, uses very large quantities of water. Using these data we can calculate exactly the amount of water that will be used in the manufacture of a given amount of any material. Because water is such a limited and precious resource, water consumption represents an important environmental impact, and can be used as a measure of such impact.

Units: litres/kg of material.

22.6.4 Material intensity factor

This index measures the amount of material that may be discarded in making a unit quantity of product. It was derived from the Ecological Rucksack proposed by Friedrich Schmidt-Bleek of the Wuppertal Institute in Germany in 1994. If a material is very scarce, we may need to process many tonnes of ore, etc. just to make one kilogram of product. Rucksack measures a kind of material efficiency. Material intensity (MI) is the reciprocal of the Rucksack value.

Units: kilogram of waste or spoil/kilogram of material.

22.6.5 Ecological footprint

This index measures the impact in terms of land area of any source of material, energy or service and it was devised by Wackernagel and Rees (1996) in Canada. For example, a modern, integrated steel plant will

Table 22.3 Water usage in the manufacture of various materials

Material	Water usage (litres/kg)
Aluminium	495.0–1,490.0
Cement	10.8–32.5
Clay brick	2.8–8.4
Concrete	1.7–5.1
Copper	269.0–298.0
Glass	6.8–37.5
Glass – toughened	26.2–28.9
Lead	1.84–2.03
Gypsum plaster	4.7–13.8
Plastics	
HDPE	194.0–215.0
LDPE	198.0–218.0
Polyurethane	256.0–283.0
PVC	155.0–172.0
Rubber – ABS (synthetic)	308.0–340.0
Steel	22.9–68.6
Stone	1.7–5.1
Zinc	160.0–525.0

Source: Granta Design, *CES Edupack 2008, Materials Selection Software*, Cambridge.

Table 22.4 MI factors of a few building materials

Material	MI factor (kg/kg)
Aluminium, virgin	85
Aluminium, recycled	3.5
Iron with zinc (small galvanised parts)	9
Copper, virgin	500
Copper, recycled	10
Lead	16
Steel (from basic oxygen furnace)	7
Steel (from electric arc furnace)	3.4
Zinc	23
Gypsum	1.8
Gravel	1.2
Quartz (for glass-making)	1.4
Sand	1.2
Insulation paper (recycled)	1.7
Concrete, structural	1.3
Concrete, lightweight aerated	2.3
Bitumen	2.6
Expanded polystyrene foam	11
Fibre-board, medium thickness	2
Glass wool	4.7
Flat glass	3
Granite slabs, polished	1.9
Hardboard	2.9
Sandstone	1.3
Polyurethane, hard foam	7.3
Plywood	2
Stoneware	2.9
Portland cement	3.2
Clay roof tile	2.1
Facing brick	2.0
Clay brick	2.1
Spruce, roundwood	1.9
Spruce, boards, beams, slats	2.2
Spruce, flooring, shuttering	2.8
Spruce, windows and doors, etc.	3.5
Glass fibre (E glass)	6.2
Rubber	5
Linoleum	2
Porcelain	10.0
Acrylic paint	2.7

Source: Schmidt-Bleek, F. (1994), *Ecological Rucksack*, Wuppertal Institut, Germany.

be located on a large plot of land, which will therefore be unavailable for any other purpose. The steel plant is said to have a 'footprint' of so many hectares. To measure the total footprint for steel we shall need to know the land area of the iron ore mine, the coal mine and the limestone quarry, as well as that of the steelworks. This index is a good way of measuring impact, because the land area of the Earth is finite, and each hectare of land can only be used for one purpose, whether it is growing crops, growing timber, supporting a road or motorway, supporting a residential building, etc.

Units: hectare/person (ha/person).

22.6.6 Other types of environmental impact

The forms of impact discussed in the preceding section are all quantifiable using the various indices outlined above. Furthermore, they can all be applied to the manufacture and use of the various construction materials. However, there are other types of impact that cannot be evaluated by any of these specific indices, and these include:

- emission of dust particles, including particles of colloidal dimensions, i.e. aerosols;
- emission of liquids, including toxic liquids;
- emission of gases other than CO_2 – these would include oxides of sulphur, i.e. SO_2 and SO_3, oxides of nitrogen, NO_x, etc.
- end of life disposal – although this is just one item in the list, it is a very important stage in the life of a material, and one that can have enormous impacts. We shall look at this later.

In Section 22.5 we mentioned certain limits (nine) which should not be exceeded: carbon emissions, biodiversity, nitrogen, land use, water purity, toxic emissions, aerosols, ocean acidification and ozone layer damage (Lynas, 2011). The four bullet-pointed impacts listed above need to be assessed against these nine limits. However, we shall not consider all of these limits here – some are more pertinent than others. One of the most important impacts is that of *aerosols* in the Earth's atmosphere (particles of colloidal dimensions). Aerosols are produced in many ways, both naturally and anthropogenically, and we need to define what they are, where they come from and what effects (impact) they can have.

22.6.7 Aerosol particles

Definition

Colloidal particles, or aerosols as they are now usually referred to, whether they are solid particles or fine liquid drops, are defined by their *small size*. To be defined as aerosols or colloids, they have to be in the size range from 1 micrometre (1 μm = 10^{-6} m or 1,000 nanometres) down to 1 nanometre (1 nm = 10^{-9} m) in diameter. Put another way, they range in diameter from one millionth of a metre down to one billionth of a metre. This small size confers a number of important properties. For example, particles in the size range 0.1 μm up to 1.0 μm have a lifetime in the atmosphere of the order of ten days, and in this time they can travel 1,000–10,000 km in the air from their point of origin. Finer particles down to 1 nm diameter will stay aloft much longer, and be able to travel further. Indeed, such particles may stay in the air so long that they can reach anywhere on the Earth's surface. Two more properties are worth mentioning here: the wavelengths of visible light (violet, λ = 300 nm to red, λ = 700 nm) falls within the size range of aerosols, so aerosols can have marked optical effects despite their small size; the other point is that particles with diameters less than 5 μm are, on inhalation, capable of penetrating far into human lungs. Once there they can deposit and eventually release certain undesirable substances into the bloodstream. Fly ash particles released into the air can carry on their surfaces elements volatilised during combustion such as arsenic, chromium, selenium, lead and certain radioactive species. This latter observation raises serious health questions, important in the achievement of sustainability.

Table 22.5 Atmospheric aerosol fluxes

Natural source	Average flux (Mt a^{-1})[a]	Flux range D < 25 μm[b]
Primary		
Windblown dust		1,000–100,000
Mineral dust	917	
Mineral dust	573	
Forest fires		3–350
Sea salt	10,100	1,000–100,000
Volcanoes	30	4–40,000
Biological	50	26–60
Secondary		
Sulphates from DMS	12.4	60–010
Sulphates from volcanic SO_2	20	10–100
From biogenic VOC	11.2	40–000
From biogenic NO_x		10–00

Note: DMS = dimethyl sulphide
VOC = volatile organic compounds
(a) Seinfeld, A & Pandis, S.N. (2006), *Atmospheric Chemistry and Physics*, 2nd edition, John Wiley, New York. (b) Brasseur, G.P. *et al.* (1999), *Atmospheric Chemistry and Global Change*, Oxford University Press, New York.

The amounts of aerosol particles circulating in the Earth's atmosphere at any one time can be considerable; Table 22.5 shows the quantities of various types flowing in mega-tonnes (millions of tonnes) per annum. Note that only some of these are anthropogenic in origin and particulates from many industrial processes are not included.

This brief discussion will serve to highlight the fact that, although aerosols are tiny, they have the potential to have major effects on, for example, how much sunlight reaches the Earth's surface, and on human health, to mention but two. Aerosol particles are not always accorded the importance that they should have in the attainment of sustainability. Close examination and analysis of past naturally occurring events give us the best guide as to their significance. These naturally occurring events include volcanic eruptions and large meteorite strikes on the Earth. There are around 1,500 active volcanoes around the world, so eruptions are more common than large meteorite impacts. Many eruptions are small in scale, but the large ones can have a major effects, and one or two of these will be discussed in the next section.

A fog is an example of an aerosol system with which we are all familiar. Fog consists of water drops that are so fine that the action of gravity is negligible compared with the lift of air currents on their surfaces, and so they can float around in the air without settling, unlike raindrops. Raindrops are larger, and fall to earth under the action of gravity. Clouds in the sky are also aerosol systems – again, they consist of very small water drops.

Clouds of dust particles, where the particles are of colloidal dimensions, will behave similarly to ordinary clouds. The particles will not settle or fall to earth; they may reach the upper atmosphere, where they can remain for months or even years. Such dust particles can arise from natural processes such as forest fires, volcanic eruptions, etc., or they may be caused anthropogenically by mining operations, blasting, quarrying, industrial crushing and grinding operations, combustion processes, etc.

Origin of aerosol particles

We have seen that particles of colloidal dimensions or aerosols can be produced both by natural causes and also by man's agency.

Examples of naturally occurring systems include the hydrological cycle, where water is evaporated from the seas and oceans and ordinary clouds are formed. Forest fires can produce very fine smoke particles (particles of carbon). The freeze–thaw weathering of rocks will produce sand particles in the long term; a small fraction of these sand particles will be of colloidal dimensions. In a sandstorm, the sand can be blown and whipped into the air, the larger particles will soon fall back to earth, while those of colloidal dimensions can remain airborne for much longer. Vulcanism can produce aerosols. There are two types of volcanoes, effusive and explosive. The effusive type emits a steady lava flow, and because the lava is hot, it may lead to widespread burning of vegetation, producing smoke. Here again, many of the smoke particles will be of colloidal sizes. Explosive volcanoes (e.g Krakatoa in 1883) literally explode, and several cubic kilometres of rock, earth and vegetation can be blown kilometres into the atmosphere. The sizes of the rock particles can range between very large (the size of a bus, or larger) down to very small, including aerosol size. If such fine particles reach the upper atmosphere, they can remain there for months or years. The other form of natural formation arises from an event which is much rarer, namely, a meteorite impact. Meteorites are lumps of rocky or metallic material of extra-terrestrial origin. They are drawn in by the Earth's gravitational field and strike the Earth while travelling at high speed. Such events are referred to as hypervelocity impact events. If the meteorite is large, it will form a large crater and bury itself in the Earth's crust, throwing out a lot of material (called *ejecta*) in the process. These ejecta particles can range in size from very large (many metres in diameter) down to aerosol sizes.

Aerosols can also be anthropogenic in origin. Examples of activities giving rise to small particles include industrial processes involving combustion of fossil fuels, such as iron-making and cement manufacture, and comminution (grinding) processes such as the grinding of cement clinker. Mining operations, including blasting, etc., will also generate dust, some of which will be very fine. Other causes can be transport operations in desert and other dusty locations.

Transport systems generate a lot of colloidal material. The world's airline fleets contain several tens of thousands of aircraft, and these craft all burn fossil fuel (kerosene). The products of combustion include carbon dioxide and water vapour, and also some tiny particulates, and these are all expelled into the atmosphere at high altitudes of 10,000–12,000 metres, approximately.

Effects of aerosols

These fine particles or liquid droplets will have various physical and chemical effects. Important physical effects include scattering and reflection of incoming solar radiation, and serving as nucleation sites for precipitation (rainfall) processes. The aerosols are also able to take part in various chemical and photo-chemical processes that occur in the Earth's upper atmosphere. These effects have been recognised and are the subject of current research by climatologists and environmentalists (e.g. Svensmark & Calder, 2007). The recent Royal Society report *Geoengineering the Climate* (Royal Society, 2009) proposes the introduction of aerosols into the Earth's atmosphere as a solar radiation management technique to reduce global warming. However, their effects have been observed (though not understood at the time) and written about for at least the last 250 years. There is a lot of information available for those prepared to do some historical research; as always, little occurs on Earth that has not been seen before.

There has not been a major meteorite impact apart from the Tunguska fireball of 1908, which occurred in a very remote region of Western Siberia. The exact nature of this event is still not fully understood, and is the object of a current investigation. However, there have been several major volcanic eruptions in the last two centuries, the effects of which have entered the historical records. These are the Tambora and

Krakatoa volcanoes of 1815 and 1883, respectively, and the Laki Craters eruption in Iceland in 1783 to 1784. Tambora and Krakatoa were both explosive events, whereas the Laki event was an effusive eruption lasting from June 1783 to early 1784. Tambora is the second most powerful explosive event ever, with Krakatoa being the fifth.

Other gaseous emissions

There are other indices that might be considered; for example, if we were concerned about acid rain, we could propose the embodied sulphur dioxide index. In fact, the operators of industrial processes take steps to 'scrub' their emissions of SO_2. Also, careful selection of fuel sources can minimise SO_2 production. However, when we burn fuel we shall always produce an amount of carbon dioxide proportional to the amount of fuel burned.

While the Laki volcano was less severe than the Siberian Traps event 250 million years ago, it involved the emission to the atmosphere of enormous quantities of sulphur dioxide. Indeed, it has been estimated that 100 million tonnes of acid rain were deposited over this period. This is approximately the same amount as one year's acid rainfall on the entire Earth today. So, one year's emission of SO_2 was released over Iceland in a matter of months in 1783. The results were catastrophic. Much vegetation was destroyed or damaged, including nearly all the crops produced by farmers that year. Because there was nothing to feed the livestock, 75 per cent of all farm animals on Iceland died. There was little or no grain with which to make bread, and so there was a desperate, country-wide famine in which the population of Iceland declined by over 25 per cent due to starvation (15,000 deaths out of a population of 50,000+). At this time, Iceland was a dependency of Denmark, and at one stage the Danish authorities were giving consideration to the evacuation of the entire population of Iceland. This is powerful evidence of the severe climatic effects that can be produced by aerosol particles in the atmosphere.

The climatic effects of the Laki volcano of 1783 have been recorded as eyewitness accounts by several authors in Iceland and in England. Gilbert White's account published in 1788–1789 is quite graphic and easily obtainable. Beside the effects in Iceland, the toxic clouds described by White caused an increase in mortality among agricultural workers in England.

Acidification

Acidification is the name given to the process whereby surface waters, rain water and sea water can be rendered acidic by the dissolution of acids, usually as a result of some kind of pollution. In the UK and many other developed countries, much legislation has come into force in recent decades to protect the environment from industrial pollution. In the early years of the Industrial Revolution there was little or no legislation to protect the environment, and industrial effluents would find their way into the rivers and streams flowing through the cities and towns where the new industries were becoming established. At the time, little or nothing was known about the adverse effects on health and the environment of these industrial by-products – such knowledge was acquired only later.

What are these acids and where do they come from? The release of amounts of liquid acids into rivers and streams by industry in the UK is now a very rare occurrence because of the environmental protection legislation and the heavy financial penalties that would be exacted. Nowadays, the usual agency is via gaseous emissions from industry, from transport systems and as a result of space heating in buildings, to name the main sources. The gases that produce the major problems include carbon dioxide, sulphur dioxide, sulphur trioxide, oxides of nitrogen, etc. These are all produced when fossil fuels are burned to generate heat and energy. A certain amount of CO_2 dissolves in sea water, and one consequence of rising CO_2 levels in the atmosphere is that more is dissolving in the seas and making them more acidic. This is having a very adverse effect on certain life-forms that live in the sea, such as crustaceans and coral reefs. When dissolved in water, carbon dioxide forms weak carbonic acid.

The extra CO_2 in the atmosphere similarly dissolves in rain water to form carbonic acid, so that rainfall becomes acidic. As well as CO_2, oxides of sulphur also dissolve in rain water, making it yet more acidic. This can damage and kill trees, and examples of this have been observed with increasing frequency in the forests of northern Europe. Since trees play a pre-eminent role in photosynthesis and the production of oxygen and removal of CO_2, the killing of trees on a large scale is a very serious matter.

22.7 Durability/longevity issues

Once a building has been constructed, it will hopefully serve its designed purpose for a long time. However, it will require some regular maintenance; most buildings do. Maintenance involves the input of further energy and materials to the original structure. In the interests of achieving sustainability, these further expenditures of energy and materials should be the minimum possible, consistent with safe and satisfactory operation of the facility. Sustainability requires that we consider the whole life of the building, rather than just the initial costs when designing and constructing buildings. Taking a financial analogy, we need to 'amortise' the initial input of energy and material over the longest possible period of time.

Taking the 'whole life' into account may well mean that the use of better-quality materials and better design (and possible more expensive) solutions at the outset is less expensive and troublesome in the long run. The building industry culture of always going for the least cost will not lead to more sustainable buildings in the long term. Sayce (2002) has surveyed and studied the factors that determine the life of a building, and which help it merit the title of a sustainable building. Demolishing and replacing buildings within a decade of their construction is not a sustainable activity. Good buildings can last for centuries, and given the impact of construction, we should be aiming for our buildings to be durable, and to retain the attractiveness and usefulness for as long as possible. Given the importance of initial cost to most clients, this is not easy to achieve. Most clients have no thought or concern that their new building will still be good in 200 years' time, and they may have a very specific use in mind which militates against a future, easy change of use.

To achieve a long life, a building needs a number of attributes:

- loose-fit – the ability to adapt successfully to meet occupier needs;
- low energy – this is important both in building operation and the construction phase;
- location, or the integration of buildings to support the range of urban activities;
- 'likeability' – the ability to provide an environment that is liked by the occupants, and so the stakeholders respond with affection.

Apart from energy use, these are factors that may not be evaluated numerically, but we all recognise their importance if we think of those buildings that we have used and which we liked. Some buildings will be designed for a specific, known, long-term use, for example, the parliament buildings in a capital city. They will be appropriately located, and because they are high-profile public buildings, cost will not be a major issue. The best quality, high-durability materials will be used to create a useful, aesthetically appealing building, requiring minimum maintenance. The life of the building will not be a major concern, as it will be expected to last for a very long time.

A building on a suburban business/light industry park will be quite different. It will have a much shorter expected life; utility and not aesthetics will be the keynote. Having said that, such buildings are frequently adaptable, and because they are all grouped together on the edges of towns and cities, the lack of aesthetic appeal does not matter. For these buildings to be considered sustainable, they should be designed to be capable of easy dismantling so that the materials from which they are made can very easily be re-used or recycled.

The building of houses falls somewhere between these two extremes of service life. A notional building life of 60–80 years may be quoted as the design ideal. However, the UK building stock contains a lot of houses older than 60–80 years. A population of over 60 million, which continues to grow, means that there is a constant demand for housing. No housing was built in the UK during the Second World War, and in an attempt to quickly alleviate this problem some of the aircraft factories were turned over to the mass production of pre-fabricated housing which could be quickly erected after minimal site preparation. These units had a design life of ten years; in some parts of the UK the idea worked, the pre-fabs were taken out of service after a decade and replaced with conventional dwellings. However, in other areas many were still in use 40–50 years later.

The other great post-war housing experiment was the high-rise accommodation block, constructed in large numbers, especially in the 1960s, in the wake of slum clearance. Large quantities of materials were committed to their construction, often using untried methods involving the quick assembly of factory-produced items such as wall panels and floor units onto a steel frame. These high-rise housing units were put up by local councils in the expectation of drawing rents from them for at least 40 years, but many were condemned and demolished after 20 years. These developments failed for a variety of reasons, some technical and some social. They were demolished using controlled explosions, and the resulting piles of rubble and waste were largely committed to land-fill. This was the very antithesis of sustainable development.

These developments had some advantages, however. Large populations could be housed close to their places of work, thus reducing the need for long commuting distances and the associated problems of traffic congestion.

22.8 End of life issues

In nature, when a plant or animal reaches the end of its life and it dies, its remains are recycled so that the materials from which it is made are returned to the environment, and are made available to the current living population for their use and support. Of course, some animals are predators, preying on other species for food. Herbivores live by eating live vegetable matter. So in many cases, animals and plants are recycled before they die. However, it remains true that nature is brilliant at recycling its materials in ways that we can only marvel at.

As examples, consider what happens when a tree dies in a forest. It becomes damp and infested with fungi. Fungi belong to a class of organisms called saprophytes (also including other plants and micro-organisms) and these live on dead or decaying organic tissues, unlike parasites, which live on living tissues. The fungi break down the cellulose-type structures in the timber and turn them into CO_2 and other hydrocarbons that return to the soil. Similarly, in the animal world there are scavengers, animals that feed on carrion, dead plant material and refuse. A pride of lions will kill deer or zebra and take a meal from the fresh corpse. When they have eaten their fill, they will leave the remains lying in the open. Buzzards and hyenas will then gather round and finish off what the lions have left. Finally, small insects and birds may well pick the bones clean so that nothing (apart from the bones) is left.

Currently, one of humanity's biggest problems is the sheer volume of waste material that our modern lifestyles give rise to. Whereas nature recycles everything very efficiently, we definitely do not. Waste material does not just represent material, it is also waste energy. If we throw away a clay brick or a steel beam, we are not just discarding the fired clay or the steel, but also the energy that we invested in making it. In nature, if fresh plants are to grow, nature uses the material from the previous generation that has been returned to the soil. Living animals use living or dead tissues to sustain themselves – no piles of dead or 'waste' materials accumulate in nature. In contrast, the landscape is littered with discarded materials, derelict buildings, abandoned vehicles, etc. By not recycling or re-using our worn out and discarded materials we put pressure on our reserves of raw materials, and needlessly consume enormous amounts of energy in extracting and producing virgin materials. To express this another way, nature 'closes the loop' and humanity does not; nature operates in a sustainable manner, humanity does not.

The 'end of life' problem is therefore one of the largest problems we have to solve if we are ever to achieve a sustainable mode of living. The technology for recycling metals, particularly iron and steel, has been in existence since the early twentieth century, together with the necessary infrastructure to collect scrap iron and steel, but it was not possible to recycle plastics until the early 1990s. Again, the necessary scrap collection facilities are currently being developed. Without the means to collect large quantities of clean, homogeneous waste material, recycling cannot be done. It is true that the technology for effective recycling of more material is currently being developed, but other trends in material use are making recycling more difficult. Chapter 18 is devoted to the subject of composite materials, and the advantages that they offer. However, they work by creating intimate mixtures of two or more materials arranged in such a way that the resulting composite offers a combination of properties not possessed by any of the constituents on their own. The way the materials are put together in many cases makes their separation at end of life a difficult proposition.

Nevertheless, methods are being developed for situations where the materials are valuable. For example, scrap electronic devices in which there are small quantities of valuable metals such as gold, silver, lead and tin make recovery an attractive proposition. One problem with building materials is the use of Portland cement mortars for bricklaying. The old lime mortars used by Victorian builders were very easy to remove from bricks; Portland cement becomes as strong or even stronger than the bricks it binds together, and its removal is very difficult. Since bricks are not inherently high-value materials, used bricks are often crushed for use as hardcore, with no attempt being made to remove the mortar.

The replacement window trade is another one where recycling is now becoming common. Increasingly, when window units are removed for replacement, the frames are uPVC rather than the traditional wooden ones. These units make removal of the glass easy, and both glass and plastic can be recycled.

One form of recycling that has always gone on is the recycling of the whole building, short of demolishing it. This can take the form of building refurbishment and rehabilitation (Gorse and Highfield, 2009). In these cases the main structure of the building is usually retained, stripped out and refurbished. During the process, the building may be extended in some way, horizontally or vertically.

However, an interesting and quite recent development was the refurbishment of Winterton House in London. This was a 26-storey residential tower block built in the 1960s, and was typical of its time, with a steel frame and concrete floor decks. It was originally clad with factory-made wall units, but it contained the usual defects inherent in this method of building. However, instead of completely demolishing the entire structure, it was decided to remove all interior fittings, wall units and floor decks and to strip the building back to its steel frame. When exposed, it was found that the steel frame was in surprisingly good condition. The decision was then taken to re-fit new floor decks, to enclose the building with a brick masonry skin, and then to refit the interior and return the building to residential use. The work was completed, and the refurbished or recycled building put back to use. The cost of doing this turned out to be only about half of what complete demolition and total replacement would have cost.

22.9 Conclusions

This chapter has discussed many important issues, issues that would easily merit a whole book to themselves. Sustainability has become a very important topic in recent years, and with the Earth's population set to increase further in the next 50 years, the need to achieve a sustainable mode of living will become ever more necessary. From these discussions, we can distil a number of principles that underlie sustainable operation:

- Materials should be sourced locally, as far as possible.
- Re-used, recycled and recyclable materials should be used, if possible.
- If such materials are not possible, then the materials should be renewable (e.g. timber), or so plentifully available as to be effectively inexhaustible.
- Labour should also be sourced locally, as far as possible.
- The building should consume as little energy as possible, when in use.
- Because water resources are limited and under such pressure, the building should use as little water as possible, when in use.
- The building should be designed with an appropriate level of durability and require the minimum of maintenance.
- If the building is not designed to have a very long life – e.g. an industrial unit – the structure should be able to be easily dismantled for recycling.
- If the building is designed for a long life, the possibility of future changes in use should be borne in mind.

Finally, although all the points mentioned above will contribute to the achievement of more sustainable buildings, it is worth emphasising two of them as being of crucial importance in determining whether we ever attain a truly sustainable way of life. The following two factors are of over-riding importance: minimal CO_2 emissions and limiting demand for water. In terms of resource limitations, water may well turn out to be the factor which puts an upper limit on the Earth's population.

22.10 Critical thinking and concept review

1. Define, and give the correct units for the following indices of environmental impact:
   (i) embodied energy for a material
   (ii) embodied carbon dioxide
   (iii) water consumption
   (iv) Ecological Rucksack.
2. Explain what is meant by the ecological footprint for a material.
3. In what important respect do the four indices above (Q.1 (i)–(iv)) differ from the ecological footprint?
4. If the embodied energy for steel is 22.5 GJ/tonne, how much energy will be consumed in manufacturing 5,500 tonnes of steel for a bridge structure?
5. What will be the total carbon footprint in the manufacture of the following materials for the construction of an industrial unit:

Material	Amount	Embodied CO_2 (tonne CO_2/tonne)
Steel	100 tonnes	1.93
Concrete	250 tonnes	0.12
Bricks	50,000 bricks	0.19
Glass	10 tonnes	1.19

Note: 1,000 bricks weigh 3.6 tonnes.

6. An industrial building is demolished, giving rise to 500 tonnes of steel sections, beams and plate. If the embodied energy of virgin steel is 23 GJ/tonne, calculate the amount of energy that would be lost by putting the scrap steel into landfill.

7. The construction of a large public building requires the following quantities of materials:

Material	Quantity	Embodied energy (GJ/tonne)	Water usage (litres/kg)
Steel	500 tonnes	23.0	40.0
Concrete	2,000 tonnes	1.5	3.0
Bricks	200,000	2.5	5.0
Glass	60 tonnes	16.0	30.0
Plastics	50 tonnes	103.0	210.0

Note: 1,000 bricks weigh 3,600 kg.

Using the data given above, calculate:
   (i) the total energy used to make these materials, and
   (ii) the water consumed in the making of the materials.

8. A factory is set up to recycle scrap uPVC window frames. It is required that no more than 1.0 GJ/tonne of energy be put into the scrap material in collecting it for recycling. If the scrap is carried using trucks that use 3 MJ of energy per tonne per kilometre, and ignoring return journeys, what is the greatest distance from the recycling plant at which scrap can be collected?

9. Which gases cause acid rain? Where do the gases come from? Explain two adverse effects of acid rain, one on the natural world, and one on the built environment.

10. What are aerosol particles? Give *two* natural and *two* anthropogenic reasons for the presence of aerosol particles in the Earth's atmosphere.

11. Give *two* important differences between environmental impact mitigation and the promotion of sustainability.

12. Which *three* classes of material may be considered to be sustainable?

22.11 References and further reading

ACORN, A. (1998), *Embodied Energy Coefficients of Building Materials*, Centre for Building Performance Research, New Zealand.

ALLWOOD, J.M. and CULLEN, J.M. (2012), *Sustainable Materials: With Both Eyes Open*, UIT Cambridge Ltd, Cambridge.

ASHBY, M.F. (2009), *Materials and the Environment: Eco-informed Material Choice*, Butterworth-Heinemann, Oxford.

DAVIES, B. (2009), 'Material flow in mining and metals', *Materials World*, 17: 8.

DIAMOND, J. (2006), *Collapse: How Societies Choose to Fail or Survive*, Penguin Books, London.

FRANCIS, P. and OPPENHEIMER, C. (2004), *Volcanoes*, 2nd edition, Oxford University Press, Oxford.

GORSE, C. and HIGHFIELD, D. (2009), *Refurbishment and Upgrading of Buildings*, 2nd edition, Spon Press, London.

GRANTA DESIGN (2008), *CES Edupack 2008, Materials Selection Software*, Cambridge.

INSTITUTE OF METALS (1987), *Materials for Construction and Building in the UK*, Report of a working party of The Materials Forum and the Institution of Civil Engineers, Institute of Metals, London.

LYNAS, M. (2011), *The God Species: How the Planet can Survive the Age of Humans*, Fourth Estate, London.

ROYAL SOCIETY (2009), *Geoengineering the Climate: Science, Governance and Uncertainty*, Report of Working Group, Policy Document 10/09, The Royal Society, London.

SAYCE, S.L. (2002), 'The Quest for Sustainable Buildings: Is Longevity the Key?', pp. 399–408, *Proceedings of the Eighth Annual International Sustainable Development Research Conference*, 8–9 April 2002, University of Manchester.

SCHMIDT-BLEEK, F. (1994), *Carnoules Declaration of the Factor Ten Club*, Wuppertal Institute, Germany.

STURGES, J.L. (1998), 'Environmentally Aware Materials Selection for Construction. Part 1. Some Problems', pp. 55–64, *The Product Champions, Proceedings of the 2nd International Conference on Detail Design in Architecture*, Ed. S Emmitt, Leeds Metropolitan University.

STURGES, J.L. (1999), 'Construction Materials Selection and Sustainability' pp. 297–304, *Proceedings of the 2nd International Conference on Construction Industry Development*, Vol. I, Singapore.

SVENSMARK, H. and CALDER, N. (2007), *The Chilling Stars: A New Theory of Climate Change*, Icon Books, Cambridge.

WACKERNAGEL, M. & REES, W.E. (1996), *Our Ecological Footprint*, New Society Publishers, Canada.

WHITE, G. (1788–1789), *The Natural History of Selborne*, republished 1987 by Penguin Books, London.

Index

Page numbers in *italics* show a figure and a table is shown in **bold**.

eBooks
from Taylor & Francis

Helping you to choose the right eBooks for your Library

Add to your library's digital collection today with Taylor & Francis eBooks. We have over 50,000 eBooks in the Humanities, Social Sciences, Behavioural Sciences, Built Environment and Law, from leading imprints, including Routledge, Focal Press and Psychology Press.

Choose from a range of subject packages or create your own!

Benefits for you
- Free MARC records
- COUNTER-compliant usage statistics
- Flexible purchase and pricing options
- 70% approx of our eBooks are now DRM-free.

Benefits for your user
- Off-site, anytime access via Athens or referring URL
- Print or copy pages or chapters
- Full content search
- Bookmark, highlight and annotate text
- Access to thousands of pages of quality research at the click of a button.

ORDER YOUR FREE INSTITUTIONAL TRIAL TODAY

Free Trials Available

We offer free trials to qualifying academic, corporate and government customers.

eCollections

Choose from 20 different subject eCollections, including:

- Asian Studies
- Economics
- Health Studies
- Law
- Middle East Studies

eFocus

We have 16 cutting-edge interdisciplinary collections, including:

- Development Studies
- The Environment
- Islam
- Korea
- Urban Studies

For more information, pricing enquiries or to order a free trial, please contact your local sales team:

UK/Rest of World: **online.sales@tandf.co.uk**
USA/Canada/Latin America: **e-reference@taylorandfrancis.com**
East/Southeast Asia: **martin.jack@tandf.com.sg**
India: **journalsales@tandfindia.com**

www.tandfebooks.com